AF302298

Modern Methods of Plant Analysis

New Series Volume 3

Editors
H. F. Linskens, Nijmegen
J. F. Jackson, Adelaide

Volumes Already Published in this Series:

Volume 1: Cell Components
1985, ISBN 3-540-15822-7

Volume 2: Nuclear Magnetic Resonance
1986, ISBN 3-540-15910-X

Volume 3: Gas Chromatography/Mass Spectrometry
1986, ISBN 3-540-15911-8

Gas Chromatography/
Mass Spectrometry

Edited by
H.F. Linskens and J.F. Jackson

Contributors
R.S. Bandurski G. Combaut A. Ehmann P. Hedden
B. Janistyn H. Kameoka H. Kodama D.V. Lynch
J.K. MacLeod H. Nyberg L.M.S. Palni L. Rivier
R.R. Selvendran B.J.H. Stevens S.A.B. Tay
G.A. Thompson Jr., L. Witte

With 98 Figures

Springer-Verlag
Berlin Heidelberg New York Tokyo

Professor Dr. HANS FERDINAND LINSKENS
Botanisch Laboratorium
Faculteit der Wiskunde en Natuurwetenschappen
Katholieke Universiteit
Toernooiveld
NL-6525 ED Nijmegen
The Netherlands

Professor Dr. JOHN F. JACKSON
Department of Biochemistry
Waite Agricultural Research Institute
University of Adelaide
Glen Osmond, S.A. 5064
Australia

ISBN-13:978-3-642-82614-6   e-ISBN-13:978-3-642-82612-2
DOI: 10.1007/978-3-642-82612-2

Library of Congress Cataloging-in-Publication Data. Gas chromatography/mass spectrometry. (Modern methods of plant analysis; new ser., v. 3) 1. Plants – Analysis. 2. Gas chromatography. 3. Mass spectrometry. I. Linskens, H. F. (Hans F.), 1921–. II. Jackson, J. F. (John F.), 1935–. III. Series. QK865.G37 1986   581.19′285   86-3875 ISBN-13:978-3-642-82614-6 (U.S.)

This work is subject to copyright. All rights are reserved, whether the whole or part of the material is concerned, specifically those of translation, reprinting, re-use of illustrations, broadcasting, reproduction by photocopying machine or similar means, and storage in data banks.

Under § 54 of the German Copyright Law, where copies are made for other than private use, a fee is payable to "Verwertungsgesellschaft Wort", Munich.

© by Springer-Verlag Berlin Heidelberg 1986
Softcover reprint of the hardcover 1st edition   1986

The use of registered names, trademarks, etc. in this publication does not imply, even in the absence of a specific statement, that such names are exempt from the relevant protective laws and regulations and therefore free for general use.

Product Liability: The publisher can give no guarantee for information about drug dosage and application thereof contained in this book. In every individual case the respective user must check its accuracy by consulting other pharmaceutical literature.

2131/3130-543210

Introduction

Modern Methods of Plant Analysis

When the handbook *Modern Methods of Plant Analysis* was first introduced in 1954 the considerations were:
1. the dependence of scientific progress in biology on the improvement of existing and the introduction of new methods;
2. the difficulty in finding many new analytical methods in specialized journals which are normally not accessible to experimental plant biologists;
3. the fact that in the methods sections of papers the description of methods is frequently so compact, or even sometimes so incomplete that it is difficult to reproduce experiments.

These considerations still stand today.

The series was highly successful, seven volumes appearing between 1956 and 1964. Since there is still today a demand for the old series, the publisher has decided to resume publication of *Modern Methods of Plant Analysis*. It is hoped that the New Series will be just as acceptable to those working in plant sciences and related fields as the early volumes undoubtedly were. It is difficult to single out the major reasons for success of any publication, but we believe that the methods published in the first series were up-to-date at the time and presented in a way that made description, as applied to plant material, complete in itself with little need to consult other publications.

Contributing authors have attempted to follow these guidelines in this New Series of volumes.

Editorial

The earlier series *Modern Methods of Plant Analysis* was initiated by Michel V. Tracey, at that time in Rothamsted, later in Sydney, and by the late Karl Paech (1910–1955), at that time at Tübingen. The New Series will be edited by Paech's successor H.F. Linskens (Nijmegen, The Netherlands) and John F. Jackson (Adelaide, South Australia). As were the earlier editors, we are convinced "that there is a real need for a collection of reliable up-to-date methods for plant analysis in large areas of applied biology ranging from agriculture and horticultural experiment stations to pharmaceutical and technical institutes concerned with raw material of plant origin". The recent developments in the fields of plant biotechnology and genetic engineering make it even more important for workers in the plant sciences to become acquainted with the more sophisticated methods, which sometimes come from biochemistry and biophysics, but which also have been developed in commercial firms, space science laboratories, non-university research institutes, and medical establishments.

Concept of the New Series

Many methods described in the biochemical, biophysical, and medical literature cannot be applied directly to plant material because of the special cell structure, surrounded by a tough cell wall, and the general lack of knowledge of the specific behavior of plant raw material during extraction procedures. Therefore all authors of this New Series have been chosen because of their special experience with handling plant material, resulting in the adaptation of methods to problems of plant metabolism. Nevertheless, each particular material from a plant species may require some modification of described methods and usual techniques. The methods are described critically, with hints as to their limitations. In general it will be possible to adapt the methods described to the specific needs of the users of this series, but nevertheless references have been made to the original papers and authors. While the editors have worked to plan in this New Series and made efforts to ensure that the aims and general layout of the contributions are within the general guidelines indicated above, we have tried not to interfere too much with the personal style of each author.

Volume Three – Gas Chromatography/Mass Spectrometry

The New Series in Modern Methods of Plant Analysis was initiated in 1985 with a volume on Cell Components, and quickly followed by a second volume on Nuclear Magnetic Resonance (NMR). Both included chapters contributed by world experts on all aspects of the chosen topics. We have now followed the principle adopted in the second volume, that of presenting the application of a relative new and powerful analytical technique, this time devoting the third volume to gas chromatography coupled to mass spectrometry (GC/MS). Here the separation power of GC is combined with the selective detection properties of MS, where compounds are ionized within the mass spectrometer and fragment ions separated and analyzed on the basis of their mass-to-charge ratio.

Gas chromatography was one of the earliest techniques used to determine plant hormones. When it is combined with mass spectrometry, a technique capable of functioning as a highly sensitive and selective gas chromatography detector, especially when it is focused on a particular ion, then we have a very sensitive analytical tool indeed. As little as 10 pg may be enough for detection with this technique. Little wonder, then, that GC/MS is often used to determine plant hormones. This is reflected in the make-up of the present volume, where we feature chapters on determination of cytokinins, auxins, gibberellins, cyclic nucleotides, and other biologically active compounds. Other plant materials are not neglected, so that chapters on the determination of tobacco constituents, essential oils, phospholipids, fatty acids, volatile flower compounds, pectic polysaccharides, and terpenoids can also be found in this volume.

As in the previous volume on NMR, we hope that this collection of chapters by world experts on plant analysis by GC/MS will encourage the further use of this highly sensitive and discriminating technique amongst scientists, students, and industrial analysts working with plant materials.

Acknowledgements. The editors express their thanks to all contributors for their efforts in keeping to production schedules, and to Dr. Dieter Czeschlik, Ms. K. Gödel and Ms. E. Schuhmacher of Springer publishers for their cooperation with this and other volumes in Modern Methods of Plant Analysis. The constant help of José Broekmans is gratefully acknowledged.

Nijmegen and Adelaide, May 1986   H. F. LINSKENS
                                                                   J. F. JACKSON

Contents

**The Use of Combined Gas Chromatography-Mass Spectrometry
in the Analysis of Plant Growth Substances**
P. HEDDEN (With 8 Figures)

1 Introduction   1
2 Identification by GC-MS   1
   2.1 Derivatisation   2
      2.1.1 Methylation   2
      2.1.2 Trimethylsilylation   2
      2.1.3 Permethylation   3
   2.2 Gas Chromatography   3
   2.3 Qualitative Mass Spectrometry   4
   2.4 Quantitative Mass Spectrometry   5
3 Gibberellins   9
   3.1 Extraction and Purification 10
   3.2 GC-MS 11
4 Abscisic Acid and Related Compounds 14
   4.1 Qualitative GC-MS 16
   4.2 Quantitative GC-MS 17
5 Brassinosteroids 18
References 19

**Applications of Mass Spectrometry for the Examination
of Pectic Polysaccharides**
R. R. SELVENDRAN and B. J. H. STEVENS (With 6 Figures)

1 Introduction 23
2 Structural Analysis of Pectic Polysaccharides 23
3 Determination of the Nature of the Glycosidic Linkages 24
4 Separation and Identification of Partially Methylated Alditol Acetates . 27
5 Extensions of Methylation Analysis 29
   5.1 Controlled Partial Acid Hydrolysis Studies 29
   5.2 β-Eliminative Degradation Studies 30
      5.2.1 Neutral Glycosyl Residues Linked to Galacturonosyl Residues 32
      5.2.2 Glycosyl Residues Linked to 0–4 of 2,4-Linked
            Rhamnopyranosyl Residues 32
6 Sequencing of Sugar Residues in Pectins 33
   6.1 Partial Acid Hydrolysis 33
      6.1.1 Characterisation of Oligosaccharides as Permethylated
            Derivatives 34

  6.2 Partial Acetolysis 37
  6.3 Enzymatic Hydrolysis 38
7 Sequencing of Pectic Polysaccharides by Partial Depolymerisation
   of Permethylated Derivatives 38
8 Experimental 40
References 43

**GC-MS Methods for Cyclic Nucleotides in Higher Plants
and for Free High Unsaturated Fatty Acids in Oils**
B. JANISTYN (With 10 Figures)

 1 Introduction 47
 2 Methods of GC-MS 48
 3 Adenosine-3′:5′-Monophosphate (cAMP) in Maize Seedlings
    (*Zea mays*) 48
 4 An Isotope Dilution GC-MS Spectrometric Assay for cAMP
    in Cultured Tobacco Tissue 51
 5 Stability of Cyclic Purine Nucleotides in the Presence of
    Hydrochloric Acid During Extraction 55
 6 Guanosine-3′:5′-Monophosphate (cGMP) in Maize Seedlings
    (*Zea mays*) 56
 7 GC-Separation of Synthetic cAMP and cGMP in a Mixture 59
 8 Cyclic Pyrimidine Nucleotides in Plants? 60
 9 Free High Unsaturated Fatty Acids in Oils 60
10 Conclusions 64
References 65

GC-MS Methods for Lower Plant Glycolipid Fatty Acids
H. NYBERG (With 4 Figures)

 1 Introduction 67
 2 Extraction of the Plant Material 67
    2.1 Handling and Storage 67
    2.2 Extraction with Organic Solvents 68
    2.3 Purification of the Extract 69
 3 Separation of Glycolipids from the Total Lipid Extract 70
    3.1 Column Chromatography 70
    3.2 Thin-Layer Chromatography (TLC) 72
    3.3 Other Applications 73
 4 Isolation of the Glycolipids 74
    4.1 Thin-Layer Chromatography 74
    4.2 Localization of the Glycolipids on TLC 75
    4.3 Removing the Spots from the TLC Plates 76
 5 Derivatization of the Glycolipid Fatty Acids for GC-MS 76
    5.1 General Features 76
    5.2 Formation of Methyl Esters 77
    5.3 Silylation and Other Methods 77

6 GLC Instrumentation for Fatty Acid Analysis 78
  6.1 General Features 78
  6.2 Carrier Gas System 79
  6.3 Injection and Injectors 79
  6.4 The Detector 82
7 Column Selection for Fatty Acid GLC Analysis 83
  7.1 Column Types 83
  7.2 Supports, Liquid Phases, and Their Characteristics 84
8 Interpretation of GC Data and Calculation of Results 87
  8.1 Identification of Peaks Using Standard Compounds 87
  8.2 Quantitation of Results 88
  8.3 External and Internal Standardization 89
9 Fatty Acid Ester Structure Determination by GC-MS 90
  9.1 Equivalent Chain Lengths (ECL) 90
  9.2 Semilogarithmic Correlations 91
  9.3 Mass Spectrometers and Their Function Principles 91
  9.4 Interpretation of Mass Spectra of Fatty Acid Esters 92
10 Abbreviations 94
References 95

**Analysis of Phospholipid Molecular Species by Gas Chromatography
and Coupled Gas Chromatography-Mass Spectrometry**
D. V. LYNCH and G. A. THOMPSON, Jr. (With 6 Figures)

1 Introduction 100
2 Lipid Preparation 101
  2.1 Lipid Extraction 101
  2.2 Purification of Phospholipids 102
    2.2.1 Column Chromatography 102
    2.2.2 Thin Layer Chromatography 103
3 Formation of Derivatives for GC or GC-MS 104
  3.1 Phospholipase C Treatment 104
  3.2 Conversion of Diacylglycerols to Silyl Derivatives 106
    3.2.1 Formation of Trimethylsilyl Derivatives 107
    3.2.2 Formation of tert-Butyldimethylsilyl Derivatives 107
4 Gas Chromatography 108
5 Mass Spectrometry 110
  5.1 Instrumentation 111
  5.2 Operating Conditions 111
  5.3 Identification of Molecular Species 111
  5.4 Quantitation of Molecular Species by GC-MS 113
  5.5 Quantitation of Molecular Species by GC-MS Following Reduction
      of Double Bonds Using Deuterium 115
  5.6 Direct MS Analysis of Underivatized Phospholipids 117
6 Determination of Positional Distribution of Acyl Chains
  Using Phospholipase A_2 117
7 Conclusion 118
References 119

Contents   XI

GC-MS of Plant Sterol Analysis
G. Combaut (With 2 Figures)

1 Introduction 122
2 Development of GC-MS Plant Sterol Analysis 123
3 Operations Before GC-MS Sterol Analysis 124
  3.1 Extraction and Isolation of Plant Sterols 124
  3.2 Free Sterols and (or) Sterols from Steryl-Esters 124
  3.3 Purification of Sterolic Fractions 125
  3.4 Derivatization 125
4 Characterization of Sterols 125
  4.1 Characterization of Sterols by GC Data 125
  4.2 Characterization of Sterols by MS Data 126
  4.3 Characterization of Sterols by GC and MS Data 127
    4.3.1 A Typical Analysis of 4-Demethyl and 4,4-Dimethyl Sterols
          from *Zea mays* 127
    4.3.2 Co-Occurrence of Δ^5- and Δ^7-Sterols in Tracheophytes . . . 128
    4.3.3 Side Chain-Hydroxylated Sterols from Red Algae 129
    4.3.4 4-Methyl Sterols of Dinoflagellates 130
5 Conclusion 130
References 131

GC-MS Methods for Terpenoids
L. Witte

1 Introduction 134
2 Isolation Methods 135
3 Prefractionation and Ancillary Reactions 136
4 Gas Chromatography 138
5 Retention Data 139
6 Mass Spectrometry 140
References 142

GC-MS of Auxins
L. Rivier (With 22 Figures)

1 Introduction 146
2 The Compounds Involved 147
3 Reference Compounds 148
4 Extraction 150
5 Purification 152
6 Columns for GC 153
7 Injection Techniques 155
8 Derivatisation 157
9 Interface Between GC and MS 161
10 Mass Spectrometer 161
11 Data Systems 163
12 Ionization 163

13 GC-MS Strategy for Auxin Analysis 165
14 Quantification 176
15 The Internal Standard 178
16 Experimental Procedure 182
17 Conclusions 185
References 185

GC-MS Methods for the Quantitative Determination and Structural Characterization of Esters of Indole-3-Acetic Acid and myo-Inositol
R. S. Bandurski and A. Ehmann (With 4 Figures)

1 Introduction 189
   1.1 Discovery of IAA-Inositols 189
   1.2 Occurrence of IAA Conjugates 190
   1.3 Importance of Measuring and Identifying Hormone Conjugates . . 191
2 Quantitative Analysis and Identification of the IAA-Inositols 191
   2.1 Analysis After Hydrolysis 191
      2.1.1 Methods for Hydrolysis of IAA Conjugates 192
      2.1.2 Use of Internal Standards 194
   2.2 Analysis Before Hydrolysis 195
      2.2.1 A Quantitative Estimation of IAA-Inositol Using [^{3}H]-IAA-
            myo-Inositol as an Internal Standard 195
3 Qualitative Analysis of IAA-Inositols 195
   3.1 The Inositol Moiety 195
   3.2 Derivitization of IAA-Inositols for GC-MS 196
   3.3 Mass Spectral Fragmentation Pattern 197
      3.3.1 1-DL-1-O-(Indole-3-Acetyl)-myo-Inositol (6 TMS, MW 769) . 201
      3.3.2 2-O-(Indole-3-Acetyl)-myo-Inositol (6 TMS MW 769) 201
      3.3.3 Di-O-[N-(Trimethylsilyl) Indole-3-Acetyl]-O-Tetra-O-
            Trimethylsilyl-myo-Inositol 201
      3.3.4 Tri-O-[N-(Trimethylsilyl) Indole-3-Acetyl]-O-Tri-O-
            Trimethylsilyl-myo-Inositol 206
      3.3.5 IAA-myo-Inositol-Arabinoside and IAA-myo-Inositol-
            Galactoside 206
   3.4 Uses of GC-MS to Identify and Characterize IAA-Esters 206
4 Conclusions 210
5 Abbreviations 211
References 211

GC-MS Methods for Cytokinins and Metabolites
L. M. S. Palni, S. A. B. Tay, and J. K. MacLeod (With 8 Figures)

1 Introduction 214
2 Gas Chromatography (GC) 220
   2.1 Instrumentation 220
      2.1.1 Liquid Stationary Phases and Columns 221
      2.1.2 Injectors 222
      2.1.3 Detectors 224

2.2 Derivatisation of Cytokinins 225
    2.2.1 Trimethylsilyl (TMSi) Derivatives 225
    2.2.2 Permethyl Derivatives 227
    2.2.3 tert.-Butyldimethylsilyl (t-BuDMSi) Derivatives 228
    2.2.4 Trifluoroacetyl (TFA) Derivatives 229
2.3 Preparative GC 230
3 Mass Spectrometry 230
    3.1 Instrumentation 230
        3.1.1 Sample Introduction 230
        3.1.2 Ionisation Methods 231
        3.1.3 Analysers 231
        3.1.4 Data Systems 233
    3.2 Combined Gas Chromatography-Mass Spectrometry (GC-MS) . . 233
4 Applications of Mass Spectrometry in Cytokinin Analysis 234
    4.1 Structural Studies 234
    4.2 Quantification of Cytokinins 236
        4.2.1 Internal Standards 237
        4.2.2 Stable Isotope Dilution Mass Spectrometry 237
        4.2.3 Quantification Using GC-MS 240
        4.2.4 Probe Analysis 241
    4.3 Metabolic Profiling 244
5 General Remarks and Conclusion 245
References 245

GC-MS Method for Volatile Flavor Components of Foods
H. KAMEOKA (With 8 Figures)

1 Introduction 254
2 GC-MS Methods 254
    2.1 Preparation Methods of Flavor Samples 255
    2.2 Operational Methods 256
3 Volatile Flavor Components 256
    3.1 Fruits 256
    3.2 Vegetables 263
    3.3 Mushrooms 268
    3.4 Tea 270
    3.5 Beans and Nuts 271
    3.6 Grains 271
    3.7 Jams 272
    3.8 Fermentation Products 273
References 274

GC-MS Methods for Tobacco Constituents
H. KODAMA (With 20 Figures)

1 Introduction 277
2 Cembranoids and Their Degraded Compounds 277

3 Labdanoids and Their Degraded Compounds 279
4 Carotenoid-Degraded Compounds 280
5 Sesquiterpenoids 289
6 Terpenoid Glycosides 291
7 Linked Scanning 294
References 298

Subject Index 299

List of Contributors

BANDURSKI, ROBERT S., Michigan State University, Botany & Plant Pathology Department, East Lansing, MI 48824-1312, USA

COMBAUT, GEORGES, Laboratoire de Biologie Végétale, Université, Avenue de Villeneuve, F-66025 Perpignan Cedex, France

EHMANN, AXEL, Shell Agricultural Chemical Company, P.O. Box 4248, Modesto, CA 95352, USA

HEDDEN, PETER, Long Ashton Research Station, Department of Agricultural Sciences, University of Bristol, Long Ashton, Bristol, BS18 9AF, United Kingdom

JANISTYN, BORIS, Gresserstraße 8, D-7800 Freiburg, FRG

KAMEOKA, HIROMU, Department of Applied Chemistry, Faculty of Science and Engineering, Kinki University, 3-4-1 Kowakae, Higashiosaka-shi, Osaka 577, Japan

KODAMA, HISASHI, Central Research Institute, Japan Tobacco Inc., 6-2 Umegaoka, Midoriku, Yokohama, Kanagawa 227, Japan

LYNCH, DANIEL V., Agronomy Department, Cornell University, Ithaca, NY 14853, USA

MacLEOD, JOHN KEITH, Research School of Chemistry, Australian National University, G.P.O. Box 1, Canberra, A.C.T. 2601, Australia

NYBERG, HARRI, Department of Botany, University of Helsinki, Unioninkatu 44, SF-00170 Helsinki, Finland

PALNI, LOK MAN SINGH, Department of Developmental Biology, Research School of Biological Sciences, The Australian National University, P.O. Box 475, Canberra City, A.C.T. 2601, Australia

RIVIER, LAURENT, Institut universitaire de Médecine légale, Laboratoire de Toxicologie analytique, CH-1005 Lausanne, Switzerland

SELVENDRAN, R.R., Chemistry and Biochemistry Division, AFRC Institute of Food Research, Norwich Laboratory, Colney Lane, Norwich NR4 7UA, United Kingdom

STEVENS, B.J.H., Chemistry and Biochemistry Division, AFRC Institute of Food Research, Norwich Laboratory, Colney Lane, Norwich, NR4 7UA, United Kingdom

Tay, Stephen Ah Boon, Regional Laboratory, Australian Government
    Analytical Laboratories, P.O. Box 385, Pymble, NSW 2073, Australia

Thompson, Jr., Guy Allen, Department of Botany, University of Texas,
    Austin, TX 78713, USA

Witte, Ludger, Gesellschaft für Biotechnologische Forschung,
    Mascheroder Weg 1, D-3300 Braunschweig, FRG

The Use of Combined Gas Chromatography-Mass Spectrometry in the Analysis of Plant Growth Substances

P. HEDDEN

1 Introduction

The extreme sensitivity and selectivity of combined gas chromatography-mass spectrometry (GC-MS) is of particular advantage for the analysis of plant growth substances, which may be present in plant tissues at ppb concentrations or below. The use of this technique has now been extended to all major classes of plant growth substance except ethylene. However, in the space available it is necessary to be highly selective. Some general principles of GC-MS analysis will be discussed and illustrated by examples from three classes of growth substance, the gibberellins (GA's), the abscisins and the brassinosteroids. Auxin and cytokinin analysis is discussed by Rivier and Palni et al. in this volume.

Analysis by GC-MS is possible only after an often long and laborious purification sequence. The extent and nature of the purification steps will depend on the type of plant tissue to be extracted, the concentration of the substance of interest in the tissue and the nature of the major contaminants. There are no universal recipes, but some examples of extraction procedures from different types of tissue will be discussed separately for each class of growth substance. Most attention, however, will be given to the GC-MS analytical procedure itself, which will be discussed in terms of both qualitative and quantitative analysis.

2 Identification by GC-MS

An identification by GC-MS may be possible from <1 ng of a compound provided its mass spectrum is relatively free from extraneous ions. The identification requires comparisons with authentic compounds, both on the basis of their mass spectra and GC retention times. This latter property can be of considerable value, especially when wall-coated open tubular (WCOT) capillary columns of very high resolving power are used. Although the resolution on such columns is not sufficient to base an identification solely on retention time, samples and standards having different retention times are clearly not identical compounds.

The value of mass spectra lies in their high information content (Reeve and Crozier 1980). It is often possible to compare two mass spectra at several hundred different points. Thus most compounds can be distinguished on the basis of their mass spectra, and good matches give a high probability of identity. The mass spectra of some compounds, however, are indistinguishable. For example, certain epimers such as GA_{48} and GA_{49} or dihydrophaseic acid and epidihydrophaseic acid can be distinguished only by their different GC retention times. Thus combining capillary GC with MS produces a powerful analytical technique.

If standards or reference mass spectra are unavailable, identification by GC-MS is not possible. Although mass spectra can give considerable structural information, full chemical characterisation by a combination of synthesis and spectroscopic methods is necessary to identify a previously uncharacterised compound. If there is insufficient material for a full characterisation, but a structure can be inferred from the mass spectrum, it may be possible to synthesise the putative structure and compare the synthetic and natural compounds by GC-MS. This method has been employed in the identification of several GA's (see for example Kirkwood and MacMillan 1982). Since a large number of GA's are now known and many of their characteristic MS fragmentations are documented, the experienced worker can often infer the structure of a new GA from its mass spectrum.

2.1 Derivatisation

The volatility of the acidic plant growth substances must be increased prior to GC, usually by converting them to methyl (Me) esters. Hydroxylated compounds may be further converted to trimethylsilyl (TMS) ethers, both to enhance GC resolution and to give mass spectra with more intense molecular ions.

2.1.1 Methylation

Carboxylic acids are converted to Me esters by dissolving the sample in a small volume of methanol and adding dropwise a solution of diazomethane (Schlenk and Gellerman 1960) in diethylether to excess, i. e., until the solution remains yellow and N_2 evolution ceases. CARE! DIAZOMETHANE IS HIGHLY TOXIC AND POTENTIALLY EXPLOSIVE. IT SHOULD BE HANDLED ONLY IN A FUME CUPBOARD. After 5–10 min at room temperature excess diazomethane is removed by evaporating the sample to dryness in a stream of N_2.

2.1.2 Trimethylsilylation

The sample is transferred to a glass ampoule (made conveniently from a Pasteur pipette) and dried thoroughly under N_2 or in a desiccator over P_2O_5. Then excess (usually 5–50 µl) BSTFA (N,O-bistrimethylsilyltrifluoroacetamide) or MSTFA (N-methyl-O-trimethylsilyltrifluoroacetamide) is added, the ampoule sealed and the sample heated to 90 °C for 30 min. The silylating reagent can be used directly as solvent for gas chromatography or can be removed under reduced pressure and replaced with an alternative solvent, which must be aprotic and thoroughly dry to avoid hydrolysis of the TMS ethers.

MSTFA and BSTFA, which are relatively bulky molecules, often do not fully trimethylsilylate vicinal diols or sterically hindered hydroxy groups. Inclusion of a small proportion (5%) of trimethylchlorosilane (TMCS) in the reagent or the use of Sweeley reagent (hexamethyldisilazane-TMCS-pyridine; 3:1:9) (Sweeley et al. 1963) is preferable in this case. Ketones, such as abscisic acid (ABA) or GA_{26} and GA_{33}, may be converted to TMS enol ethers. This reaction is usually incomplete and results in a mixture of derivatives.

2.1.3 Permethylation

This is a potentially useful alternative to trimethylsilylation that has been applied to GA's and GA-glucosides (Rivier et al. 1981). It produces methyl ether esters, which are more stable and of lower molecular weight than the corresponding TMS derivatives. The procedure itself, however, is more involved. The sample compounds are first converted to methyl esters with diazomethane as described above. The dried sample is dissolved in dimethylformamide (0.5 ml) and methyl iodide (0.5 ml), both freshly distilled, and treated with NaH (5 mg) in a reaction vial that is flushed with dry N_2 until H_2 evolution ceases. The vial is closed, shaken for 15 min, and then left for 2 h. Excess reagent is destroyed by adding methanol (0.5 ml) and the solvents evaporated to dryness under N_2 at 50 °C. Water (1.0 ml) is then added and the permethylated products extracted with ethyl acetate (3×1 ml).

2.2 Gas Chromatography

Most modern GC-MS instruments utilise wall-coated open tubular (WCOT) capillary GC columns with efficiencies of around 100,000 effective theoretical plates or higher. Fused silica flexible columns are being employed increasingly because of their ease of handling. These columns can be connected directly to the MS ion source rather than via a transfer line, thereby eliminating a potential source of adsorption losses. The high peak capacity of WCOT columns increases the value of retention time as a parameter in identification. Ideally retention times should be compared by running the sample and standards sequentially on the same column and then co-injecting them. However, if a standard is not available, relative retention indices, or Kovats Indices, are useful parameters for comparison.

For a linear temperature program the Kovats Index can be obtained from a linear plot of straight-chain hydrocarbon carbon number against retention time (Van den Dool and Kratz 1963). The retention index (I) is related to the retention time by:

$$I = 100\,i\,\frac{R_x - R_n}{R_{n+i} - R_n} + 100\,n\,,$$

where R_x is the retention time of the sample, and R_n and R_{n+i} are the retention times of hydrocarbons with n and n+i carbon numbers respectively. In order to obtain a retention index, the sample is co-injected with hydrocarbons, of which Parafilm is a convenient source for the range C_{23} to C_{31} (Gaskin et al. 1971). If necessary the Parafilm can be supplemented with lower and/or higher molecular weight hydrocarbons. Columns with liquid phases of comparable polarity will give similar retention indices, but compounds with only slightly differing retention times will need to be co-injected with authentic standards to distinguish them.

2.3 Qualitative Mass Spectrometry

Electron impact (EI) positive ion mass spectra are probably the most useful for identifying previously characterised compounds. A certain degree of fragmentation is necessary to produce a spectrum with sufficient information content to be useful for identification. Less energetic ionisation processes such as chemical ionisation (CI) are valuable for determining the molecular weight of unknown compounds that give no molecular ion by EI mass spectrometry.

The use of dedicated computers in combination with GC-MS instruments has revolutionised the procedure as an analytical technique. Data systems have become especially valuable with the advent of capillary GC columns, which require rapid scanning rates to take full advantage of their high resolving power. Total cycle times of 1 s or less are now routine. The vast amount of data collected in a typical GC run of 30 min or longer can be processed practically only by computer. Figures 1 and 2 illustrate the use of such data systems in detecting and identifying minor components such as growth substances in a complex plant extract. It is often impractical to examine every GC peak in such an extract, especially when the compounds of interest are very minor components. In such circumstances it is possible to search for these compounds using mass chromatography, also known as mass fragmentography or cross-scanning. The computer is instructed to search every scan in a run for ions characteristic of potential components. Scans containing these ions are then displayed in full and compared with standard spectra. It is common practice to subtract a background scan containing extraneous ions due to column bleed etc. from the scan of interest in order to produce an enhanced spectrum that is easier to interpret. Subtraction of one scan from another may also assist in interpreting mixed spectra resulting from components that are incompletely separated. If the components are slightly separated, it should be possible to obtain a clean spectrum of each. However, this technique should be used with caution as the resulting subtracted spectra are sometimes atypical of the compound.

The matching of sample and reference spectra often tends to be subjective, especially if the sample spectrum is weak and contaminated. Reeve and Crozier (1980) suggested a means of quantifying the correlation between spectra based on their information content. This suggestion has, however, rarely been followed. In practice, spectra are compared on the basis of relatively few characteristic ions, usually in the higher mass region where ions are common to fewer compounds.

GC-MS data systems usually obtain spectrum libraries which may be searched by the computer for close matches with sample spectra. Although the mass spectra of most growth substances are not included in the available libraries, users may construct their own library if they have a large number of authentic standards. In practice, spectra need to be fairly intense and free from extraneous ions for a library search to be successful.

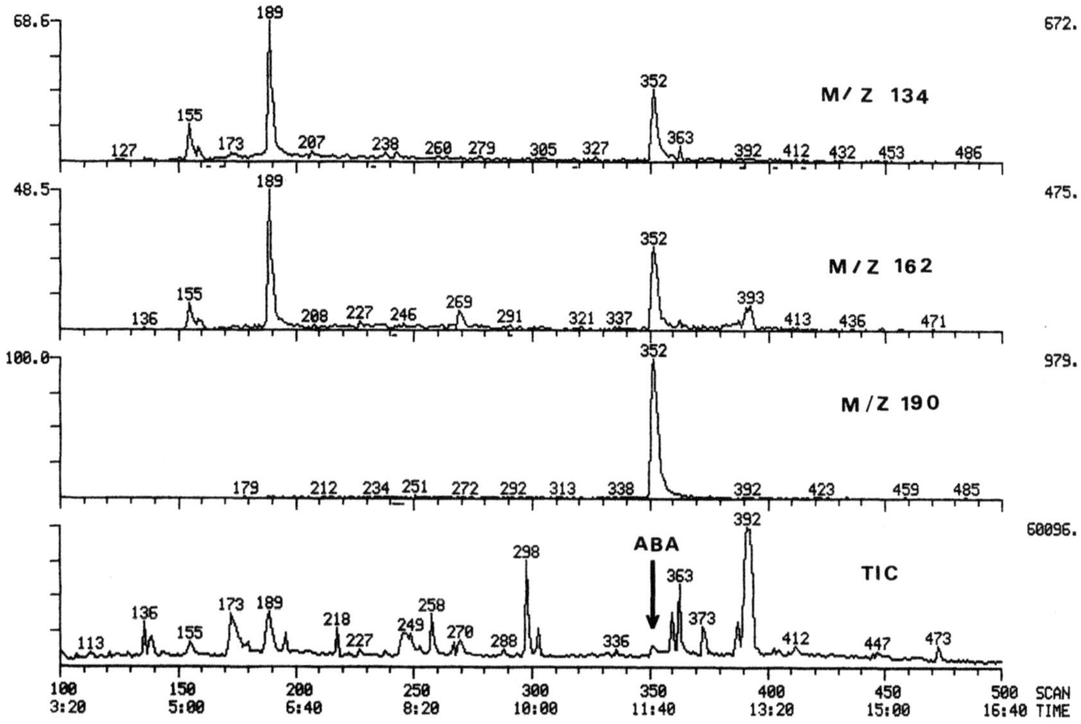

Fig. 1. TIC trace and mass chromatograms from a purified and methylated extract of rye shoots (Taylor and Hedden, unpublished results). GC-MS was carried out using a Finnigan 4000 system. The sample (1 μl in methanol) was injected onto a BP-1 quartz silica WCOT capillary column (25 m × 0.2 mm) at 55 °C with no split. After 0.5 min a 50:1 split was opened and after 1 min the temperature was programmed to 140 °C at 15 °C per minute and then at 7.5 °C per min to 250 °C. The carrier gas (helium) inlet pressure was 1.2 mbar. Spectra were acquired from 140 °C at 2 s per cycle using an electron energy of 40 eV. The mass chromatograms were chosen for ions in the mass spectrum of ABA (1 in Fig. 7) methyl ester

2.4 Quantitative Mass Spectrometry

Quantitative MS is based on isotope dilution of a heavy-isotope labelled internal standard. The standard, which ideally is a labelled analogue of the compound being analysed, is added to the plant extract immediately after homogenisation, so that purification losses are accounted for. The relative amounts of endogenous compound and internal standard are then determined by measuring the intensities of equivalent ions in the mass spectra of the two compounds.

The ion intensities in quantitative GC-MS are usually determined using selective ion monitoring (SIM). Since only a limited number of ions are monitored, this technique is more sensitive than full scanning, although it produces less information. The traces obtained from SIM are equivalent to mass chromatograms, which can also be used for quantitation, albeit with lower sensitivity. When SIM

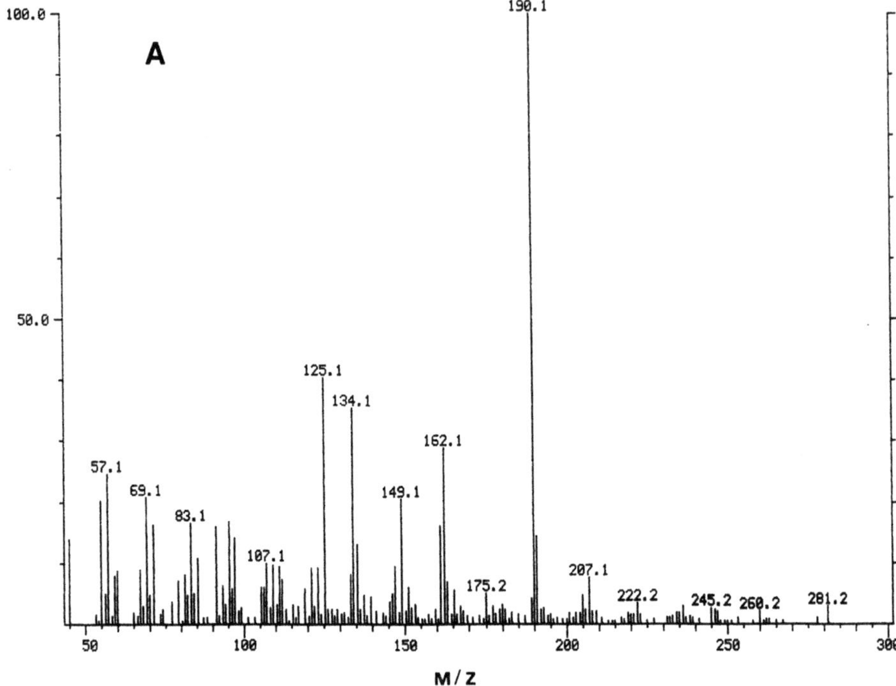

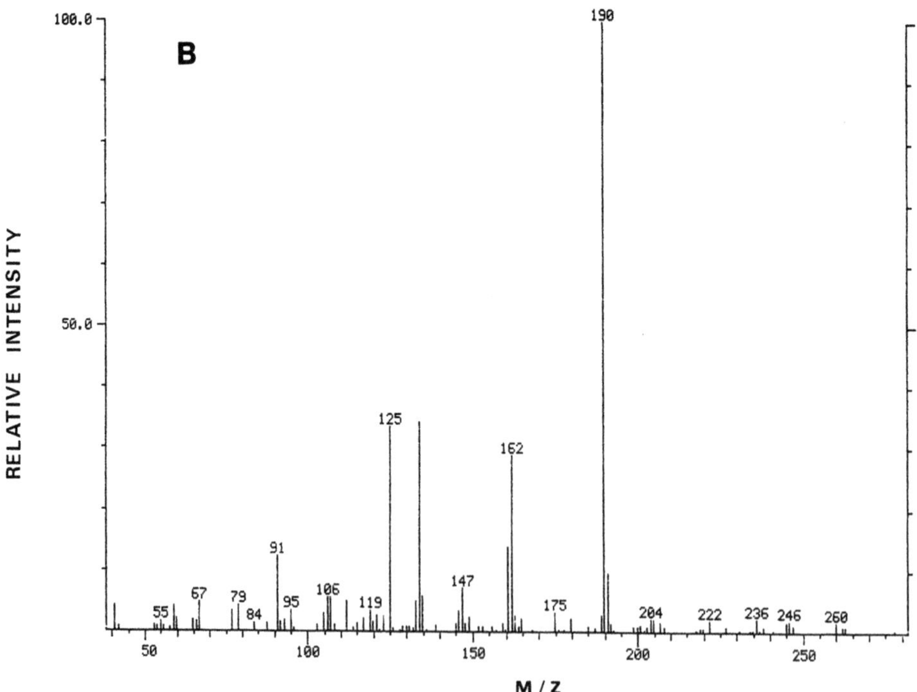

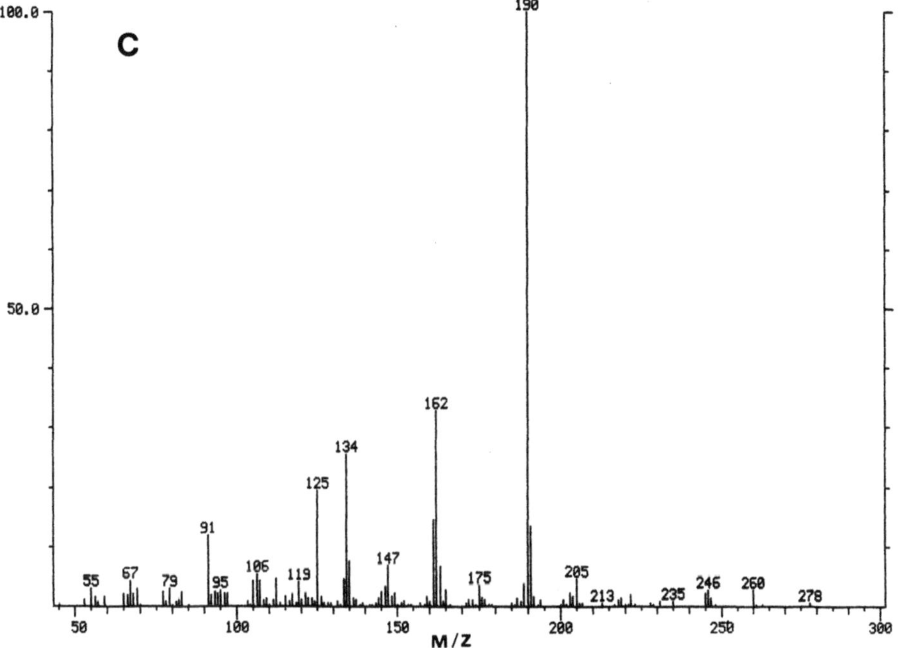

Fig. 2 A–C. A The mass spectrum from scan 352 in Fig. 1. **B** The background subtracted spectrum (scan 352 – scan 354). **C** An authentic spectrum of ABA Me ester. GC-MS conditions are as described for Fig. 1

is used it is advisable to monitor several pairs of ions to ascertain that the correct compound is being monitored.

The technique is illustrated by the quantitative analysis of GA_4 in immature apple embryos (Fig. 3). The internal standard was prepared from GA_4 by labelling with two deuterium atoms on C-17 using the Wittig reaction. A calibration curve was constructed by mixing different proportions of deuterated and non-deuterated GA_4. The mixtures as MeTMS derivative were subjected to GC-MS in SIM mode, and the peak areas for the ions at m/z 284 (GA_4 MeTMS) and 286 ($[^2H]GA_4$ MeTMS) were measured from the SIM traces. These fragment ions were chosen since the molecular ion for GA_4 MeTMS is relatively weak. The molar ratios were plotted against the peak area ratios corrected for the contribution from GA_4 MeTMS to the intensity of the ion at m/z 286 due mainly to the ^{30}Si isotope and estimated at 5%. The uncorrected curve is indicated in Fig. 3 by the dotted line. A straight line was drawn through the corrected curve by the method of least squares.

An 80% aqueous methanol extract of embryos (1.5 g) dissected from apple (cv. Sunset) seeds 80 days after anthesis was spiked with $[^2H_2]GA_4$ (1.5 µg) from the solution used for the calibration. After purification by solvent partition and high performance liquid chromatography (HPLC), the fraction containing GA_4

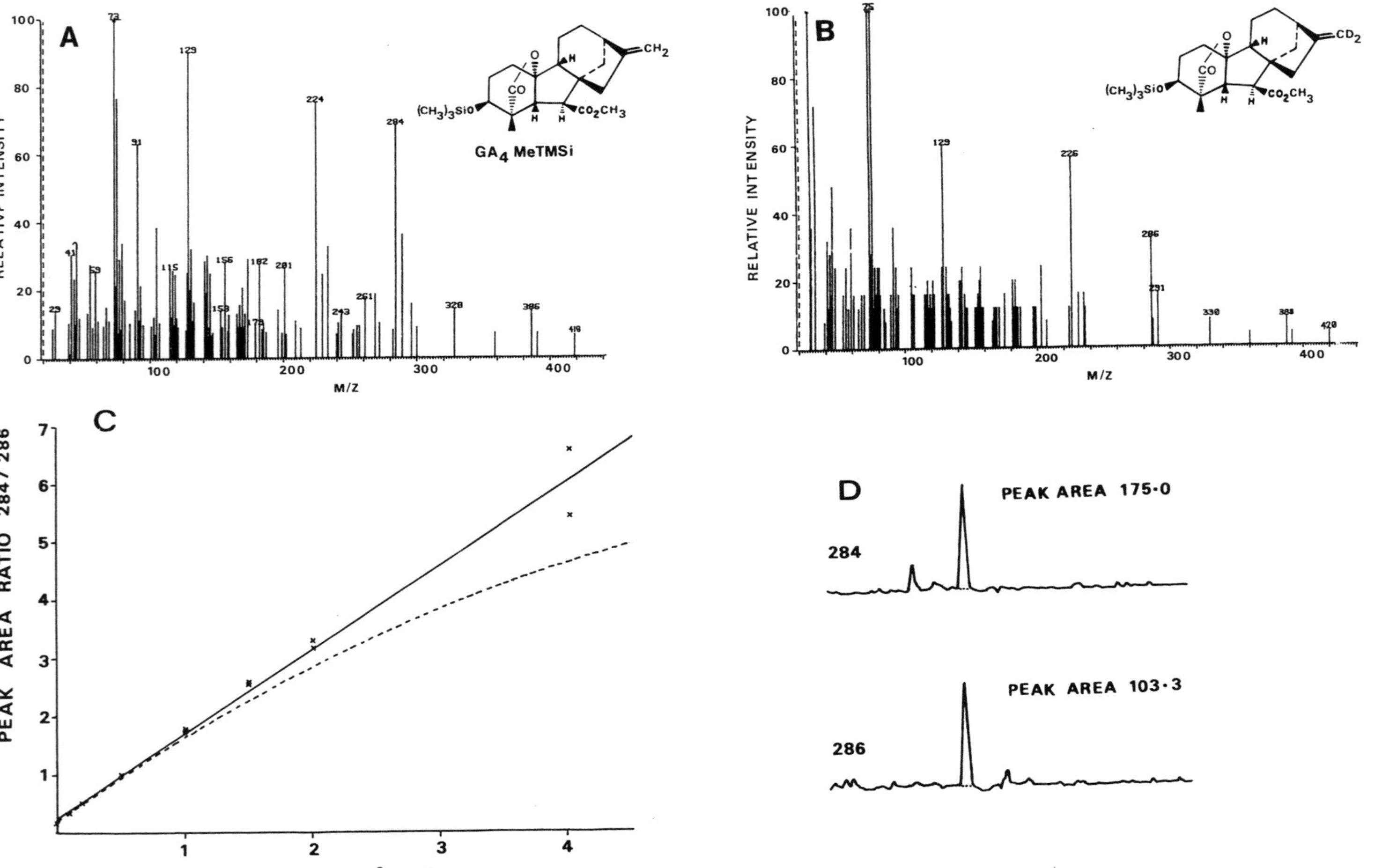

Fig. 3 A–D. Quantitation of GA$_4$ in immature apple embryos. **A** Mass spectrum of GA$_4$ Me ester TMS ether. **B** Mass spectrum of [^{2}H$_2$]GA$_4$ used as internal standard. **C** Calibration curve (the corrected curve is depicted by the *continuous line* and the uncorrected curve by the *dotted line*. **D** SIM traces for a purified and derivatised extract of apple embryos. GC-MS was carried out using a VG 1212 instrument

was derivatised and analysed by GC-MS SIM. The corrected peak area ratio m/z 284/286 was 1.85, which by reference to the calibration curve corresponds to a molar ratio of 1.1. Therefore the extract contained 1.65 µg GA_4, corresponding to a tissue concentration of 1.1 µg GA_4 g^{-1} fresh weight.

It is obviously important that the internal standard is chemically very similar to the analyte. If it is not possible to obtain a labelled analogue of the analyte, a related compound may be used, in which case ions of similar mass should be monitored. It is essential to ensure that the analyte and internal standard do not separate during purification. If a labelled internal standard is used, the label must be stable during extraction and purification. The possible isotopes for use with plant growth substances are ^2H, ^{13}C, ^{14}C, ^{15}N, ^{17}O, and ^{18}O. The radioactivity of ^{14}C can also serve as a convenient means of detection after purification. Combinations of ^2H and ^3H have been used for the same purpose (Sponsel and MacMillan 1978).

If a labelled analogue of the analyte is used as internal standard, it is desirable that the ions monitored are sufficiently different in mass to prevent cross-interference by isotope-containing ions. This requires that at least two atoms of ^2H or ^{13}C are incorporated into the standard when TMS ethers are analysed, since the silicon isotopes, ^{29}Si and ^{30}Si, make a substantial contribution (natural abundance 4.7% and 3.1% respectively). If interference is unavoidable it can be corrected for empirically as indicated above, but for high accuracy the amounts of internal standard and analyte in the sample should not differ greatly.

Deuterated compounds may separate slightly from their protonated equivalents on GC and HPLC, the degree of separation depending on the extent and position of substitution. Isotope dilutions are better determined from SIM or mass chromatogram profiles than from ion intensities in individual mass spectra.

3 Gibberellins

There are over 70 fully characterised GA's. They can be divided into five groups on the basis of structure as illustrated in Fig. 4. The C_{19}-GA's such as GA_1 are monocarboxylic acids and include the physiologically active compounds. The four groups of C_{20}-GA's differ in the extent of oxidation at C-20 and include mono-, di-, and tricarboxylic acids. The biosynthetic relationship of the GA's in Fig. 4 is indicated by the arrows. The GA's also differ in the extent of their hydroxylation, giving a wide range of polarities. It is therefore difficult to devise a single purification method. If it is necessary to analyse for all the GA's that may be present in an extract several different fractions may have to be examined by GC-MS.

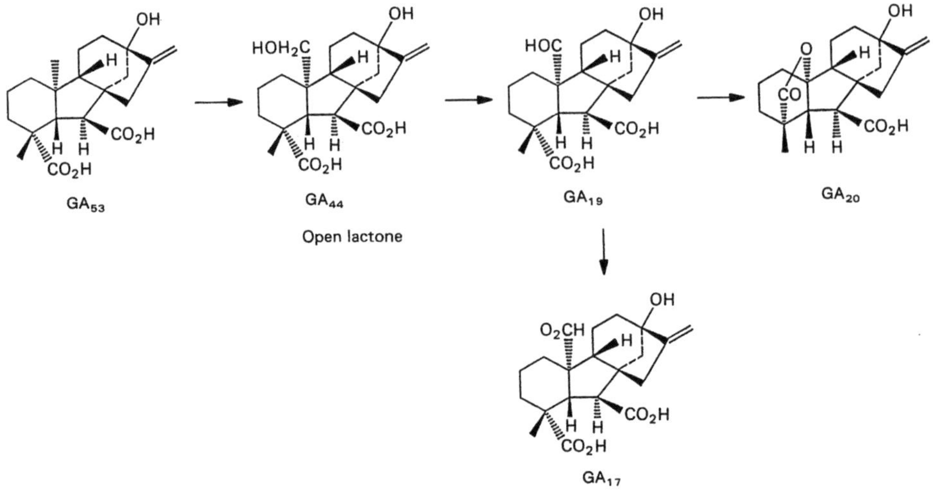

GA₅₃ → GA₄₄ (Open lactone) → GA₁₉ → GA₂₀ → GA₁₇

Fig. 4. Examples of the five basic GA structural types and their biosynthetic relationship

3.1 Extraction and Purification

The method based on aqueous methanol extraction followed by solvent partition has changed little in recent years and has been described in detail elsewhere (Yokota et al. 1980; Hedden 1985). A basic procedure is outlined in Fig. 5. Several group separation techniques have been described for the GA's, including columns of polyvinylpyrrolidone (PVP) (Glenn et al. 1972; Sandberg et al. 1981) and charcoal-celite (Durley et al. 1971), anion exchange chromatography on DEAE Sephadex G-25 (Crozier and Durley 1983) or Dowex 1X1-100 (Browning and Saunders 1977), and reversed-phase cartridges (Hedden 1985).

An alternative to solvent partition as a purification method has been described for extracts from small quantities of plant tissues (<1 g fresh weight) (Koshioka et al. 1983b). The method entails passing the aqueous methanol extract through two C_{18} reversed-phase Bondapak columns, the first to remove highly lipophilic material and the second to separate most GA's and GA-glucosyl conjugates (eluted in 50% aqueous methanol) from less polar material including GA biosynthetic precursors (eluted with methanol). Separation of GA's from GA conjugates was accomplished on a SiO_2 gel partition column. This column can be eluted stepwise or by a continuous gradient of formic acid-saturated hexane-ethyl acetate to remove the free GA's as a group or to accomplish some separation. GA-glucosyl conjugates were eluted with methanol.

Preparative separation of GA's and GA conjugates prior to GC-MS analysis is achieved most commonly by reversed-phase HPLC. Retention volumes for several GA's and GA-conjugates on Bondapak C_{18} analytical columns have been published (Jones et al. 1980; Koshioka et al. 1983a). Separation of GA's with a broad range of polarities requires a solvent gradient, most commonly of water-methanol. Acetic acid at low concentration (50 μl l^{-1}) is added to the solvents to suppress ionisation.

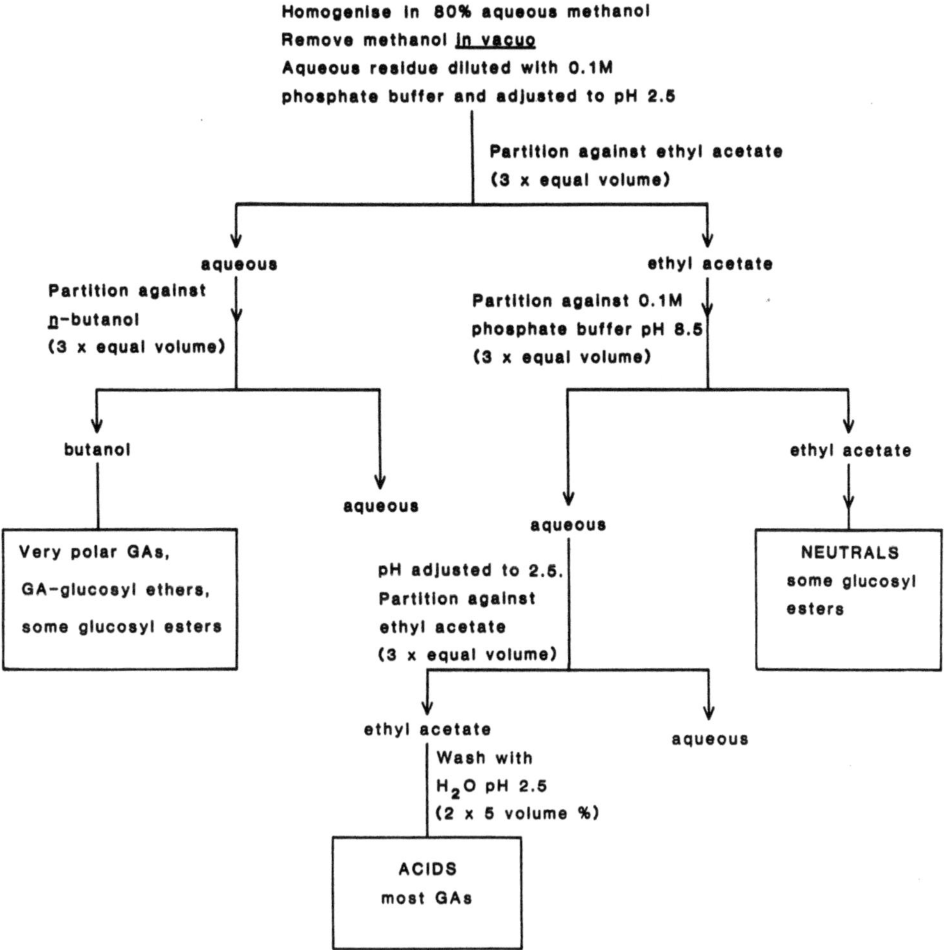

Fig. 5. Scheme for the extraction and initial purification of GA's and GA-conjugates by solvent-solvent partition

3.2 GC-MS

Since GA's contain no chemical features that allow specific detection by most physicochemical methods, apart from the recently introduced immunoassays, GC-MS is the only reliable method available for detecting and quantifying these compounds. They are generally derivatised to methyl esters TMS ethers (see Sect. 2.1) although they can also be gas chromatographed as Me esters or TMS ether esters. The EI mass spectra of GA's 1–24 as Me and MeTMS derivatives were published nearly 20 years ago (Binks et al. 1969), but there is no equivalent compendium for GA's discovered since that time. Abbreviated mass spectral data for most of the characterised GA's have been published recently (Crozier and Durley 1983; Hedden 1985). Significant ions in the EI MS of the more commonly

Table 1. Ten significant ions in the mass spectra of the more common C_{19}-GA's as MeTMS derivatives (see Fig. 6 for their structures). The spectra are normalised on the most intense ion other than those at m/z 73 or 75

GA$_1$ M^+ 506 (100%), 491 (13), 448 (22), 377 (15), 313 (17), 208 (15), 207 (37), 193 (24), 131 (20), 129 (21).

GA$_3$ M^+ 504 (100), 489 (11), 477 (4), 445 (7), 388 (7), 370 (11), 347 (12), 208 (43), 207 (20), 193 (26).

GA$_4$ M^+ 418 (27), 400 (9), 386 (20), 358 (11), 328 (32), 289 (73), 284 (100), 225 (84), 224 (75), 129 (73).

GA$_7$ M^+ 416 (18), 401 (4), 384 (49), 356 (67), 312 (27), 298 (44), 282 (27), 281 (40), 223 (91), 222 (100).

GA$_8$ M^+ 594 (100), 579 (9), 535 (9), 504 (4), 448 (4), 379 (22), 238 (29), 208 (18), 207 (40), 147 (60).

GA$_9$ M^+ 330 (6), 298 (100), 286 (14), 270 (78), 243 (43), 227 (48), 226 (47), 217 (27), 184 (27), 148 (36).

GA$_{20}$ M^+ 418 (100), 403 (16), 390 (5), 375 (45), 359 (14), 301 (16), 235 (9), 208 (16), 207 (38), 193 (14).

GA$_{29}$ M^+ 506 (100), 491 (10), 447 (8), 389 (8), 375 (15), 303 (15), 291 (6), 235 (8), 208 (12), 207 (32).

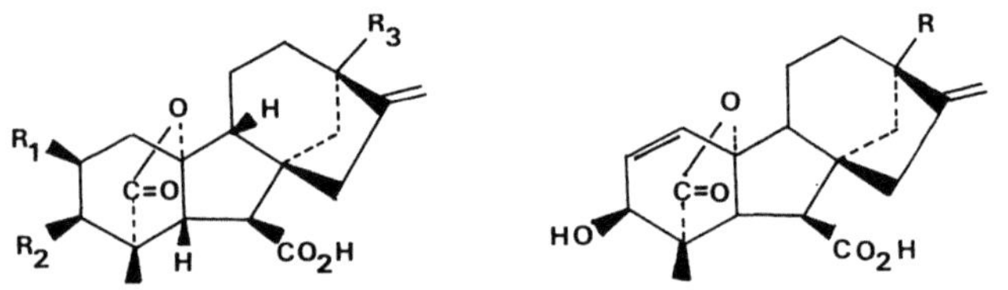

	R$_1$	R$_2$	R$_3$			R
GA$_1$	H	OH	OH		GA$_3$	OH
GA$_4$	H	OH	H		GA$_7$	H
GA$_8$	OH	OH	OH			
GA$_9$	H	H	H			
GA$_{20}$	H	H	OH			
GA$_{29}$	OH	H	OH			

Fig. 6. Structures of commonly occurring C_{19}-GA's for which mass spectral data are listed in Table 1

occurring GA's are listed in Table 1. The structures of these GA's are given in Fig. 6.

Despite the large number of known GA structures, there are doubtless many new GA's still to be characterised. If a new structure is discovered, an identification is not possible by MS alone. However, there are characteristic fragmentations that indicate particular structural features. Some of these are listed in Table 2. If a possible structure can be inferred from the mass spectrum, the putative compound can be produced by partial synthesis or biochemical conversion.

Table 2. Characteristic ions or losses in the MS of GA MeTMS derivatives

Fragment ion or loss m/z	Structural feature
-103	Primary alcohol e.g. GA_{22}
-116	1,3 Dihydroxy grouping on ring A e.g. GA_{16}, GA_{54}
129	Single hydroxy group at C-1 or C-3 e.g. GA_1, GA_4
130	16-Hydroxy group e.g. GA_2, GA_{10}
147	Vicinal diol, e.g. GA_8, GA_{34}
156	Single hydroxy group on rings C and D at C-15, e.g. GA_{45}, GA_{63}
207/8	Single hydroxy group on rings C and D at C-13, e.g. GA_1, GA_{20}

Gibberellins may occur as conjugates, the most important of which are the glucosyl ethers or esters. Their extraction and analysis has been described in some detail (Schneider 1983). Although the highly involatile conjugates can be gas chromatographed as TMS derivatives (Schneider et al. 1975), the high temperatures required often lead to decomposition. The use of permethylated derivatives with much lower molecular weights than the TMS compounds would seem a more promising technique (Rivier et al. 1981). The EI mass spectra of both TMS and permethylated derivatives contain very weak molecular ions and fragmentation patterns dominated by the glucose moiety (Yokota et al. 1975; Rivier et al. 1981). However, negative ion EI MS at low electron energy gives more intense molecular ions and may prove useful for GA-conjugates (Schneider 1983).

Due to the difficulty of analyzing GA conjugates directly by GC-MS, conjugates are usually first hydrolysed enzymatically and the aglycone analysed. This method gives no information about the nature of the glycosidic linkage. Furthermore, since the efficiency of enzyme hydrolysis varies considerably and is dependent on the position of glucose attachment (Schneider and Schliemann 1979), quantitative measurements are not possible unless the particular GA conjugate is available in labelled form and its spectrum compared with that of the natural product.

Two examples of the use of GC-MS for GA identification are given below.

1. Identification of GA_9 in Norway Spruce Seedlings (Oden et al. 1982). An acidic ethyl acetate fraction was produced from a methanol extract of 100 g tissue by dissolving the extract in 0.5 M sodium phosphate at pH 8.0, washing with petroleum ether, acidifying to pH 2.7 and extracting with ethyl acetate. After group purification on a PVP column eluted with 0.1 M sodium phosphate at pH 8, initial separation was achieved by normal phase HPLC (stationary phase – 0.5 M formic acid on silica support, mobile phase – gradient of 0.5 M formic acid saturated n-hexane-ethyl acetate). Bioassay of the fractions revealed two regions of biological activity, the less polar of which was chromatographed a second time by reversed-phase HPLC on 10 µm Nucleosil C_{18} eluted isocratically with methanol-water-acetic acid (50.0 : 49.5 : 0.5).

An active component in the second separation was located by bioassay and shown to have the same retention volume as $[^3H]GA_9$. The active fractions were pooled, methylated with diazomethane, dissolved in dichloromethane and analysed by GC-MS using a quadrupole mass spectrometer. Gas chromatography was carried out on an 25 m × 0.25 mm OV-101 quartz capillary column using Grob splitless injection at 70 °C and then increasing to 250 °C at 25 °C per minute. The mass spectra were obtained at an electron energy of 20 eV. A compound eluting after 13 min gave a mass spectrum that matched very closely that of GA_9 Me ester, which also co-chromatographed with the sample compound. The concentration of the compound in the tissue was estimated at ca. 1 ng g^{-1} fresh weight.

2. Identification of GA's in Pumpkin Endosperm (Blechschmidt et al. 1984). Pumpkin (*Cucurbita maxima*) endosperm was homogenised and adjusted to pH 3. After adding an equal volume of acetone to half the homogenate, acids were extracted with ethyl acetate, the ethyl acetate phase washed with water and reduced to dryness. The extract was dissolved in phoshate buffer at pH 8, washed with petroleum ether, and again adjusted to pH 3 and partitioned against acetone-ethyl acetate to give an acidic ethyl acetate fraction. The other half of the homogenate was adjusted to pH 4.5 and treated with cellulase from *Aspergillus niger* at 37 °C for 16 h to hydrolyse GA-conjugates. The GA-aglycones were extracted into ethyl acetate as before to give a fraction containing GA's that were originally both free and conjugated.

Both fractions were examined by GC-MS as Me (diazomethane) TMS(MSTFA) derivatives. The samples were co-injected with a mixture of C_{22} to C_{32} alkanes in order to determine Kovats Retention Indices. Several GA's, including GA_{13}, GA_{25}, GA_{43}, and GA_{49}, were identified in both fractions by comparison of mass spectra and Kovats Indices with those of authentic samples. In addition a number of unfamiliar mass spectra were obtained which, it was suspected, were due to a series of previously unknown 12-hydroxylated GA's. This was confirmed by preparing several 12α- and 12β-hydroxylated GA's from incubations of ent-12β- and ent-12α-hydroxykaurenoic acid with cultures of the fungus *Gibberella fujikuroi* (Gaskin et al. 1984). From comparisons of mass spectra and Kovats Indices the following GA's were identified in the extracts:

12α-hydroxyGA_4, -GA_{12}, -GA_{14}, -GA_{25}, and -GA_{37}.

12α-HydroxyGA_4 was isolated and its structure confirmed by NMR spectroscopy. Consequently, it was given the new designation, GA_{58}. The 12-hydroxylated GA's were found predominantly in the second extract and are thus present in the endosperm mainly as conjugates.

4 Abscisic Acid and Related Compounds

Abscisic acid (ABA) (**1** in Fig. 7) and its metabolites, 6-hydroxy ABA (**2**), phaseic acid (PA) (**3**), dihydrophaseic acid (DPA) (**4**) and epidihydrophaseic acid (epiDPA) (**5**) can be extracted and purifed essentially as outlined for GA's

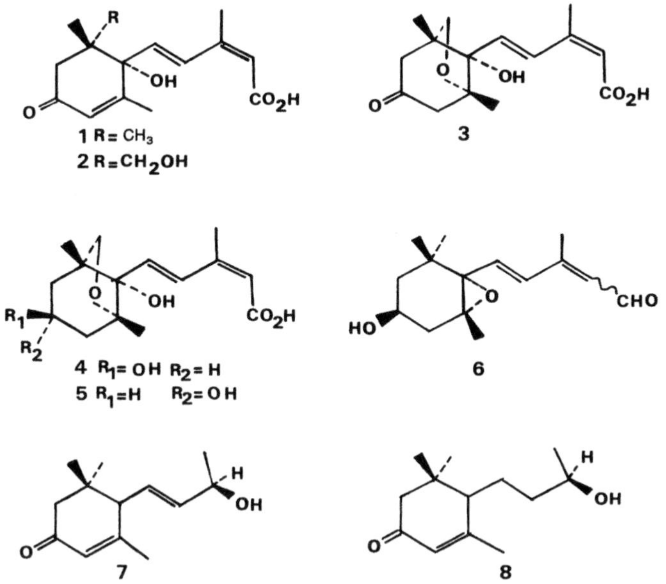

Fig. 7. Structures of ABA and related compounds

(Fig. 5). Indeed, purification schemes have been devised so that all the acidic growth substances can be analysed in a single plant extract (Rademacher and Graebe 1984). However, for extraction of ABA and its glucosyl ester, aqueous acetone is preferred to aqueous methanol, since under basic conditions in methanol the glucosyl ester is readily transmethylated to ABA methyl ester (Milborrow and Mallaby 1975). For ABA, but not its more polar metabolites, ether may be substituted for ethyl acetate in (Fig. 5) to give a cleaner final acidic extract.

The major glucose conjugates of ABA are ABA glucosyl ester (ABAGE) (Koshimizu et al. 1968) and ABA 1'-O-glucoside (ABAGS) (Loveys and Milborrow 1981). Other polar metabolites are DPA 3'-O-glucoside (Milborrow and Vaughan 1982), -hydroxy and -methylglutarylhydroxy ABA (Hirai et al. 1978). These compounds are not analysed as such by GC-MS, but first converted to the unconjugated metabolite by base hydrolysis. ABA glucosyl ester is hydrolysed at pH 11 and 65 °C in 3 min (Niell et al. 1983), but the glucosyl ether requires more vigorous conditions (Loveys and Milborrow 1981).

As with the GA's, reversed-phase HPLC is now the preferred method of purification for ABA and its metabolites (Vaughan and Milborrow 1984). In contrast to the GA's, ABA is easily detected by UV absorption ($\lambda_{max}260$ nm, $\varepsilon = 21,400$). It can also be detected with high sensitivity and reasonable selectivity by electron capture after GC separation. However, MS is the most reliable method and if possible the results obtained by other methods should be validated by GC-MS.

4.1 Qualitative GC-MS

ABA and metabolites are usually analysed as methyl esters. They do not trimethylsilylate easily because the tertiary hydroxy group is sterically hindered. In addition, the ketone function incompletely trimethylsilylates as the enol, so that a mixture of products results. The significant ions in the EI and CIMS of ABA and its metabolites as Me esters are listes in Table 3. In the EIMS the molecular ions are small. However, in CIMS there is considerably less fragmentation, and ABA and 2-trans-ABA give significantly different spectra (Netting et al. 1982). These isomers give quite similar EI mass spectra.

Several neutral ABA-related compounds have biological activity similar to ABA in some bioassays. These include xanthoxin (**6** in Fig. 7) (Taylor and Burden 1982), vomifoliol (**7**) and 3,4-dihydrovomifoliol (**8**) (Bercht et al, 1976). Significant ions in the EIMS of the underivatised compounds are listed in Table 4. Extraction and purification procedures for each of these compounds have been published and will not be discussed here in detail. In the procedure outlined in Fig. 5

Table 3. MS data for ABA and metabolites as methyl esters. (Numbers in heavy type refer to structures in Fig. 7)

ABA (**1**)
EI M^+ 278 (1%), 260 (4), 246 (3), 222 (3), 190 (100), 162 (39), 147 (12), 134 (37), 125 (40), 91 (38).
CI (isobutane)[a] $[MH]^+$ 279 (100), 261 (88), 247 (13)
CI (ammonia)[a] $[MNH4]^+$ 296 (100), 278 (2), 279 (8), 261 (8)
CI (methane)[b] $[MH]^+$ 279 (14), 261 (100), 247 (43), 229 (12), 219 (9)

2-trans-ABA
EI M^+ 278 (1), 260 (5), 246 (3), 222 (9), 190 (100), 162 (30), 147 (10), 134 (36), 125 (15), 91 (25)
CI (methane)[b] $[MH]^+$ 279 (100), 261 (14), 247 (13), 229 (4), 223 (12)

PA (**3**)
EI M^+ 294 (15), 276 (15), 217 (11), 177 (25), 167 (34), 139 (38), 135 (46), 125 (100), 122 (83), 94 (55)
CI (isobutane)[a] $[MH]^+$ 295 (28), 277 (100), 263 (12)
CI (ammonia)[a] $[MNH4]^+$ 312 (100), 294 (8), 295 (6), 277 (47)

DPA (**4**)
EI[c] M^+ 296 (7), 278 (18), 246 (11), 220 (17), 188 (23), 154 (53), 125 (49), 122 (100), 109 (68), 94 (64)
CI (isobutane)[a] $[MH]^+$ 297 (26), 279 (100), 265 (6), 261 (4), 247 (10)
CI (ammonia)[a] $[MNH4]^+$ 314 (100), 296 (34), 297 (16), 279 (34)

epi-DPA (**5**)
EI[d] M^+ 296 (6), 278 (9), 246 (8), 220 (9), 188 (18), 154 (61), 125 (66), 122 (100), 109 (64), 94 (61)
CI (isobutane)[a] $[MH]^+$ 297 (28), 279 (100), 265 (4), 261 (5), 247 (5)
CI (ammonia)[a] $[MNH4)^+$ 314 (100), 296 (22), 297 (18), 279 (55)

Source of data: [a] Takeda et al. (1984).
                [b] Netting et al. (1982).
                [c] Tietz et al. (1979).
                [d] Zeevaart and Milborrow (1976)
                Otherwise mass spectra were obtained on a Kratos MS80 RFA instrument at 60 eV.

they would occur in the neutral ethyl acetate fraction. Xanthoxin is extracted as a mixture of the cis,trans and trans,trans isomers. Prior to gas chromatography it has usually been esterified to the acetate with acetic anhydride in pyridine (Firn et al. 1972) or to the perfluorobutyrate with perfluorobutyric anhydride in dry benzene (Böttger 1978). However, capillary gas chromatography of the free alcohols is possible and they are suitable for GC-MS in this form.

4.2 Quantitative GC-MS

There are many published examples of the use of deuterated ABA as internal standard for isotope dilution GC-SIM measurements. Rivier et al. (1977) prepared $[^2H_6]ABA$ by base-catalysed exchange in deuterated water, and, using EIMS, monitored the base peak at m/z 190 for ABA Me and at m/z 194 for the internal standard. The disadvantage of this standard is that the label is exchangable and cannot be used at pH values above 8.

Other workers have used $[^2H_3]ABA$, which has a base stable label on the C-3 methyl group (Netting et al. 1982; Niell et al. 1983). Niell et al. (1983) converted this standard to the glucosyl ester for estimating ABA-glucose ester levels after base hydrolysis and GC-SIM on the liberated hormone. Netting et al. (1982) used CI GC-SIM with methane as reagent gas to estimate ABA and 2-trans-ABA in *Eucalyptus haemastoma* leaves. Their method is outlined here.

The leaves were spiked with $[3\text{-Me-}^2H_3]ABA$ and $2\text{-trans-}[3\text{-Me-}^2H_3]ABA$ and extracted in 1% acetic acid in acetone containing 10 mg l^{-1} 2,6-di-t-butyl-4-methylphenol as an antioxidant. The extract was evaporated to the aqueous phase and partitioned against diethyl ether. The ether phases were extracted with saturated sodium bicarbonate solution, which was then adjusted to pH 3 and partitioned against ether to give an acid fraction. The aqueous phase from the first partition was spiked a second time with the internal standards and saponified with 95% ethanol-60% KOH (2:1) to hydrolyse ABA glucosyl conjugates. After hydrolysis and acidification with sulphuric acid, the free ABA was then extracted into ether. The acid extracts were purified initially on C_{18} SepPak cartridges and then on a reversed-phase C_{18} HPLC column eluted with 95% ethanol-0.2% acetic

Table 4. EIMS data for some neutral ABA-related growth inhibitors

---

Xanthoxin (**6**; Fig. 7)
cis, trans- M^+250 (2), 232 (4), 221 (2), 199 (10), 168 (10), 149 (21), 121 (17), 107 (19), 95 (41), 43 (100)
trans, trans- M^+250 (2), 232 (2), 221 (2), 199 (1), 168 (10), 149 (19), 121 (15), 107 (18), 95 (29), 43 (100)
(+)-Vomifoliol (**7**)[a] M^+224 (0.5), 206 (4), 168 (9), 151 (4), 150 (6), 135 (5), 124 (100), 122 (7), 111 (7), 107 (3)
Dihydrovomifoliol (**8**)[a] M^+226 (2), 193 (3), 183 (4), 171 (72), 170 (73), 153 (60), 152 (77), 125 (25), 111 (64), 110 (100)

---

[a] Data taken from Bercht et al. (1976)

acid (2:3), which separated ABA and 2-trans-ABA. The hydrolyzed conjugate fraction required prior separation on a C_8 HPLC column.

The purified ABA and 2-trans ABA were examined as methyl esters by GC-CIMS using methane as GC carrier and reagent gas. Quantitation was carried out by SIM of the base peaks for the two analytes and the corresponding deuterated standards, i.e. m/z 261 and 264 for ABA and [$^2H_3$]ABA, and m/z 279 and 282 for the 2-trans isomers. Peak height ratios for each pair of ions were related to the relative concentrations of analytes and internal standards by reference to calibration curves.

5 Brassinosteroids

Brassinolide (**9** in Fig. 8), first isolated from pollen of *Brassica napus* (Grove et al. 1979), stimulates cell elongation and division, mimicking the effect of auxin (Yopp et al. 1981), GA's or cytokinins (Mandava et al. 1981) in several bioassays. At least nine brassinolide-related steroids, collectively known as brassinosteroids, have been isolated from several plant sources. They can be divided into two groups on the basis of structure. The first includes brassinolide itself and its members possess a seven-membered B-ring lactone. Members of the second group contain the ergostan-6-one skeleton as in castasterone (**10**), which is presumed to be biosynthetic precursor of brassinolide (Abe et al. 1984). Some examples are given in Fig. 8.

The brassinosteroids are purified from the neutral fraction after extraction of the plant material with aqueous acetone or methanol. Since they are present in plant tissues at extremely low concentrations extensive purification is necessary

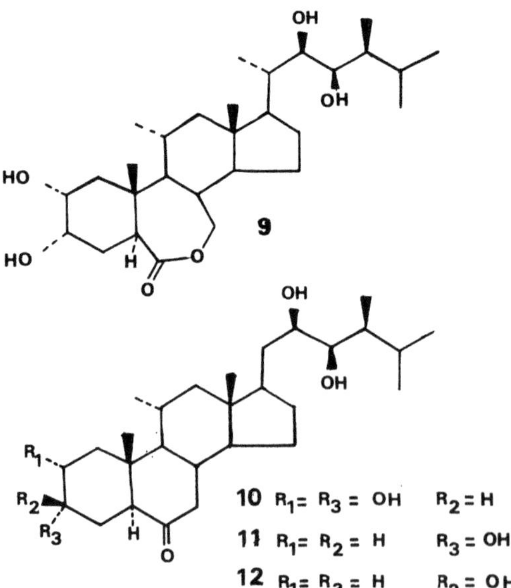

Fig. 8. Structures of some brassinosteroids

Table 5. MS data for brassinosteroids. (Bold numbers refer to structures in Fig. 8)

Brassinolide (**9**)
EI (bismethaneboronate)[a] M$^+$528 (2%), 457 (4), 374 (21), 345 (10), 332 (12), 177 (52), 155 (100)
CI (isobutane) (bismethaneboronate)[a] [MH]$^+$529 (100), 511 (11), 469 (17), 373 (5), 345 (11), 155 (9)

Castasterone (**10**)
EI (bismethaneboronate)[b] [M$^+$] 512 (66), 441 (13), 399 (11), 358 (40), 357 (30), 329 (16), 302 (17), 287 (36), 282 (9), 155 (100)

Typhasterol (**11**)
EI (methaneboronate TMS)[c] M$^+$544 (100), 529 (68), 526 (44), 525 (95), 454 (86), 155 (54)

Teasterone (**12**)
EI (methaneboronate TMS)[c] M$^+$544 (36), 529 (84), 515 (100), 454 (10), 155 (20)

Source of data: [a] Takatsuto et al. (1982).
[b] Suzuki et al. (1985).
[c] Abe et al. (1984).

before GC-MS can be attempted. For example Suzuki et al. (1985) estimated the concentration of castasterone and brassinone in immature seeds of *Pharbitis purpurea* as 1.1 ng g^{-1} and 0.2 ng g^{-1} fresh weight respectively. They extracted 27 kg plant material and used sequentially solvent partition, countercurrent distribution, two silica gel adsorption columns and hplc on gel permeation, normal phase and reversed-phase columns before GC-MS analysis. The rice lamina inclination bioassay (Wada et al. 1981) was used to detect active components at each purification step.

A GC-MS analytical technique has been developed by Takatsuto et al. (1982) using the methaneboronate derivative. Brassinolide contains two pairs of vicinal hydroxy groups and thus forms a bismethaneboronate. The sample is dissolved in dry pyridine containing methaneboronic acid (100 µg acid per 50 µl pyridine) and heated at 60 °C for 30 min. Brassinolides containing only a single hydroxy group on ring A are converted to their methaneboronate TMS ethers by adding trimethylsilylimidazole (30 µl) to the above reagents after the boronation reaction is complete and allowing the mixture to stand at room temperature for 30 min. The solutions can be used directly for GC-MS analysis. Table 5 lists significant ions in the EI and CI (isobutane) MS of several brassinolides as methaneboronate TMS derivatives. Recently CI GC-SIM of the [MH]$^+$ ion has been used to detect brassinosteroids in plant extracts (see for example Morishita et al. 1983). Such evidence based only on retention time and a single ion must be treated with extreme caution.

References

Abe H, Morishita T, Uchiyama M, Takatsuto S, Ikekawa N (1984) A new brassinolide-related steroid in the leaves of *Thea sinensis*. Agric Biol Chem 48:2171–2172
Bercht CA, Samrak HM, Lousberg RJJCh, Theuns H, Salemink CA (1976) Isolation of vomifoliol and dihydrovomifoliol from *Cannabis*. Phytochemistry 15:830–831

Binks R, MacMillan J, Pryce RJ (1969) Plant hormones VII. Combined gas chromatography-mass spectrometry of the methyl esters of gibberellins A_1 to A_{24} and their trimethylsilyl ethers. Phytochemistry 8:271–284

Blechschmidt S, Castel U, Gaskin P, Hedden P, Graebe JE, MacMillan J (1984) GC/MS analysis of the plant hormones in seeds of *Cucurbita maxima*. Phytochemistry 23:553–558

Böttger M (1978) The occurrence of cis, trans-, and trans,trans-xanthoxin in pea roots. Z Pflanzenphysiol 86:265–268

Browning G, Saunders PF (1977) Membrane localised gibberellins A_9 and A_4 in wheat chloroplasts. Nature 265:375–377

Crozier A, Durley RC (1983) Modern methods of analysis of gibberellins. In: Crozier A (ed) The biochemistry and physiology of gibberellins, vol I. Praeger, New York, pp 485–560

Durley RC, MacMillan J, Pryce RJ (1971) Investigation of gibberellins and other growth substances in seed of *Phaseolus multifloris* and *Phaseolus vulgaris* by gas chromatography and combined gas chromatography-mass spectrometry. Phytochemistry 11:317–326

Firn RD, Burden RS, Taylor HF (1972) The detection and estimation of the growth inhibitor xanthoxin in plants. Planta 102:115–126

Gaskin P, MacMillan J, Firn RD, Pryce RJ (1971) "Parafilm": A convenient source of n-alkane standards for the determination of gas chromatographic retention indices. Phytochemistry 10:1155–1157

Gaskin P, Hutchison M, Lewis N, MacMillan J, Phinney BO (1984) Microbiological conversion of 12-oxygenated and other derivatives of *ent*-kaur-16-en-19-oic acid by *Gibberella fujikuroi*, mutant B1-41a. Phytochemistry 23:559–564

Glenn JL, Kuo CC, Durley RC, Pharis RP (1972) Use of insoluble polyvinylpyrrolidone for purification of plant extracts and chromatography of plant hormones. Phytochemistry 11:345–351

Grove MD, Spencer GF, Rohwedder WK, Mandava N, Worley JF, Warthen JD Jr, Steffens GL, Flippen-Anderson JL, Cook JC Jr (1979) Brassinolide, a plant growth-promoting steroid isolated from *Brassica napus* pollen. Nature 281:216–217

Hedden P (1985) Gibberellins. In: Crozier A (ed) Methods in plant growth substance analysis (in press)

Hirai N, Fukui H, Koshimizu K (1978) A novel abscisic acid metabolite from seeds of *Robinia pseudacacia*. Phytochemistry 17:1625–1627

Jones MG, Metzger JD, Zeevaart JAD (1980) Fractionation of gibberellins in plant extracts by reverse-phase high performance liquid chromatography. Plant Physiol 65:218–221

Kirkwood PS, MacMillan J (1982) Gibberellins A_{60}, A_{61}, and A_{62}: Partial synthesis and natural occurrence. J Chem Soc Perkin Trans I:689–697

Koshimizu K, Inui M, Fukui H, Mitsui T (1968) Isolation of $(+)$-abscisyl-β-D-glucopyranoside from immature fruits of *Lupinus luteus*. Agric Biol Chem 32:788–791

Koshioka M, Harada J, Takeno K, Noma M, Sassa T, Ogiyama K, Taylor JS, Rood SB, Legge RL, Pharis RP (1983 a) Reverse-phase C18 high performance liquid chromatography of acidic and conjugated gibberellins. J Chromatogr 256:101–115

Koshioka M, Takeno K, Beall FD, Pharis RP (1983 b) Purification and separation of plant gibberellins from their precursors and glucosyl conjugates. Plant Physiol 73:398–406

Loveys BR, Milborrow BV (1981) Isolation and characterization of 1'-O-abscisic acid-β-D-glucopyranoside from vegetative tomato tissue. Aust J Plant Physiol 8:571–589

Mandava NB, Sasse JM, Yopp JH (1981) Brassinolide, a growth-promoting steroidal lactone. II. Activity in selected gibberellin and cytokinin bioassays. Physiol Plant 53:453–461

Milborrow BV, Mallaby R (1975) Occurrence of methyl-$(+)$-abscisiate as an artefact of extraction. J Exp Bot 26:741–748

Milborrow BV, Vaughan GT (1982) Characterization of dihydrophaseic acid 4'-O-β-D-glucopyranoside as a major metabolite of abscisic acid. Aust J Plant Physiol 9:361–372

Morishita T, Abe H, Uchiyama M, Marumo S, Takatsuto S, Ikekawa N (1983) Evidence for plant growth promoting brassinolides in leaves of *Thea sinensis*. Phytochemistry 22:1051–1053

Netting AG, Milborrow BV, Duffield AM (1982) Determination of abscisic acid in *Eucalyptus haemastoma* leaves using gas chromatography/mass spectrometry and deuterated internal standards. Phytochemistry 21:385–389

Niell SJ, Horgan R, Heald JK (1983) Determination of the levels of abscisic acid-glucose ester in plants. Planta 157:371–375

Oden P-C, Andersson B, Gref R (1982) Identification of gibberellin A_9 in extracts of Norway spruce [*Picea abies* (L.) Karst.] by combined gas chromatography-mass spectrometry. J Chromatogr 247:133–140

Rademacher W, Graebe JE (1984) Isolation and analysis by gas-liquid chromatography of auxins, gibberellins, cytokinins and abscisic acid from a single sample of plant material. Ber Dtsch Bot Ges 97:75–85

Reeve DR, Crozier A (1980) Quantitative analysis of plant hormones. In: MacMillan J (ed) Hormonal regulation of development I. Molecular aspects of plant hormones. Encyclopedia of plant physiology. New series, vol 9. Springer, Berlin Heidelberg New York, pp 203–280

Rivier L, Milon H, Pilet P-E (1977) Gas chromatography-mass spectrometric determinations of abscisic acid levels in the cap and the apex of maize roots. Planta 134:23–27

Rivier L, Gaskin P, Albone KS, MacMillan J (1981) GC-MS identification of endogenous gibberellins and gibberellin conjugates as their permethylated derivatives. Phytochemistry 20:687–692

Sandberg G, Dunberg A, Oden PE (1981) Chromatography of acid phytohormones on columns of Sephadex LH-20 and insoluble poly-N-vinylpyrrolidone, and application to the analysis of conifer extracts. Physiol Plant 53:219–224

Schlenk H, Gellerman JL (1960) Esterification of fatty acids with diazomethane on a small scale. Anal Chem 32:1433–1440

Schneider G (1983) Gibberellin conjugates. In: Crozier A (ed) The biochemistry and physiology of gibberellins, vol I. Praeger, New York, pp 389–456

Schneider G, Schliemann W (1979) Untersuchungen zur enzymatischen Hydrolyse von Gibberellin-O-glucosiden. II. Hydrolysegeschwindigkeit von Gibberellin-2-O- und Gibberellin-3-O-glucosiden. Biochem Physiol Pflanz 174:746–751

Schneider G, Janicke S, Sembdner G (1975) Gibberelline XXXIV Mitt. Beitrag zur Gaschromatographie von Gibberellinen und Gibberellin-O-glucosiden -N,O-Bis(trimethylsilyl)acetamid als Silylierungsreagens. J Chromatogr 109:409–412

Sponsel VM, MacMillan J (1978) Metabolism of gibberellin A_{29} in seeds of *Pisum sativum* cv. Progress No. 9; use of [^{2}H] and [^{3}H]GAs, and the identification of a new GA catabolite. Planta 144:69–78

Suzuki Y, Yamaguchi I, Takahashi N (1985) Identification of castasterone and brassinone from immature seeds of Pharbitis purpurea. Agric Biol Chem 49:49–54

Sweeley CC, Bentley R, Makita M, Wells WW (1963) Gas-liquid chromatography of trimethylsilyl derivatives of sugars and related substances. J Am Chem Soc 85:2497–2507

Takatsuto S, Ying B, Morisaki M, Ikekawa N (1982) Microanalysis of brassinolide and its analogues by gas chromatography and gas chromatography-mass spectrometry. J Chromatogr 239:233–241

Takeda N, Harada K, Suzuki M, Tatematsu A, Hirai N, Koshimizu K (1984) Structural characterization of abscisic acid and related metabolites by chemical ionization mass spectrometry. Agric Biol Chem 48:685–694

Taylor HF, Burden RS (1972) Xanthoxin, a recently discovered plant growth inhibitor. Proc R Soc Lond B 180:317–346

Tietz D, Dorffling K, Wohrle D, Erxleben I, Liemann F (1979) Identification by combined gas chromatography-mass spectrometry of phaseic acid and dihydrophaseic acid and characterization of further abscisic acid metabolites in pea seedlings. Planta 147:168–173

Van den Dool H, Kratz PD (1963) A generalization of the retention index system including linear temperature programmed gas-liquid chromatography. J Chromatogr 11:463–471

Vaughan GT, Milborrow BV (1984) The resolution by HPLC of RS-[2-^{14}C]Me 1',4'-*cis*-diol of abscisic acid and the metabolism of (−)-*R*- and (+)-*S*-abscisic acid. J Exp Bot 35:111–120

Wada K, Marumo S, Ikekawa N, Morisaki M, Mori K (1981) Brassinolide and homobrassinolide promotion of lamina inclination of rice seedlings. Plant Cell Physiol 22:323–325

Yokota T, Hiraga K, Yamane H, Takahashi N (1975) Mass spectrometry of trimethylsilyl derivatives of gibberellin glucosides and glycosyl esters. Phytochemistry 14:1569–1574

Yokota T, Murofushi N, Takahashi N (1980) Extraction, purification and identification. In: MacMillan J (ed) Hormonal regulation of plant development. I. Molecular aspects of plant hormones. Encyclopedia of plant physiology. New series, vol 9. Springer, Berlin Heidelberg New York, pp 113–201

Yopp JH, Mandava NB, Sasse JM (1981) Brassinolde, a growth-promoting steroidal lactone. I. Activity in selected auxin bioassays. Physiol Plant 53:445–452

Zeevaart JAD, Milborrow BV (1976) Metabolism of abscisic acid and the occurrence of *epi*-dihydrophaseic acid in *Phaseolus vulgaris*. Phytochemistry 15:493–500

Applications of Mass Spectrometry
for the Examination of Pectic Polysaccharides

R. R. SELVENDRAN and B. J. H. STEVENS

1 Introduction

Mass spectrometry has proved to be an important technique for the study of carbohydrate derivatives from a range of tissues, and its potential applications for
the elucidation of the structural features of polysaccharides have been discussed
in a number of reviews (Lindberg 1972; Lönngren and Svensson 1974; McNeil et
al. 1982a). As the basic concepts of interpretation of mass spectra from carbohydrate derivatives have been adequately reviewed (Kochetkov and Chizhov 1966;
Hanessian 1971; Lönngren and Svensson 1974; Kováčik et al. 1978), no attempt
will be made to interpret the spectra in detail. However, attention will be drawn
to some of the distinctive and characteristic features associated with the fragmentation pathways used for the identification of derivatives from pectic polysaccharides.

The main purpose of this article is to show the utility of mass spectrometry
(MS) for the examination of pectic polysaccharides, some of which have un-
usually complex structural features (Aspinall 1980; Stephen 1983; McNeill et al.
1984). It is imperative that the pectic polysaccharides undergo minimal degradation during chemical extraction and subsequent purification, and some of the recent developments in these areas have been reviewed (Selvendran et al. 1985;
Selvendran 1985). Alternatively fragments of pectic polysaccharides could be
solubilised from cell walls by using highly purified pectin degrading enzymes (Talmadge et al. 1973; Darvill et al. 1978; McNeill et al. 1980). For the purpose of
this article it is assumed that the pectic polysaccharide preparations are of sufficient purity, i.e., they are free of non-covalently associated non-carbohydrate
components, and are reasonably homogeneous.

2 Structural Analysis of Pectic Polysaccharides

Investigations on the structure of a polysaccharide can provide information on
(a) the nature and proportions of glycosyl residues, (b) the absolute configuration, D or L, of each glycosyl residue, (c) the nature of the glycosidic linkages, i.e.,
the substitution patterns of the glycosyl residues, (d) the ring form, pyranose or
furanose, of each glycosyl residue, (e) the anomeric configuration of the glycosidic linkages, (f) the sequence of glycosyl residues along the chain, and (g) the
determination of the points of attachment of non-carbohydrate substituents. The
main emphasis of this article will be on points c, d, and f, but some pertinent com-

ments will be made on some of the other points, particularly point a, because an accurate estimate of the glycosyl residue composition is a prerequisite for detailed structural studies.

Although the carbohydrate residues of neutral pectic polysaccharides can be hydrolysed quantitatively with dil. H_2SO_4 or dil. trifluoroacetic acid, the quantitative release of neutral sugars, particularly rhamnose, from pectins has posed problems (Albersheim et al. 1967; Selvendran et al. 1979; Selvendran and DuPont 1984). This is partly due to the unusual stability of the aldobiouronic acid 2-O-(α-D-galactopyranosyluronic acid)-L-rhamnose, i.e., Gal$p$A-(1→2)-Rha$p$, and partly due to the fact that the pectic material tends to precipitate in acidic medium, and can only be dissolved after protracted heating. Advantage is, however, taken of the stability of the glycosiduronic linkage in pectins to isolate higher acidic oligossaccharides (Aspinall et al. 1967, 1968 a, b).

The stability of the glycosiduronic linkage also prevents the quantitative estimation of Gal$p$A in pectins colorimetrically (Selvendran et al. 1979; Selvendran and DuPont 1984). An alternative method for water-soluble pectic material is to convert the galacturonosyl residues to the corresponding galactosyl residues by reduction of the carbodiimide-activated carboxyl groups with $NaBH_4$ or $NaBD_4$ (–COOH→–CH_2OH or –CD_2OH, respectively) (Taylor and Conrad 1972). This procedure replaces the acid-resistant glycosiduronic bonds by the more acid-labile glycosyl bonds, and the neutral polysaccharide can then be analysed, after hydrolysis, by formation of the alditol acetates (Darvill et al. 1978; McNeil et al. 1980). By performing the reduction with $NaBD_4$ (preferably in D_2O) a dideutero label is introduced on C-6 facilitating the distinction between galactosyl residues derived from Gal$p$A from those native to the pectin, by gas chromatography-mass spectrometry (GC-MS). This method has been successfully applied to soluble pectins (Aspinall and Jiang 1974), but in our experience requires relatively large amounts of material (~ 40 mg), and the reduction has to be done twice or thrice to ensure about 90% conversion of galacturonosyl) residues to galactosyl residues.

In most studies on the structure of pectic substances the absolute configuration of the sugar residues is assumed – rhamnose and arabinose are assigned the L-configuration, whereas galacturonic acid, galactose, xylose, mannose, glucose, and glucuronic acid are assigned the D-configuration. However, for complete characterisation, it is preferable to determine the configuration of the sugar residues experimentally. Recently GC methods have been developed to determine the enantiomeric nature of sugar residues (Gerwig et al. 1978; Leontein et al. 1978), and these methods have been used to determine the configuration of the sugars from rhamnogalacturonan-1 (RG-1), isolated from the cell walls of suspension-cultured sycamore cells (McNeil et al. 1980).

3 Determination of the Nature of the Glycosidic Linkages

Methylation analysis is by far the most widely used procedure for linkage analysis in carbohydrate polymers. The method is based on the complete etherification of free sugar hydroxyl groups in a polysaccharide. The fully methylated polysaccharide is then hydrolysed to cleave the glycosidic linkages to release partially methylated sugars, which are then reduced to yield partially methylated alditols; the reduction is usually performed with $NaBD_4$ – the merits of this procedure will be discussed later. The newly formed hydroxyl groups are usually acetylated to give partially methylated alditol acetates (PMAA), which are then analysed by GC and identified by GC-MS. The mass spectra of PMAA are normally relatively easy to interpret, with fragmentation patterns characteristic of structure, and especially of substitution pattern (Björndal et al. 1970; Lindberg 1972; Lönngren and Svensson 1974). The method is illustrated by the results expected from a segment of a hypothetical pectic polysaccharide, after reduction of the carboxyl group of Gal$p$A to $-CD_2OH$ by the carbodiimide method, using $NaBD_4$ as the reducing agent (Fig. 1; the major ions used for the identification of the PMAA are given in Table 1). This procedure can provide information on the following: (a) the nature of the glycosidic linkages, (b) the relative proportions of non-reducing end groups and degree of branching, (c) distinction between 4-linked Gal$p$A and 4-linked Gal$p$, and (d) the ring size of the monomers (pyranose or furanose). However, no distinction can be made between 4-linked and 5-linked hexose residues. In Fig. 1 the Gal$p$ residues are assumed to be 4-linked.

For most methylations, the method of choice is that of Hakomori (1964), as described by Sandford and Conrad (1966). As we have used the above method

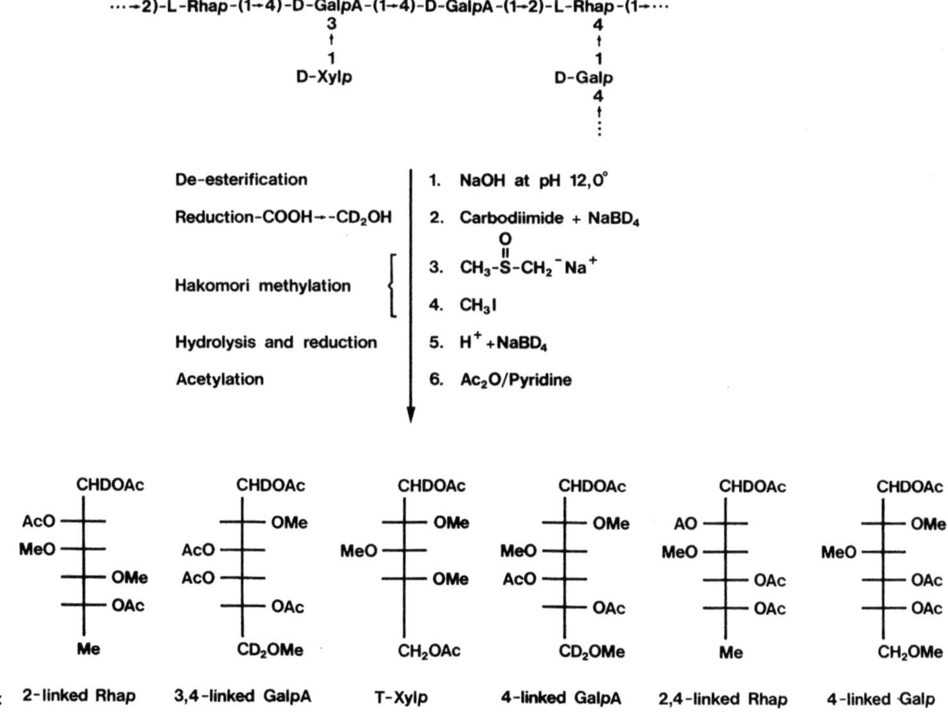

Fig. 1. Main stages in the preparation of PMAA from a hypothetical pectic polysaccharide

Table 1. Relative retention times and main diagnostic ions of PMAA from some pectins

RT[a]	Alditol acetate	Main diagnostic fragment ions with typical relative intensities (in brackets)
0.44*	2,3,5-Me$_3$Ara[b]	118(100), 129(56), 102(34), 161(28), 101(26), 162(4)
0.57*	2,3,4-Me$_3$Xyl	117(100), 101(99), 118(94), 102(82), 161(25), 162(19)
0.80	3,5-Me$_2$Ara	129(100), 130(88), 101(64), 161(57), 190(35)
0.87	3,4-Me$_2$Rha	131(100), 89(68), 130(49), 190(19)
1.07	2,3-Me$_2$Ara	118(100), 129(51), 102(20), 189(14)
1.19	2,3,4,6-Me$_4$Gal[c]	118(100), 113(73), 102(51), 233(45), 173(14), 162(14)
	2,3-Me$_2$Xyl	118(100), 129(92), 102(38), 189(18)
	3,4-Me$_2$Xyl	117(100), 130(90), 101(40), 190(20)
1.37	2-MeRha	118(100), 129(18), 275(5), 173(7)
1.57	4-MeRha	131(100), 89(40), 262(7)
1.67	3-MeRha	130(100), 143(71), 190(39), 203(32)
1.83*	2-MeRha	118(100), 261(14)
2.01*	3-MeXyl	129(100), 130(77), 189(34), 190(25)
2.15	3,4,6-Me$_3$Gal	129(100), 130(84), 161(42), 190(22)
2.22	2,3,6-Me$_3$Gal	118(100), 113(73), 102(51), 233(45), 162(14), 173(14)
2.6*	Arabinitol	115(100), 116(90), 103(82), 145(72), 146(43), 188(37), 187(34)
2.89	2,3,4-Me$_3$Gal	118(100), 102(71), 129(61), 189(18), 162(18), 233(14)
3.14	2,6-Me$_2$Gal	118(100), 129(20), 305(4)
3.57*	Xylitol	115(100), 103(82), 116(80), 145(67), 146(39)
3.6*	3,6Me$_2$Gal	130(100), 113(51), 190(48), 233(22), 173(10)
4.7	2,3-Me$_2$Gal	118(100), 127(76), 102(28), 261(17), 201(7)
	2-MeGal	118(100), 139(12), 157(1), 231(Tr)[d], 333(Tr)
	3-MeGal	130(100), 190(30), 261(15), 201(6)

[a] Retention time relative to 1,5-di-O-acetyl-2,3,4,6-tetra-O-methyl glucitol on OV-225 at 170°.
* values, found by authors, which differ from published values of Jansson et al. (1976).
[b] 2,3,5-Me$_3$Ara = 4-O-acetyl-2,3,5-tri-O-methylarabinitol etc.
[c] The galactose and xylose derivatives can be separated on ECNSS-M at 170° (see text).
[d] Tr = Trace

extensively, with minor modifications (Ring and Selvendran 1978; O'Neill and Selvendran 1980), our comments will be based on our experience with the above procedure. In the Hakomori methylation, the polysaccharide dissolved (or dispersed) in dimethylsulphoxide (DMSO) is first treated with sodium methylsulphinylmethanide (sodium dimsyl) and then allowed to react with methyl iodide. The strong base ensures complete alkoxide formation, and efficient methylation is usually achieved in one step. With some polymers, particularly those not fully dispersable in DMSO or those which are highly branched, a certain degree of undermethylation may occur. Recently potassium dimsyl has been shown to be more effective for some methylations (Darvill et al. 1978; Valent et al. 1980), and the time of treatment with base could be varied to achieve optimum methylation of some polysaccharides (Jansson et al. 1976).

Efficient methylation in a single operation is essential for pectins, because the esterified Gal$p$A residues which are native to the pectins, and those formed from the Gal$p$A residues during methylation are susceptible to base-catalysed β-elimination. Base-catalysed degradation could account for the poor recovery of Gal$p$A residues of cabbage pectins (Stevens and Selvendran 1984). In subsequent work with pectins from onions (R. J. Redgwell and R. R. Selvendran, unpublished results), potatoes and cabbage (B. J. H. Stevens and R. R. Selvendran, unpublished results) we have considerably minimised base-catalysed β-elimination and obtained much better recoveries of Gal$p$A by taking the following precautions: (1) De-esterified pectins were used as the starting material. The pectins were de-esterified with dil. NaOH at pH 12 for 2–8 h at 0 ° (Aspinall et al. 1970) to minimize as much as possible β-eliminative degradation during de-esterification. (2) The methylated pectins were reduced by refluxing for 4 h with LiAlH$_4$ or LiAlD$_4$ in a mixture of dichloromethane and ether (Lindberg and Lönngren 1978) instead of tetrahydrofuran (Lindberg 1972). Reduction with LiAlD$_4$ labels the galactose residues arising from 4-linked Gal$p$A residues, and thus helps to distinguish these from those arising from 4,6-linked Gal$p$ residues. The improved yields of Gal$p$A are not only due to the more efficient reduction, but also to the fact that the alkoxides are much more strongly nucleophilic towards the alkylating agent than the carboxylate ions, so that the net base concentration may be markedly diminished before esterification occurs with consequent degradation. Repetition of the operation to achieve full methylation must be avoided since extensive degradation would take place on addition of fresh base. However, the degradative reactions could be used to advantage under certain circumstances (see Sect. 5.2). As stated before, for water-soluble pectins, provided sufficient material is available, the problems of β-eliminative degradation may be avoided by performing the methylation on the carboxyl-reduced pectin. This technique has been effectively used by Aspinall and Jiang (1974).

4 Separation and Identification of Partially Methylated Alditol Acetates

The mass spectrum of the PMAA is characteristic of the pattern of substitution of the alditol by methoxyl and acetoxyl groups. The ambiguity in the identification of a methylated derivative can be overcome by performing the reduction with NaBD$_4$ instead of NaBH$_4$ (Step 5, Fig. 1). Incorporation of deuterium at C-1 increases by 1 mass unit every fragment containing C-1. This enables identification of pairs of derivatives e.g., 2,3- and 3,4-di-O-methyl xylitol derivatives which co-elute on GC. Mass spectral information in conjunction with retention times on a couple of GC columns should enable positive identifications of most components in a mixture. Derivatives which co-elute can be inferred from the occurrence and relative abundance of pertinent ions in the mass spectrum. Molecular ions are not seen in the electron impact (EI) spectra at 70 eV, but molecular ions can frequently be observed in the chemical ionization (CI) spectra. In Table 1 a range of PMAA which we obtained from pectic substances of cabbage, apples, potatoes and onions are listed, along with their RRt on OV-225, and major fragment ions, mostly primary, and their relative abundance; note that C-1 carries a deuterium atom and the spectra were obtained in the EI mode at 70 eV.

Primary fragment ions from PMAA are formed by α-cleavage with preferred formation of ions of lower molecular weight. The scission between carbon atoms in the alditol chain occurs more readily between two methoxylated carbon atoms than between a methoxylated and an acetoxylated carbon atom with marked preference for the methoxyl-bearing species to carry the positive charge, and least likely to occur between two acetoxylated carbon atoms. Primary fragment ions undergo a series of subsequent eliminations to give secondary fragments by single or consecutive loss of acetic acid (m/z 60), methanol (m/z 32), ketene (m/z 42) or formaldehyde (m/z 30); acetic acid and methanol being generally lost by β-elimination, although under certain circumstances acetic acid but not methanol, could be lost by α-elimination. In this connection, it is useful to note that there is little tendency for scission to take place adjacent to a deoxygenated carbon atom, but the presence of a deoxysugar is easily recognised from the primary fragments. For example, the 3,4-di-O-Me rhamnitol derivative, from 2-linked Rha$p$, has an intense ion at m/z 131 derived from cleavage between carbons 3 and 4. A very useful collection of mass spectra, together with retention time data for a range of PMAA, has been published by Jansson et al. (1976), and the mass spectra of these and a variety of other carbohydrate derivatives have been reviewed by Lindberg

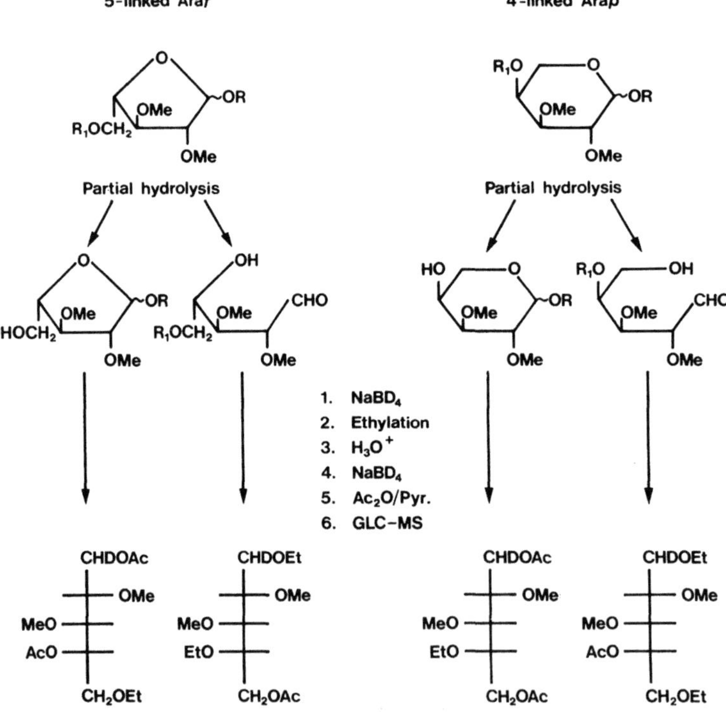

Fig. 2. Partially ethylated methylated alditol acetates from 5-linked arabinofuranosyl and 4-linked arabinopyranosyl residues after partial hydrolysis of the methylated polysaccharide and derivization to yield products which allow GLC-MS analysis. R and R_1 are glycosyl residues

(1972), Lönngren and Svensson (1974) and Aspinall (1982). As an alternative to PMAA, partially ethylated alditol acetates may be used when separation difficulties are encountered with particular combinations of the corresponding methylated derivatives (Sweet et al. 1975, 1974).

The ring size of terminal non-reducing sugars of complex carbohydrates can easily be determined from the positions of O-methyl substitution. For example terminal Ara$f$ would give 1,4-di-O-acetyl-2,3,5-tri-O-methyl arabinitol, whilst T-Ara$p$ would give 1,5-di-O-acetyl-2,3,4-tri-O-methyl arabinitol. Likewise the occurrence of T-Gal$f$ and T-Gal$p$ residues in the cell wall polymers of Chlamydomonas reinhardii was obtained from GC-MS of the corresponding PMAA (O'Neill and Roberts 1981). However, methylation analysis does not distinguish between 5-linked Ara$f$ and 4-linked Ara$p$ residues. Darvill et al. (1980) have published a general procedure for the unambiguous determination of ring size of interchain sugars. The method has been used in structural studies on RG-1 (McNeil et al. 1980). The application of the method for 5-linked Ara$f$ and 4-linked Ara$p$ residues is shown in Fig. 2. Partial hydrolysis of a methylated polysaccharide is followed by reduction with NaBD$_4$, ethylation, total hydrolysis, and GC-MS analysis of the derived partially alkylated alditol acetates. The occurrence of 5-linked Ara$f$ could be inferred from the formation of 5-O-acetyl-1,4-di-O-ethyl-2,3-di-O-methyl arabinitol and 1,4-di-O-acetyl-5-O-ethyl-2,3-di-O-methyl arabinitol. Using this procedure Darvill et al. have shown that beet pectic arabinan contains 5-linked, 2,5-linked, 3,5-linked, and 2,3,5-linked Ara$f$ residues and not the corresponding pyranosyl residues.

5 Extensions of Methylation Analysis

5.1 Controlled Partial Acid Hydrolysis Studies

Controlled partial acid hydrolysis in combination with methylation analysis has been used for determining the point of attachment of acid-labile sugars (Lindberg and Lönngren 1978). The method is particularly valuable when a polysaccharide contains a limited number of acid-labile glycosidic linkages, which may be cleaved without significant hydrolysis of other glycosidic linkages. However, preliminary experiments are required to determine the optimum conditions of hydrolysis to obtain meaningful results. Likewise, the methylated polysaccharide may be partially hydrolysed and remethylated with trideuteriomethyl iodide (or ethylated). This method yields similar but more detailed information on the positions of the acid-labile sugars. Two examples from our work on cell wall polysaccharides will be cited to illustrate the applications of the technique.

The xyloglucan from potatoes contains acid-labile T-Ara$f$ residues linked to 0-2 of some of the 2-linked Xyl$p$ residues. The fully methylated xyloglucan was subjected to mild acid hydrolysis (90% HCOOH for 40 min at 70 °), and the degraded material was reduced with NABD$_4$ and remethylated with CD$_3$I. The product was hydrolysed, and the mixture of methylated sugars was examined as their PMAA by GC-MS. The most notable results were the absence of 1,4-di-O-acetyl-2,3,5-tri-O-methyl arabinitol and a large increase in 1,5-di-O-acetyl-2,3,4-tri-O-methylxylitol, which was trideuteriomethylated to a considerable extent at C-2 (Ring and Selvendran 1981; Selvendran 1983). From these results it was inferred that a large proportion of the T-Ara$f$ residues were linked to 0-2 of the non-terminal Xyl$p$ residues.

A similar procedure was used on purified cabbage pectin to determine (a) the mode of attachment of the arabinan side-chains to the rhamnogalacturonan backbone, and (b) the mode of occurrence of 2,4-linked Xyl$p$ residues (Stevens and Selvendran 1984). The methylated pectin (which underwent appreciable degradation during methylation) was treated with CH_2N_2 to ensure complete esterification and then treated with $LiBH_4$ to reduce $-COOCH_3$ to $-CH_2OH$. The reduced product was partially hydrolysed, remethylated with CD_3I, and analysed as for potato xyloglucan. The results showed the following: (1) An appreciable proportion of the 3,4-di-O-methylrhamnitol derivative was labelled at 0-4. This clearly showed the removal of acid labile substituents from 0-4 of rhamnose and their replacement with CD_3 groups. However, a significant proportion of the 3-O-methylrhamnitol derivative (unlabelled) remained, showing that the substituents on 0-4 of the 2,4-linked Rha$p$ residues were not equally susceptible to mild acid hydrolysis; presumably 5-linked Ara$f$ (easily hydrolysable) and 4-linked Gal$p$ residues were involved. (2) The yield of 2,3,4,6-tetra-O-methylgalactitol derivative, which was labelled at 0-4 increased. This derivative could only arise from a 4-linked Gal$p$ residue which has an easily hydrolysable group on 0-4. (3) Mild acid hydrolysis resulted in the loss of all the 2,4-linked Xyl$p$ residues, and a corresponding increase in 2,3,4-tri-O-methylxylitol derivative, which was labelled at 0-2 and 0-4. This finding, in addition to confirming the occurrence of 2,4-linked Xyl$p$ residues in cabbage pectin, suggested that the residues linked to 0-2 and 0-4 are easily hydrolysable. From the combined results it could be inferred that the bulk of the arabinan-rich pectic moieties are linked to the rhamnogalacturonan backbone via 0-4 of 2,4-linked Rha$p$ residues, either directly or through short 4-linked Gal$p$ residues. Unambiguous evidence about the nature of the sugar residues linked to 0-4 of 2,4-linked Rha$p$ residues can be obtained by using base catalysed β-elimination reactions described below.

5.2 β-Eliminative Degradation Studies

Base-catalysed β-elimination reactions were developed by Lindberg et al. and applied effectively to elucidate the structural features of bacterial polysaccharides (Lindberg et al. 1973; Lindberg and Lönngren 1976; Lindberg and Lönngren 1978). For applications of base-catalysed β-elimination reactions see Aspinall (1982). Using base-catalysed reactions McNeil and co-workers determined (a) the neutral glycosyl residues to which the galacturonosyl residues are linked in RG-1 (McNeil et al. 1980; Fig. 3, Scheme 1), and (b) the glycosyl residues attached to 0-4 of 2,4-linked Rha$p$ residues of RG-1 (McNeil et al. 1982b; Fig. 3, Scheme 2). In Fig. 3, R_1 and R_2 are unspecified neutral glycosyl residues, but R_2 could be CH_3. The latter would arise by β-elimination during Hakomori methyl-

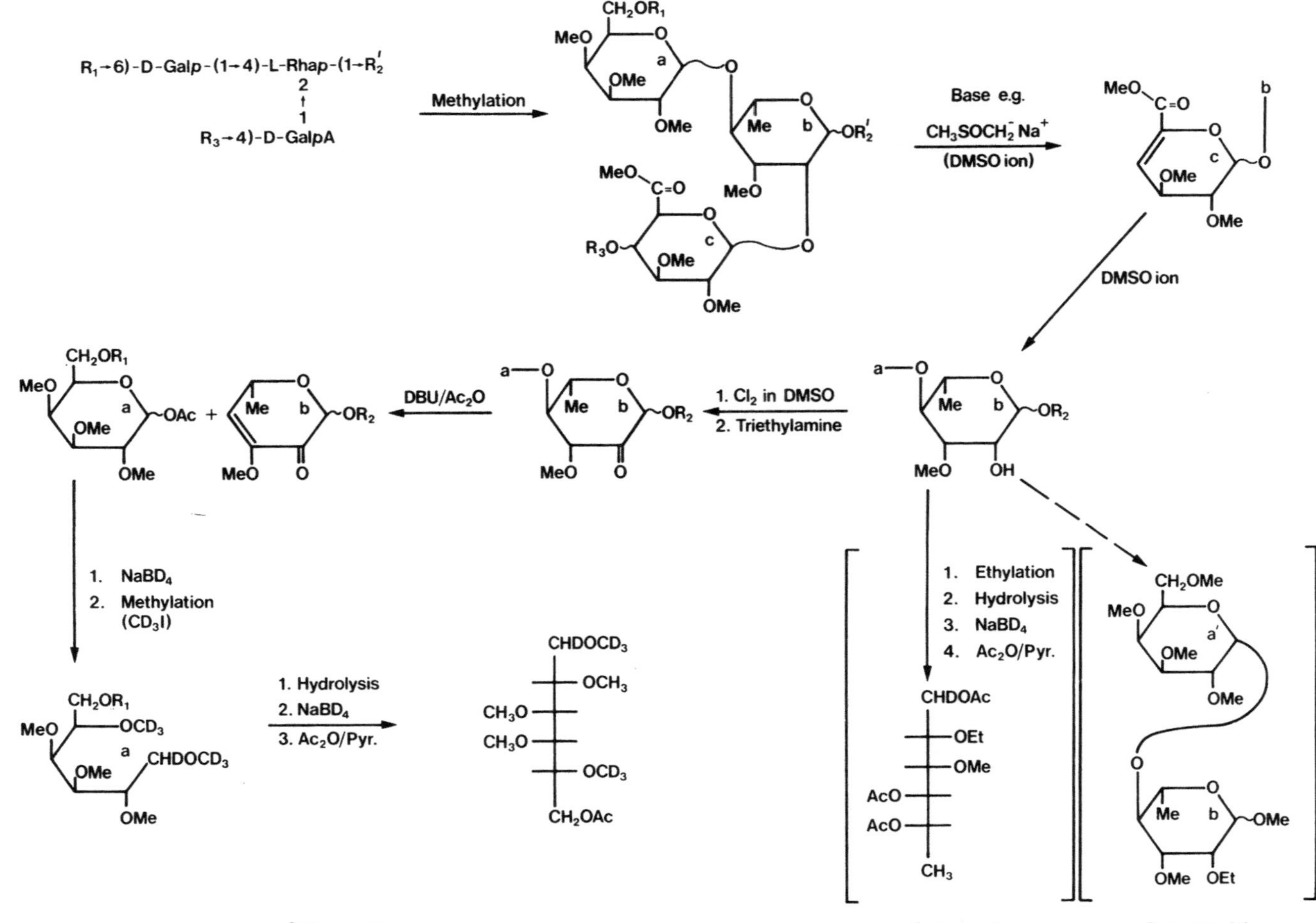

Fig. 3. Main steps in the β-eliminative degradation studies of a pectic polysaccharide. R_1 and R_3 are unspecified glycosyl residues, but in Scheme 1.1 $R_1 = CH_3$. R_2' is an unspecified moiety from the rhamnogalacturonan backbone; $R_2 = $ Rha or CH_3

ation of RG-1 if C-1 of the 2,4-linked Rha*p* residue is linked to 0-4 of 4-linked Gal*p*A; for more details on this point see McNeil et al. (1982b).

5.2.1 Neutral Glycosyl Residues Linked to Galacturonosyl Residues (Fig. 3, Scheme 1)

In this procedure RG-1 is first methylated and the permethylated product is treated with dimsyl anion in DMSO when the substituents linked to 0-4 of the 4-linked galacturonosyl residues are liberated with the formation of Δ 4:5 unsaturated galacturonosyl residues. The unsaturated galacturonosyl residues undergo a second reaction in the presence of the dimsyl anion in which the glycosiduronic bond is cleaved with the formation of an unsaturated hydroxyl group on the glycosyl residue to which the galacturonosyl residue had been attached. These unsubstituted hydroxyl groups are labelled by ethylation, and the partially methylated partially ethylated RG-1 is hydrolysed, reduced with $NaBD_4$ and acetylated (Fig. 3, Scheme 1). The resulting partially methylated partially ethylated alditol acetates, in which the position of the OEt group marks the point of attachment of the galacturonosyl residue, are identified by GC-MS. McNeil et al. (1980) detected O-ethyl groups on several derivatives but predominantly on C-2 of the derivative from 2,4-linked Rha*p* residue, which is shown in Scheme 1. RG-1 also contains D-Gal*p*A residues linked to 0-2 or 0-3 of 2- or 3-linked Ara*f* residues, and 0-4 or 0-6 of 4- or 6-linked Gal*p* residues.

Note. In the above scheme it should be noted that the addition of dimsyl anion to methylated RG-1 containing 2-linked Rha*p* residues as in the sequence →4)-Gal*p*A-(1→2)-Rha*p*-(1→4)-Gal*p*A-(1→ results in almost complete loss of the Rha*p* residues. This is (apparently) due to the fact that the elimination of methylated Gal*p*A linked to C-1 of Rha*p* results in the formation of a rhamnopyranosyl residue having an aldehyde group on C-1, which is relatively unstable in the presence of the base. The reasons for the preferential formation of –CHO group on C-1, instead of CHOCH$_3$, in the case of 2-linked Rha*p*, but not necessarily in the case of 2,4-linked Rha*p*, are not clear. The reducing group formed in the presence of the strong base eliminates the substituent on C-3, which is β to the –CHO group, producing an unsaturated glycose derivative. The unsaturated glycose residue does not form a derivative detectable by GC and thus is lost. The preferential loss of 90% of the 2-linked Rha*p* residues, when the base eliminated RG-1 is subjected to methylation analysis, has been taken as evidence that 90% of the 2-linked Rha*p* residues are linked to C-4 of Gal*p*A as shown in the above sequence.

5.2.2 Glycosyl Residues Linked to 0-4 of 2,4-Linked Rhamnopyranosyl Residues (Fig. 3, Scheme 2)

During the base-catalysed elimination of methylated RG-1, the free hydroxyl groups formed at C-2 of the 2,4-linked Rha*p* residues are subjected to oxidation with chlorine (in dichloromethane) dissolved in DMSO at − 45 ° for 5 h, and the reaction is terminated by the addition of trimethylamine. The newly formed keto group at C-2 allows base-catalysed elimination of the glycosyl residues attached to 0-4 of the (oxidised) rhamnosyl residues. This elimination should not be performed with dimsyl anion, as the base is too strong and would destroy the exposed aldehyde groups of the glycose residues eliminated from 0-4. Therefore the oxidised compound from RG-1 is subjected to the milder base-catalysed reaction of Aspinall and Chaudhari (1975), which involves treatment with the organic base 1,5-diazabicyclo[5.4.0]undec-5-ene (DBU) and acetic anhydride. In this procedure the reducing group of the eliminated product is simultaneously protected by acetylation. The resulting product is de-O-acetylated and reduced in one step by treatment with M ammonia in ethanol containing $NaBD_4$. The partially methylated oligosaccharide alditols formed are methylated using CD_3I, to label the hydroxyl groups produced by de-O-acetylation and reduction. The resulting permethylated oligosaccharide alditols are hydrolysed, reduced, acetylated and analysed by GC-MS. The labelled PMAA formed from 6-linked Gal$p$ residue is shown in Fig. 3, Scheme 2.

Using this procedure McNeil et al. (1982 b) showed that seven differently linked glycosyl residues were attached to 0-4 of 2,4-linked Rha$p$ residues of RG-1: these were 5-linked L-Ara$f$, terminal D-Gal$p$, and 3-, 4-, 6-, and 2,6-linked, and 3,6-linked D-Gal$p$ residues. Confirmatory evidence for the occurrence of T-Gal$p$ residues was obtained by isolating and characterising the derivative derived from D-Gal$p$ (1→4)-L-Rha$p$. The formation of this derivative could be deduced from Scheme 1, Step 3, where $R_1 = R_2 = CH_3$. Ethylation of such a product would give a permethylated ethylated derivative carrying an –OEt group on C-2 of the rhamnose residue (Fig. 3, Scheme 1.1).

6 Sequencing of Sugar Residues in Pectins

The main feature of the methods used to determine the sequence of sugar residues in the rhamnogalacturonan backbone, or those carbohydrate moieties linked to the backbone, involve arresting the hydrolysis of the polysaccharide (or its derivative) at some convenient stage before complete depolymerisation has occurred, and then characterising the products formed. The isolated degraded polysaccharide(s) and oligosaccharides on characterisation give detailed information on sequences and also provide confirmation of linkage types. By carrying out the hydrolysis in a stepwise manner followed by dialysis at the end of each period of hydrolysis, the progressive degradation of very acid-labile carbohydrate moieties to yield monosaccharides can largely be avoided (Aspinall 1982; Aspinall 1969; Aspinall and Whitehead 1970 a, b).

6.1 Partial Acid Hydrolysis

Partial acid hydrolysis has been used to study the structural features of pectic material from soybean cotyledons (Aspinall et al. 1967), lemon peel (Aspinall et al. 1968 a), leaf and stem of lucerne (Aspinall et al. 1968 b), and rape seed hull (Aspinall and Jiang 1974). The hydrolysis was effected in 0.5 M H_2SO_4 at 100 °, but various times of hydrolysis, from 2 to 6 h, were used. The oligosaccharides from lemon peel and rape seed hulls were fractionated by chromatography on DEAE-Sephadex A-25 or A-50 (formate form) to yield a neutral component and several acidic ones. The neutral component was fractionated on a charcoal-celite column and both neutral and acidic oligosaccharides were further purified by paper chromatography and characterised by conventional chemical methods. As GC-MS was not used in the above studies, we shall not discuss them any further, but the reader is advised to consult the above and related papers on plant gums before embarking on partial acid hydrolysis of pectic substances.

Using basically the same methodology Toman et al. (1975, 1976) isolated and characterised a range of acidic and neutral oligosaccharides from willow bark pectin. These workers used mass spectrometry, in the EI mode, to confirm the identity of the permethylated oligosaccharides or permethylated oligosaccharide alditols. Some of the oligosaccharides identified are as follows: (1) D-Gal*p*A-(1→2)-L-Rha*p*, (2) D-Glc*p*A-(1→6)-D-Gal*p*, (3) D-Glc*p*A-(1→4)-L-Fuc*p*, (4) D-Gal*p*A-(1→2)-L-Rha*p*-(1→2)-L-Rha*p*, (5) D-Gal*p*A-(1→2)-L-Rha*p*-(1→4)-D-Gal*p*A and (6) D-Gal*p*-(1→3)-D-Gal*p*. The oligosaccharide (4) is relatively unstable to hot dilute acids and is usually obtained only on acetolysis of pectic polysaccharides (Aspinall et al. 1967, 1968 b). The first three oligosaccharides were identified by mass spectrometry as the permethylated derivatives, whereas (4), (5), and (6) were identified as permethylated oligosaccharide alditols. In general, the permethylated oligosaccharide alditol is the preferred derivative, as significant degradation of some reducing oligosaccharides may occur during Hakomori methylation. Before discussing the mass spectral evidence on which the structures of a couple of oligosaccharide derivatives from pectins are based, some general comments on mass spectral fragmentation of permethylated oligosaccharide derivatives will be made.

6.1.1 Characterisation of Oligosaccharides as Permethylated Derivatives

The permethylated derivatives are well suited for mass spectrometric studies because of their volatility, ease of preparation, and fragmentation patterns in the EI and CI modes. Because most of the oligosaccharides released from pectic substances (and related polysaccharides) by partial acid hydrolysis have been identified by mass spectrometry as the permethylated (or permethylated ethylated) derivatives, this section will deal almost exclusively with such derivatives. At present, the separation and identification of permethylated derivatives by combined GC-MS are widely used for di- and tri-saccharides (Kärkkäinen 1970, 1971; Moor and Waight 1975; Ring and Selvendran 1981), although the separation of mixtures, particularly in the tri-saccharide range, is not always good. Identification of higher oligosaccharides is difficult due to poor resolution. In general, therefore, higher oligosaccharides are separated as parent compounds and the individually prepared derivatives are analysed by direct insertion mass spectrometry (Åman et al. 1981; Ashford et al. 1982). Alternatively, the oligosaccharide derivatives can be separated by HPLC and the separated derivatives can be identified by direct insertion mass spectrometry or by combined HPLC-MS (Åman et al. 1981; McNeil et al. 1982a; O'Neill and Selvendran 1985).

The most commonly used ionization mode is EI, which gives valuable fragment ions but rarely gives molecular ions. However, the molecular weights can often by inferred from the fragment ions. Molecular or pseudomolecular ions can be obtained by chemical ionisation using reagent gases such as ammonia (O'Neill and Selvendran 1983; Geyer et al. 1983), isobutane (Chizhov et al. 1976; McNeil and Albersheim 1977), methane (Fales and Milne 1969) or acetonitrile (Kochetkov and Chizhov 1966). CI gives much more abundant ions in the high-mass range but provides less information on structure. Further adduct formation between fragments is more likely in the CI mode, so that the results have to be interpreted with caution (McNeil 1983). Later on we shall discuss the applications of both EI and CI modes of fragmentation to the permethylated alditol from the aldobiouronic acid Gal$p$A-(1→2)-Rha$p$. Molecular or pseudomolecular ions can also be obtained by fast atom bombardment-mass spectrometry (FAB-MS). In this technique the ionised sample molecules are generated by bombardment with activated beams of argon or xenon, and the method has the advantage that underivatised molecules can be used. The simpler spectra obtained by FAB-MS provide less information on structure. For the applications of FAB-MS to determine the molecular weights of oligosaccharides and some of their structural features see Dell et al. (1983 a, b).

The interpretations of the mass spectra of permethylated oligosaccharide derivatives, particularly in the EI mode, have been discussed in a number of reviews and research papers (Kochetkov and Chizhov 1966; Lönngren and Svensson 1974; Kováčik et al. 1978); a knowledge of the basic principles involved is assumed. For the applications of mass spectrometry to characterise oligosaccharides, containing two to four sugar residues, from cell-wall polysaccharides see Selvendran (1983). Recently, Harada et al. (1983) published a paper on the sequencing of permethylated oligosaccharides by CI-MS.

In general, reduction of reducing oligosaccharides, prior to methylation, is best performed with $NaBD_4$ to avoid ambiguities arising from structurally symmetric terminal residues. The mass spectra may be analysed by considering, in turn, fragment ions derived from the non-reducing and reducing terminal residues, and then fragments arising from internal residues. The nomenclature used here is that developed by Kochetkov and Chizhov (1966), and later modified by Kováčik et al. (1968 a, b), with the exception that the alditol fragment will be referred to as "ald", to distinguish it from the terminal non-reducing sugar, and also the reducing sugar in the glycoside form. It should be noted that the above nomenclature is different from that of Albersheim and co-workers (McNeil et al. 1982 a; Lau et al. 1985), and as we shall use some of their results later on the reader should bear this distinction in mind.

The most easily recognizable fragment ions in the mass spectra of permethylated oligosaccharide alditols are those arising from non-reducing sugars i.e., the A_1 ions at m/z 233 (permethylated hexuronic ester), 189 (permethylated deoxyhexose), 219 (permethylated hexose) and 175 (permethylated pentose). Likewise the alditol residues ($NaBD_4$ reduction) give fragment ions at m/z 250, 206, 236, and 192 for ions derived from hexuronic acid, deoxyhexose, hexose, and pentose. The linkage of the internal sugars could be deduced from the appropriate ions.

                                                    R. R. Selvendran and B. J. H. Stevens

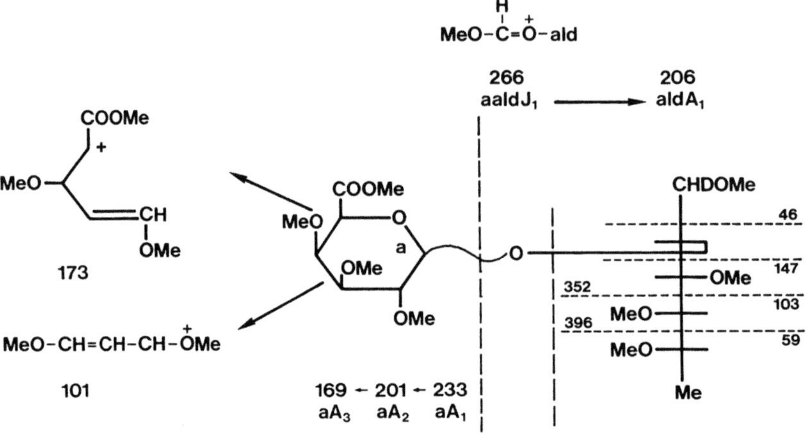

Fig. 4. MS fragmentation pattern of methylated, reduced Gal$p$A-(1→2)-Rha$p$

Mass Spectral Evidence for the Structure of D-Gal$p$A-(1→2)-L-Rha$p$. The aldobiouronic acid from sugar beet pectin was reduced with $NaBD_4$, methylated by the Hakomori procedure, and the permethylated derivative was separated by GC on OV-1 and analysed by mass spectrometry in both EI and CI (NH_3 gas) modes (B. J. H. Stevens and R. R. Selvendran, unpublished results). The pseudomolecular ions at m/z 473 $[M+18]^+$(5) and 456$[M+1]^+$(0.8) were observed only in the CI spectra. The diagnostic ions from the EI spectra at m/z 396[M-59](0.1), 352[M-103](0.1), 266[a ald J₁](3), 233[aA₁](48), 206[ald A₁](40), 201[aA₂](56) and some of the smaller ions are shown in Fig. 4; the values within brackets represent the relative intensities. The ions at m/z 473 and 456 show that the parent disaccharide alditol contains one hexuronic acid and one deoxyhexitol residue. The intense ions at m/z 233 and 206 are consistent with a hexuronosyl-dexoxyhexitol derivative. The ion at m/z 266 confirms the disaccharide nature of the derivative. The abundance ratio aA₂/aA₁ ≃ 1, suggests that the uronic acid is Gal$p$A; for Glc$p$A the ratio >4 (Kováčik et al. 1968a; Ring and Selvendran 1980). The ion at m/z 173 could only arise from a uronic acid containing derivative. The nature of the linkage between Gal$p$A and Rha$p$ could be deduced from the relatively intense ions at m/z 147 and 115, which are diagnostic of a methylated rhamnitol derivative substituted at 0-2, and hence a (1→2)-linkage. Thus the parent disaccharide could be deduced to be Gal$p$A(1→2)-Rha$p$. In the above deduction not all the ions in the spectra were considered. It should be noted, however, that in order to confirm the identity of the parent oligosaccharide the sugars constituting it and the nature of the linkage between the sugars should be established. Aspinall and Jiang (1974) have performed GC of the methanolysis products from the methylated derivative. We have converted the aldobiouronic acid to the methylester methylglycoside, reduced it with $LiAlD_4$, and analysed the resulting neutral product for the constituent sugars, and by methylation analysis (B. J. H. Stevens and R. R. Selvendran, unpublished results). A similar approach has been used to deduce the structures of higher oligosaccharides derived from cell-wall polysaccharides (Selvendran 1983; O'Neill and Selvendran 1983 and 1985; Fichtinger-Schepman et al. 1980).

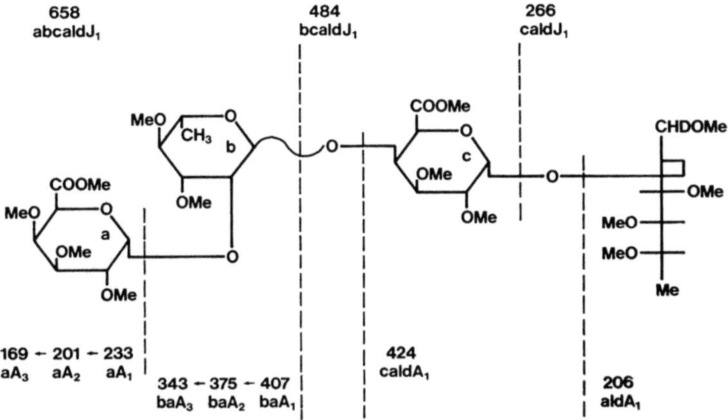

Fig. 5. MS fragmentation pattern of methylated, reduced GalpA-(1→2)-Rhap-(1→4)-GalpA-(1→2)-Rhap

Mass Spectral Evidence for the Structure of D-GalpA(1→2)-L-Rhap-(1→4)-D-GalpA-(1→2)-L-Rhap. The above oligosaccharide is released on partial acid hydrolysis of pectins (Aspinall et al. 1968 b), and has also been isolated from a pectic type of polysaccharide from Coccoliths (Fichtinger-Schepman 1980; Fichtinger-Schepman et al. 1980). The structure of the permethylated oligosaccharide alditol from the tetrasaccharide was deduced from the ions obtained in the EI mode by Fichtinger-Schepman et al. (1980). To avoid unnecessary confusion, the methylation is assumed to be performed with CH_3I and not with CD_3I as reported in their paper. The main ions used for the identification of the derivative are shown in Fig. 5. The ions of the aA series at m/z 233 and 201 and the ratio of $aA_2 : aA_1$ showed that the non-reducing terminal sugar derivative is permethylated GalpA. The ions of the baA series at m/z 407 and 375 showed that the sugar internal to the T-GalpA is a deoxyhexose. Likewise from the cbaA series of ions it can be deduced that a hexuronic acid is linked to the deoxyhexose. The ions at m/z 206 (*ald* A_1) and 147 showed that the reducing sugar in the parent oligosaccharide is a deoxyhexose and is linked to the internal sugar via 0-2 (very small intensity of ion at m/z 134). The ion at m/z 266 (c *ald* J_1) showed that the sugar internal to the terminal reducing sugar is not 3-linked. The above data in conjunction with the ions at m/z 484 (bc *ald* J_1), 658 (abc *ald* J_1) and 598 (bc *ald* A_1) define the sequence GalpA-Deoxyhexose-Hexuronic acid-Deoxyhexitol for the parent oligosaccharide alditol. Coupled with the results of sugar analysis and methylation analysis of the carboxy-reduced parent oligosaccharide the sequence GalpA-(1→2)-Rhap-(1→4)-GalpA-(1→2)-Rhap could be deduced for the parent oligosaccharide.

6.2 Partial Acetolysis

Acetolysis of a polysaccharide involves treatment of the acetylated polysaccharide under acid-catalysed conditions, with acetic anhydride-sulphuric acid mixtures, frequently with added acetic acid. The resulting mixtures of sugars are then de-acetylated with sodium or barium methoxide, reduced with $NaBD_4$, and identified as the permethylated alditols by GC-MS. For details of the acetolysis procedure as applied to pectic substances see Aspinall et al. (1968a). The rates of hydrolysis of glycosidic linkages during acetolysis are often different from those observed during partial acid hydrolysis. Thus whereas $(1\rightarrow6)$-linkages between hexose residues are the most resistant to acid hydrolysis (Peat et al. 1961), this type of linkage is readily cleaved by acetolysis.

Acetolysis has been used to characterise the side chains attached to $(1\rightarrow6)$-linked core mannans from yeasts and fungi (Lee and Ballou 1965). Acetolysis has been used to determine the side-chains of the xyloglucans from runner beans and apples (O'Neill and Selvendran 1983; Aspinall and Fanous 1984). Further, 6-deoxy-hexopyranosyl linkages, especially those of L-rhamnose (Aspinall et al. 1963) and L-fucose (Aspinall and Baillie 1963) in terminal positions, are rather readily hydrolysed by partial acid hydrolysis but are sufficiently stable to acetolysis to be isolated as oligosaccharide constituents [e.g., Gal$p$A-$(1\rightarrow2)$-Rha$p$-$(1\rightarrow2)$-Rha$p$ from pectins]. In addition, as a method of analysis acetolysis may have advantages over partial acid hydrolysis for those polysaccharides which are insoluble in aqueous medium since the acetylated polysaccharide will almost certainly be soluble in the acetolysis medium.

6.3 Enzymatic Hydrolysis

Enzymatic hydrolysis of (partially degraded) pectins have proved useful for the determination of the nature of the sugar residues linked to the rhamnogalacturonan backbone of pectins. Aspinall and co-workers (1968a) isolated the pseudoaldobiouronic acid, D-Xyl$p$-$(1\rightarrow3)$-D-Gal$p$A, from the "Pectinase"-treated extract of lemon peel pectin. Albersheim and co-workers have made extensive use of a highly purified polygalacturonanase to obtain fragments of pectic polysaccharides from suspension-cultured sycamore cell walls (Talmadge et al. 1973). More recently, the same group have isolated rhamnogalacturonan-1 (RG-1) and rhamnogalacturonan-2 (RG-2) from the endopolygalacturonanase treated sycamore pectic polysaccharides using a combination of ion-exchange chromatography and gel filtration (McNeill et al. 1980; Darvill et al. 1978). Glycosyl residue composition of RG-2 showed that it contains ten different monosaccharides including apiose, 2-O-methylxylose and O-methylfucose. In addition methylation analysis showed that RG-2 contains the rarely observed glycosyl linkages of 2-linked Glc$p$A, 3,4-linked Fuc$p$, and 3-linked Rha$p$ residues.

7 Sequencing of Pectic Polysaccharides
by Partial Depolymerisation of Permethylated Derivatives

The sequencing of complex polysaccharides by identifying the partially methylated oligosaccharides released on partial hydrolysis of the methylated polysaccharides has been extensively used by Albersheim and co-workers (Valent et al.

1980; Åman et al. 1981; McNeil et al. 1982a). The random fragmentation of the methylated polysaccharide to give a series of overlapping fragments does not require specific structures. The separation of the fragments after suitable derivatisation has been greatly facilitated by HPLC on reversed-phase columns. Recently Lau et al. (1985) have applied the glycosyl-sequencing technique of Valent et al. (1980) to determine the structure of the backbone of RG-1; the basic steps are shown in Fig. 6. The following operations were involved: (a) The D-Gal*p*A residues of RG-1 were first reduced to the corresponding, 6,6′-dideuterio-D-galactosyl residues by the carbodiimide method, using $NaBD_4$ and D_2O, thus eliminating the troublesome carboxyl groups. (b) The carboxyl-reduced RG-1 was methylated, partially hydrolysed with 88% formic acid (80 min, 70 °) and reduced with $NaBD_4$. The resulting partially methylated oligosaccharide alditols contained free hydroxyl groups wherever they were linked in the unfragmented polysaccharide, and also at C-1 and C-5 of the reducing ends; these hydroxyl groups were then labelled by ethylation. The mixture of totally alkylated oligosaccharides was

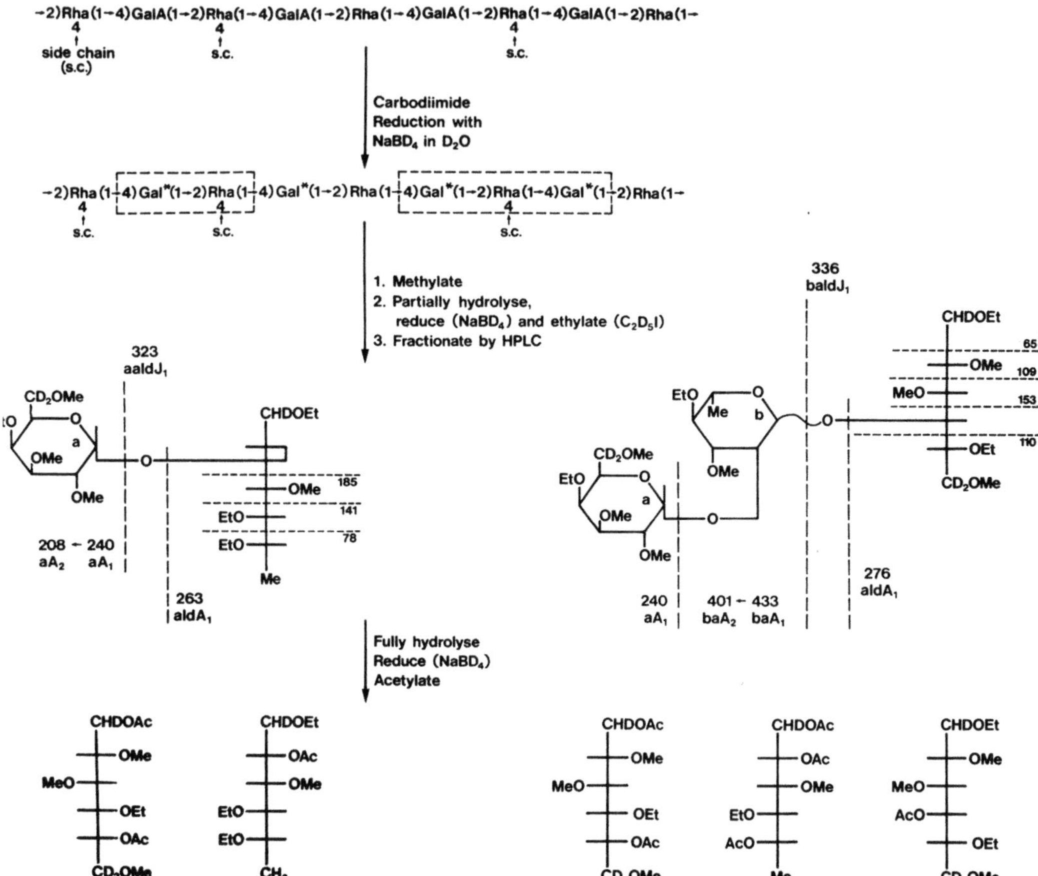

Fig. 6. Main stages in the sequencing of the backbone of rhamnogalacturonan-1

separated by reversed-phase HPLC. The separated peralkylated alditols can be examined directly by GC-MS to reveal their essential structural features. The main ions derived from two of the permethylated oligosaccharide alditols are shown in Fig. 6. (c) The structures of the peralkylated oligosaccharide alditols were confirmed by identifying the partially methylated partially ethylated alditol acetates formed on full hydrolysis, reduction with $NaBD_4$ and acetylation.

By this technique the structures of di-, tri-, and tetra-saccharides released from the backbone may be determined. As these oligosaccharides overlap within the backbone they can be pieced together to determine the sequence of the backbone of RG-1. It should be noted that in Fig. 6 the oligosaccharide derivatives that would be derived from the side-chains of RG-1 are not shown, because the main objective of this study was to determine the sequence of L-Rha*p* and D-Gal*p*A residues in the backbone. Using this technique Lau et al. (1985) showed that the bulk of the L-Rha*p* and D-Gal*p*A residues of RG-1 are present in strictly alternating sequence. The length of the alternating sequence is, however, not known, but could be as long as 400 residues. About half the rhamnosyl residues in this alternating sequence carry side-chains on 0-4 as indicated by the presence of an –OEt group on C-4 of the partially alkylated alditol acetates derived from 2,4-linked Rha*p* residues.

8 Experimental

In this section some pertinent comments on some of the experimental procedures are given. We hope these may be useful as a guide, but the reader should consult the original references for details.

De-Esterification. Pectic polysaccharides were de-esterified at 0 ° and at pH 12 by the addition of N NaOH (Aspinall et al. 1970). After 2 to 16 h the pH was adjusted to 5 with HCl and the polysaccharides dialysed. The longer times for de-esterification may be required for some highly esterified pectins.

Reduction with Carbodiimide/$NaBH_4$. The method of Taylor and Conrad (1972) was used.

Preparation of Sodium Dimsyl. The procedure used was similar to that of Sandford and Conrad (1966) and Jannson et al. (1976) except that argon was used in place of nitrogen to minimise back diffusion of air (Ring and Selvendran 1978). All apparatus was dried in a vacuum oven at 40 ° over P_2O_5 for at least 6 h beforehand. NaH (2.1 g), washed free of mineral oil with n-pentane while under a flow of argon, was heated with DMSO (30 ml spectroscopic grade, dried over molecular sieve type 4 A) at 50°–60° for 6 h.

Methylation. Finely divided polysaccharide or reduced oligosaccharide (typically 5 mg but up to 20 mg if subsequent partial hydrolysis or carboxyl reduction was to be carried out) was dried in a test tube under vacuum at 40 ° over P_2O_5 overnight, a serum cap fitted and dry DMSO added (1 ml to 5 mg sample). The tube was evacuated and flushed with argon several times. Treatment in an ultrasonic bath facilitated solution of the sample. A volume of Na dimsyl equal to that of the DMSO was added and the tube placed in a container under a flow of argon and left overnight at room temperature for alkoxide formation to take place. For oligosaccharides some workers have used a shorter treatment with base and have reported that potassium dimsyl gave more complete methylation (Valent et al. 1980). For methylation the alkoxide was cooled in ice and cooled MeI (volume equal to that of the Na dimsyl added) was slowly added with a hypodermic syringe, excess pressure being released through a second hypodermic needle. The solution was allowed to warm to room temperature with occasional agitation in an ultrasonic bath and excess MeI was removed by flashing with argon. After 1 h or longer, when the solution was clear the methylated material was treated by one of the following procedures according to whether the material was polymeric or oligomeric.

Polymeric Material. An equal volume of $CHCl_3$/MeOH (1 : 1 v/v) was added, any insoluble material removed by filtration through glass fibre paper and the soluble methylated material was dialysed against aq. EtOH (50% v/v; at least 2 changes) evaporated to dryness and re-dissolved in $CHCl_3$/MeOH. We have found dialysis (Visking size 2, Medicell International Ltd.) to be a suitable procedure for polysaccharides with molecular weights above 10,000. Alternatively the following procedures can be used: 1. The methylated polysaccharide can be extracted with CH_2Cl_2 as for oligosaccharides (see below) which ensures that no low molecular weight polysaccharides are lost but high molecular weight material may not be quantitatively extracted.

2. The methylated material may be separated by gel-filtration through Sephadex LH-20 using $CHCl_3$/MeOH as the eluant (Talmadge et al. 1973) and the fractions monitored for carbohydrate.

Oligomeric Material. The methylated material was poured into water (2 to 5 ml) and extracted four times with CH_2Cl_2 (or $CHCl_3$) (3 ml) and the combined extracts were washed three times with water (3 ml), dried with anhydrous Na_2SO_4 and concentrated to a small volume.

Reduction with $LiALD_4$. Carboxyl groups in the methylated material were reduced by refluxing with $LiALD_4$ in Et_2O/CH_2Cl_2 (Lindberg and Lönngren 1978). The methylated material and apparatus were dried beforehand as for the methylation procedure. In our experience (R.J. Redgwell, M.A. O'Neill, and R.R. Selvendran, unpublished results) a large excess (40 to 50 mg for 5 mg methylated material) of fresh reagent is required to ensure almost complete reduction. The solvents should be peroxide free but we have found that freshly opened bottles were usually satisfactory. A suspension of the reagent in the solvent (3 to 4 ml) was prepared and added to the flask containing the methylated sample dissolved in solvent ($\sim$2 ml). Agitation for a few seconds in an ultrasonic bath helped to disperse the reagent before refluxing. At the end of the reaction, after excess reagent had been destroyed, the copious precipitate of hydroxide had to be extensively washed with $CHCl_3$/MeOH to remove adsorbed polymer. This was achieved by alternate filtration and resuspension on a sintered filter (Porosity 3, ~ 25 mm diam.). The combined polymer solution and washings were evaporated to dryness, re-dissolved in $CHCl_3$/MeOH, filtered through glass fibre paper (Whatman GF/C) and evaporated to dryness for hydrolysis.

Hydrolysis of Methylated Material and Preparation of PMAA. The methylated material was sequentially hydrolysed with formic acid (90%, 1 to 3 ml) for 2 h at 100 ° and H_2SO_4 (0.25 M, 1 to 3 ml) for 12 h at 100 ° as described by Lindberg (1972). Following hydrolysis the H_2SO_4 was neutralised and the partially methylated sugars reduced and acetylated as described by Selvendran et al. (1979) except that $NaBD_4$ was used in place of $NaBH_4$. For references on some of the improved methods of work-up of the methylated material see Selvendran et al. (1985).

Gas Chromatography. PMAA were separated on 3 m × 3 mm columns of OV-225 on J. J.'s diatomite CQ or on Chromosorb W. HP, 100/120 mesh. For determination of relative retention times isothermal runs at 180 ° were performed to compare with the published data of Jansson et al. (1976) but, owing to peak broadening, retention times of low levels of the late running components were difficult to determine. For better resolution, and for GC-MS, temperature programmed (150 ° to 220 ° at 1 ° min^{-1}) runs were used. GC on ECNSS-M (3% on Gas Chrom. Q, 3 m × 3 mm column) was carried out isothermally at 170 °. Methylated oligosaccharide alditols were separated on columns (45 cm × 4 mm) of 1% OV-1 on J.J's diatomite CQ temperature programmed from 150 ° to 310 ° at 2 ° min^{-1} (O'Neill and Selvendran 1983) and on 3% OV-1 (2.8 m × 2 mm column) temperature programmed (190 °, 5 min then at 1 ° min^{-1} to 320 °) (Stevens and Selvendran 1984). The longer column gave better separation, but retention times were too long for oligomers of more than three units.

Capillary columns give much better resolution than packed columns but in our experience their lower sample capacity sometimes results in a loss of sensitivity with minor components in the mass spectrometer. Stationary phases used have included OV-225 (O'Neill and Selvendran 1983) OV-275 (Klok et al. 1982), Silar 10C (Harris et al. 1984), SE-30 (Valent et al. 1980) and SP1000 (Jansson et al. 1976; Lomax and Conchie 1982).

Owing to their flexibility and greater chemical stability, chemically bonded fused silica capillary columns are very useful where interchange between gas chromatographs alone and those connected to a mass spectrometer often necessitate altering the configuration of a glass column. Wider bore (up to 0.5 mm) columns of this type are becoming available with the advantage of increased loading capacity. On column injection techniques have given the most reproducible quantitative results (Grob and Grob 1978; Conchie et al. 1982).

Mass Spectrometry. The results of our studies reported here were obtained on either an AEI MS30 or an AEI MS80 mass spectrometer (capillary GC-MS) with an inlet temperature of 250 ° and, source temperature of 200 °. In the E.I. mode, the ionisation potential was 70 eV and ion source temperature of 220 ° and in CI mode the reagent gas was NH_3. Data were acquired with continuous scanning (10 s/decade) on an AEI DS50SM computer system.

Acknowledgements. We thank Mr. J. Eagles and Mr. K. Parsley of the AFRC Mass Spectrometry Group at the Food Research Institute, Norwich, for their help with the mass spectrometry.

References

Albersheim P, Nevins DJ, English PD, Karr A (1967) A method for the analysis of sugars in plant cell wall polysaccharides by gas-liquid chromatography. Carbohydr Res 5:340–345

Åman P, McNeil M, Franzén L-E, Darvill AG, Albersheim P (1981) Structural elucidation, using HPLC-MS and GLC-MS, of the acidic polysaccharide secreted by Rhizobium meliloti strain 1021. Carbohydr Res 95:263–282

Ashford D, Desai NN, Allen AK, Neuberger A, O'Neill MA, Selvendran RR (1982) Structural studies of the carbohydrate moieties of lectins from potato (*Solanum tuberosum*) tubers and thorn-apple (*Datura stramonium*) seeds. Biochem J 201:199–208

Aspinall GO (1969) Gums and mucilages. In: Tipson RS, Horton D (eds) Advances in carbohydrate chemistry and biochemistry, vol 24. Academic, New York, pp 333–379

Aspinall GO (1980) Chemistry of cell wall polysaccharides. In: Preiss J (ed) Carbohydrates – structure and function. Academic, New York, pp 473–500 (The Biochemistry of Plants, vol 3)

Aspinall GO (1982) Chemical characterization and structure determination of polysaccharides. In: Aspinall GO (ed) The polysaccharides, vol 1. Academic, New York, pp 35–131

Aspinall GO, Baillie J (1963) Gum tragacanth. Part I. Fractionation of the gum and the structure of tragacanthic acid. J Chem Soc 1702–1714

Aspinall GO, Chaudhari AS (1975) Base-catalysed degradations of carbohydrates. X. Degradation by β-elimination of methylated degraded leocarpan A. Can J Chem 53:2189–2193

Aspinall GO, Fanous HK (1984) Structural investigations on the non-starchy polysaccharides of apples. Carbohydr Polymers 4:193–214

Aspinall GO, Jiang K-S (1974) Rapeseed hull pectin. Carbohydr Res 38:247–255

Aspinall GO, Whitehead CC (1970a) Mesquite gum. I. The 4-O-methylglucuronogalactan core. Can J Chem 48:3840–3849

Aspinall GO, Whitehead CC (1970b) Mesquite gum. II. The arabinan peripheral chains. Can J Chem 48:3850–3855

Aspinall GO, Charlson AJ, Hirst EL, Young R (1963). The location of L-Rhamnopyranose residues in gum arabic. J Chem Soc, pp 1696–1702

Aspinall GO, Cottrell IW, Egan SV, Morrison IM, Whyte JNC (1967) Polysaccharides of soy-beans. Part IV. Partial hydrolysis of the acidic polysaccharide complex from cotyledon meal. J Chem Soc (C), pp 1071–1080

Aspinall GO, Craig JWT, Whyte JL (1968a) Lemon-peel pectin. Part 1. Fractionation and partial hydrolysis of water-soluble pectin. Carbohydr Res 7:442–452

Aspinall GO, Gestetner B, Molloy JA, Uddin M (1968b) Pectic substances from Lucerne (*Medicago Sativa*). Part II. Acidic oligosaccharides from partial hydrolysis of leaf and stem pectic acids. J Chem Soc (C), pp 2554–2559

Aspinall GO, Cottrell JW, Molloy JA, Uddin M (1970) Lemon-peel pectin. III. Fractionation of pectic acids from lemon-peel and lucerne. Can J Chem 48:1290–1295

Björndal H, Hellerqvist CG, Lindberg B, Svensson S (1970) Gas liquid chromatography and mass spectrometry in methylation analysis of polysaccharides. Angew Chem Int Ed 9:610–619

Chizhov OS, Kadentsev VI, Solov'yov AA, Levonowitch PF, Dougherty RC (1976) Polysaccharide sequencing by mass spectrometry: Chemical ionisation spectra of permethyl glycosylalditols. J Org Chem 41:3425–3428

Conchie J, Hay AJ, Lomax JA (1982) A comparison of some hydrolytic and gas chromatographic procedures used in methylation analysis of the carbohydrate units of glycopeptides. Carbohydr Res 103:129–132

Darvill AG, McNeil M, Albersheim P (1978) Structure of plant cell walls VIII. A new pectic polysaccharide. Plant Physiol 62:418–422

Darvill AG, McNeil M, Albersheim P (1980) General and facile method for distinguishing 4-linked aldopyranosyl residues from 5-linked aldofuranosyl residues. Carbohydr Res 86:309–315

Dell A, Morris HR, Egge HE, von Nicolai H, Strecker G (1983 a) Fast-atom-bombardment mass-spectrometry for carbohydrate-structure determination. Carbohydr Res 115:41–52

Dell A, Oates JE, Morris HR, Egge H (1983 b) Structure determination of carbohydrates and glycosphino-lipids by fast-atom-bombardment mass spectrometry. Int J Mass Spectrom Ion Phys 46:415–418

Fales HM, Milne GWA (1969). Chemical ionization mass spectrometry of complex molecules. J Am Chem Soc 91:3682–3685

Fichtinger-Schepman AMJ (1980) Structural studies on the polysaccharide associated with the coccoliths of the alga *Emiliania huxleyi* (Lohmann) Kamptner. PhD Thesis, University of Utrecht

Fichtinger-Schepman AMJ, Kamerling JP, Versluis C, Vliegenthart JFG (1980) Structural analysis of acidic polysaccharides derived from the methylated, acidic polysaccharide associated with coccoliths of *Emiliania huxleyi* (Lohmann) Kamptner. Carbohydr Res 86:215–225

Gerwig GJ, Kamerling JP, Vliegenthart JFG (1978) Determination of the D and L configuration of neutral monosaccharides by high-resolution capillary G.L.C. Carbohydr Res 62:349–357

Geyer R, Geyer H, Kunnhardt S, Mink W, Stirm S (1983) Methylation analysis of complex carbohydrates in small amounts: Capillary gas chromatography-mass fragmentography of methylalditol acetates obtained from *N*-glycosidically linked glycoprotein oligosaccharides. Anal Biochem 133:197–207

Grob K, Grob K Jr (1978) On-column injection on to glass capillary columns. J Chromatogr 151:311–320

Hakomori S (1964) A rapid permethylation of glycolipid, and polysaccharide catalyzed by methylsulfinyl carbanion in dimethylsulfoxide. J Biochem (Tokyo) 55:205–207

Hanessian S (1971) Mass spectrometry in the determination of structure of certain natural products containing sugars. In: Glick D (ed) Methods of biochemical analysis, vol 19, Interscience Publishers, John Wiley, New York, pp 105–228

Harada K, Ito S, Takeda N, Suzuki M, Tatematsu A (1983) Sequence determination of permethylated oligosaccharides by chemical ionization mass spectrometry. Biomed Mass Spectrom 10:5–12

Harris PJ, Henry RJ, Blakeney AB, Stone BA (1984). An improved procedure for the methylation analysis of oligosaccharides and polysaccharides. Carbohydr Res 127:59–73

Jansson PE, Kenne L, Liedgren H, Lindberg B, Lönngren J (1976) A practical guide to the methylation analysis of carbohydrates. Chem Commun Univ Stockholm, No 8

Kärkkäinen J (1970) Analysis of disaccharides as permethylated disaccharide alditols by gas-liquid chromatography-mass spectrometry. Carbohydr Res 14:27–33

Kärkkäinen J (1971) Structural analysis of trisaccharides as permethylated methyl glycosides by gas-liquid chromatography-mass spectrometry. Carbohydr Res 17:1–10

Klok J, Cox HC, De Leeuw JW, Schenck PA (1982) Analysis of synthetic mixtures of partially methylated alditol acetates by capillary gas chromatography, gas chromatography-electron impact mass spectrometry and gas chromatography-chemical ionization mass spectrometry. J Chromatogr 253:55–64

Kochetkov NK, Chizhov OS (1966) Mass spectrometry of carbohydrate derivatives. In: Adv Carbohydr Chem Biochem 21:29–93

Ková čik V, Bauer Š, Rosík J, Kováč P (1968 a) Mass spectrometry of uronic acid derivatives. III. The fragmentation of methyl ester methyl glycosides of methylated uronic and aldobiouronic acids. Carbohydr Res 8:282–290

Ková čik V, Bauer Š, Rosík J (1968 b) Mass spectrometry of uronic acid derivatives. IV. The fragmentation of methyl ester methyl glycosides of methylated aldotriouronic acids. Carbohydr Res 8:291–294

Ková čik V, Mihálov V, Hirsch J, Kováč P (1978). Mass spectrometry of uronic acid derivatives. Biomed Mass Spectrom 5:136–145

Lau JM, McNeil M, Darvill AG, Albersheim P (1985) Structure of the backbone of rhamnogalacturonan-1, a pectic polysaccharide in the primary cell walls of plants. Carbohydr Res 137:111–125

Lee Y-C, Ballou CE (1965) Preparation of mamanobiose, mannotriose, and a new mannotetraose from Saccharomyces cerevisiae mannan. Biochemistry 4:257–264

Leontein KB, Lindberg B, Lönngren J (1978). Assignment of absolute configuration of sugars by GLC of their acetylated glycosides formed from chiral alcohols. Carbohydr Res 62:359–362

Lindberg B (1972) Methylation analysis of polysaccharides. In: Ginsburg V (ed) Methods in enzymology, vol XXVIII. Academic, New York, pp 178–195

Lindberg B, Lönngren J (1976) Specific degradation of polysaccharides containing uronic acid residues. Methods Carbohydr Chem 7:142–148

Lindberg B, Lönngren J (1978) Methylation analysis of complex carbohydrates: General procedure and application for sequence analysis. In: Ginsburg V (ed) Methods in enzymology, vol L, complex carbohydrates, part C. Academic, New York, pp 3–33

Lindberg B, Lönngren J, Thompson JL (1973) Degradation of polysaccharides containing uronic acid residues. Carbohydr Res 28:351–357

Lomax JA, Conchie J (1982) Separation of methylated alditol acetates by glass capillary gas chromatography and their identification by computer. J Chromatogr 236:385–394

Lönngren J, Svensson S (1974) Mass spectrometry in structural analysis of natural carbohydrates. In: Tipson RS, Horton D (eds) Advances in carbohydrate chemistry and biochemistry, vol 29. Academic, New York, pp 41–106

McNeil M (1983) Elimination of internal glycosyl residues during chemical ionization – mass spectrometry of per-O-alkylated oligosaccharide alditols. Carbohydr Res 123:31–40

McNeil M, Albersheim P (1977) Chemical-ionization mass spectrometry of methylated hexitol acetates. Carbohydr Res 56:239–248

McNeil M, Darvill AG, Albersheim P (1980) Structure of plant cell walls. X. Rhamnogalacturonan-1, a structurally complex pectic polysaccharide in the walls of suspension-cultured sycamore cells. Plant Physiol 66:1128–1134

McNeil M, Darvill AG, Åman , Franzén L-E, Albersheim P (1982 a) Structural analysis of complex carbohydrates using high-performance liquid chromatography, and mass spectrometry. In: Gingsburg V (ed) Methods in enzymology, Complex carbohydrates, vol 83. Academic, New York, pp 3–45

McNeil M, Darvill AG, Albersheim P (1982 b) Structure of plant cell walls. XII. Identification of seven differently linked glycosyl residues attached to 0-4 of the 2,4-linked L-rhamnosyl residues of rhamnogalacturonan 1. Plant Physiol 70:1586–1591

McNeil M, Darvill AG, Fry SC, Albersheim P (1984). Structure and function of the primary cell wall of plants. Annu Rev Biochem 53:623–663

Moor J, Waight ES (1975) The mass spectra of permethylated oligosaccharides. Biomed Mass Spectrom 2:36–45

O'Neill MA, Roberts K (1981) Methylation analysis of cell wall glycoproteins and glycopeptides from Chlamydomonas reinhardii. Phytochemistry 20:25–28

O'Neill MA, Selvendran RR (1980) Methylation analysis of cell wall material from Phaseolus vulgaris and Phaseolus coccineus. Carbohydr Res 79:115–124

O'Neill MA, Selvendran RR (1983) Isolation and partial characterisation of a xyloglucan from the cell walls of Phaseolus coccineus. Carbohydr Res 111:239–255

O'Neill MA, Selvendran RR (1985) Structural analysis of the xyloglucan from Phaseolus coccineus cell walls using cellulase derived oligosaccharides. Carbohydr Res 145:45–58

Peat S, Whelan WJ, Edwards TE (1961) Polysaccharides of baker's yeast. Part IV. Mannan. J Chem Soc, pp 29–34

Ring SG, Selvendran RR (1978) Purification and methylation analysis of cell wall material from *Solanum tuberosum*. Phytochemistry 17:745–752

Ring SG, Selvendran RR (1980) Isolation and analysis of cell wall material from beeswing wheat bran. Phytochemistry 19:1723–1730

Ring SG, Selvendran RR (1981) An arabinogalactoxyloglucan from the cell wall of *Solanum Tuberosum*. Phytochemistry 20:2511–2519

Sandford PA, Conrad HE (1966) The structure of the Aerobacter aerogenes A3 (S1) polysaccharide. I. A reexamination using improved procedures for methylation analysis. Biochemistry 5:1508–1517

Selvendran RR (1983) Applications of gas chromatography-mass spectrometry for cell wall analysis. In: Frigerio A (ed) Recent developments in mass spectrometry in biochemistry, medicine and environmental research 8. Elsevier Scientific Publishing, Amsterdam, pp 159–176

Selvendran RR (1985) Developments in the chemistry and biochemistry of pectic and hemicellulosic polymers. J Cell Sci Suppl 2:51–88

Selvendran RR, DuPont MS (1984) Problems associated with the analysis of dietary fibre and some recent developments. In: King RD (ed) Developments in food analysis techniques 3. Elsevier Applied Science Publishers, London, pp 1–68

Selvendran RR, March JF, Ring SG (1979) Determination of aldoses and uronic acid content of vegetable fibre. Anal Biochem 96:282–92

Selvendran RR, Stevens BJH, O'Neill MA (1985) In: Brett CT, Hillman JR (eds) Biochemistry of plant cell walls, Society for experimental biology seminar series, vol 28. Cambridge University Press, pp 37–78

Stephen AM (1983) Other plant polysaccharides. In: Aspinall GO (ed) The polysaccharides, vol 2. Academic, pp 98–193

Stevens BJH, Selvendran RR (1984) Pectic polysaccharides of cabbage (*Brassica oleracea* var. Decema) Phytochemistry 23:109–115

Sweet DP, Shapiro RH, Albersheim P (1974) The mass spectral fragmentation of partially ethylated alditol acetates, a derivative used in determining the glycosyl linkage composition of polysaccharides. Biomed Mass Spectrom 1:263–268

Sweet DP, Albersheim P, Shapiro RH (1975) Partially ethylated alditol acetates as derivatives for elucidation of the glycosyl linkage-composition of polysaccharides. Carbohydr Res 40:199–216

Talmadge K, Keegstra K, Bauer WD, Albersheim P (1973). The structure of plant cell walls. 1. The macromolecular components of the walls of suspension-cultured sycamore cells with a detailed analysis of the pectic polysaccharides. Plant Physiol 51:158–173

Taylor RL, Conrad HE (1972) Stoichiometric depolymerization of polyuronides and glycosaminoglycuronans to monosaccharides following reduction of their carbodiimide-activated carboxyl groups. Biochemistry 11:1383–1388

Toman R, Karácsonyi Š, Kubačková M (1975) Studies on the pectin present in the bark of white willow (*Salix alba* L.): Fractionation and acidic depolymerisation of the water-soluble pectin. Carbohydr Res 43:111–116

Toman R, Karácsonyi Š, Kubačková M (1976) Studies on pectin present in the bark of white willow (*Salix alba* L.) Structure of the acidic and neutral oligosaccharides obtained by partial acid hydrolysis. Cellul Chem Technol 10:561–565

Valent BS, Darvill AG, McNeil M, Robertson BK, Albersheim P (1980) A general and sensitive chemical method for sequencing the glycosyl residues of complex carbohydrates. Carbohydr Res 79:165–192

GC-MS Methods for Cyclic Nucleotides in Higher Plants and for Free High Unsaturated Fatty Acids in Oils

B. Janistyn

1 Introduction

Vegetable substances can be categorized in a multitude of chemical compound classes. Within each class they vary considerably in concentration.

The differentiation hitherto made between primary vegetable substances of low concentration (e.g., phytohormones) and secondary vegetable substances, usually occurring in high concentrations (e.g., flavonoids, alkaloids, terpenes etc.), appears to be questionable from this point of view.

All phytohormones known so far develop their physiological efficacy in concentrations which are also effective in animal systems (Cleland 1972; Galston 1961; Black et al. 1974; Mohr and Schopfer 1978; Grove et al. 1979).

The compounds not yet defined as of phytohormones, such as arachidonic acid and prostaglandines, lie in similarly low ranges of concentration (Janistyn 1982 a, b).

By definition, the concentrations of the secondary vegetable substances lie several orders of magnitude above those of the primary substances. However, the above-mentioned differences in concentration in secondary vegetable substances (e.g., mentha oil: 50–70% menthol and $<1\%$ piperitone) (Hefendehl 1962) do not preclude the possibility that the compounds, which are present in small amounts in this case, may have primary regulatory physiological effects in plants.

On the other hand, high concentration of secondary vegetable substances also appear to develop physiological effects. For example, the ubiquitous occurrence of flavonoides in leaves as protection against ultraviolet radiation is presently under discussion (Wellmann 1983).

This complex functional pattern of primary and secondary vegetable substances has been further complicated by the fact that the "second messengers" adenosine-3′:5′-monophosphate (cAMP) and guanosine-3′:5′-monophosphate (cGMP), known from animal organisms, are also found in plants (Greengard and Costa 1970; Goldberg and Haddox 1977; Brown and Newton 1981; Janistyn 1983).

In essence, there are two reasons for the problems of analysis of vegetable substances:

First, the small concentrations of vegetable substances. They are in the range of pmol g^{-1} fresh weight ($\hat{=} 0.1$ ppb) and hence they require either large amounts of plant material for their isolation, or correspondingly sensitive methods of detection, e.g., bio-assays. In contrast to animal organisms, plants contain a variety of compounds which, in their amount, lie several orders of magnitude above the primary vegetable substances. Therefore, they substantially impair bio-assays.

The second reason for problems within vegetable substance analysis lies in the fact that the incorporation of radioactively labeled precursor compounds, even into dominant secondary vegetable substances, occurs only with low apparent rates. Subsequent purification is consequently difficult and time-consuming. Furthermore, a large pool of outlined components cannot be expected, since they are subject to a correspondingly high turn-over due to the high plant-physiological activity.

2 Methods of GC-MS

In proportion to the physicochemical nature of the GC-MS analysis and to its high detection sensitivity, the above-mentioned problems become less in significant.

In the case of compounds lying within the range of pmol g^{-1} fresh weight of the material to be investigated, the enrichment and purification methods are, however, necessary.

Within the limits of this chapter, the mode of operation and the fundamental methodology of the GC-MS analysis cannot be dealt with. For this purpose, there are a large number of monographs and other publications (Budde and Eichelberger 1979; Waller and Denver 1980; Spiteller 1970).

3 Adenosine-3′:5′-Monophosphate (cAMP) in Maize Seedlings (*Zea mays*)

This cyclic nucleotide is well known as a mediator in the action of several hormones and of changes induced by the environment in mammals and bacteria (Greengard and Costa 1970; Robison et al. 1968; Rasmussen 1970; Jost and Rickenberg 1971; Pastan et al. 1975). Earlier attempts to detect cAMP with certainty in higher plants failed for the above-mentioned reasons.

Maize seeds (*Zea mays* L., Golden Bantam 8 row and Bear Hybrid WF 9/38 Vaughan's Seed Co.) were kept under running water for 24 h and were spread out on steam sterilized moist filter paper in plastic boxes. The seeds were sprayed with a 0.1% aqueous solution of penicillin G, Na-salt and kept at 20 °C in the dark. Pieces of seedling were controlled periodically for sterility according to Berlin et al. (1971). The seedlings were harvested after 5 days under green safelight and a fresh weight of 20,450 g was determined. They were frozen in liquid nitrogen and

---

Freeze—dried tissue—extract 1 (petrolether, 40—60°C;
discarded), extract 2 (ethylacetate, discarded), extract
3 (ethanol—water, 80 : 20, v/v) containing the nucleotides,
—charcoal—anion—exchange—column—1. TLC—2. TLC—PC—
silylation—GC—MS

Fig. 1

Table 1

Purification	Tracer radioactivity		Sample dry weight (tracer added)		Sample dry weight (without tracer)	
	cpm $\times 10^5$	[%]	[mg]	[%]	[mg]	[%]
Extract 3			4.15×10^4	100	4.15×10^5	100
Tracer added: adenine-U-^{14}C-cAMP; 0.5 µCi	6.97	100				
Charcoal	6.67	95.7	5420	1.3	5411	1.3
Anion-exchange-column	5.81	83.5	510	1.2×10^{-1}	505	1.2×10^{-1}
1. TLC	4.86	69.8	120.3	3.0×10^{-2}	118.8	2.8×10^{-2}
2. TLC	3.47	49.8	9.04	2.7×10^{-3}	8.98	2.7×10^{-3}
PC	3.15	45.2	1.16	2.7×10^{-4}	1.15	2.7×10^{-4}

lyophilized at -20 °C; 1640 g of dry weight, representing 8% of the fresh weight, was thus obtained. The dry material was submitted to a series of extraction steps on a Soxhelet-Apparatus, each over 24 h. The procedures for extraction and purification of cAMP are summarized in Fig. 1.

Extract 3 was evaporated under vacuum and freeze-dried. A brown residue of 830 g (50.6%) was obtained. The residue was dissolved in 1 l redist. water.

The one half of the solution [adenine-U-^{14}C]adenosine-3′ : 5′-monophosphate (0.5 µCi; specific radioactivity: 238 mCi mmol^{-1}; Radiochemical Center, Amersham Buchler, Braunschweig) was added as a marker in the subsequent purification steps. Both solutions, with and without marker, were treated identically (Table 1). The solutions were stirred with 50 g of activated charcoal (Merck) overnight and centrifuged at 35,000 g for 30 min. The supernatants were discarded and the sediments were resuspended and repelleted until the supernatants were colorless. The sediments were extracted by water-ethanol-ammonia 25% (48 : 50 : 2, v/v) and dried under vacuum to give light brown residues of 5.420 g (^{14}C tracer) and 5.411 g. The residues were dissolved in redist. water and applied to a weak basic anion exchange column (MN-2100, ECTEOLA Cellulose, Machery, Nagel & Co, D-5160 Düren; 1.5 cm/100 cm). Fractions of 10 ml in redist. water were collected and 1 ml aliquots were assayed for radioactivity in 10 ml of a dioxan scintillation cocktail. The dioxan-cocktail contained 100 g naphthalene and 5 g PPO (2,5-diphenyloxazole) in 1 l dioxane. The fractions containing radioactivity were pooled and freeze-dried. Residues of 510 mg (^{14}C tracer) and 505 mg were obtained. In the next step the samples were submitted to a twofold preparative thin-layer chromatography on pre-coated silica gel plates (Merck, F, 354). For the first chromatography a mixture of n-butanol-methanol-ethylacetate-ammonia 25% (7 : 3 : 4 : 4, v/v) was used. The cAMP zones were detected in UV-light and by radioactive scanning (Bertold LB 2722).

The zones were scraped off, eluted with methanol and dried under vacuum to yield residues of 120.3 mg (^{14}C tracer) and 119.8 mg. For the second chromatography a mixture of isopropanol-methanol-water-ammonia 25% (10 : 1 : 2 : 5, v/v)

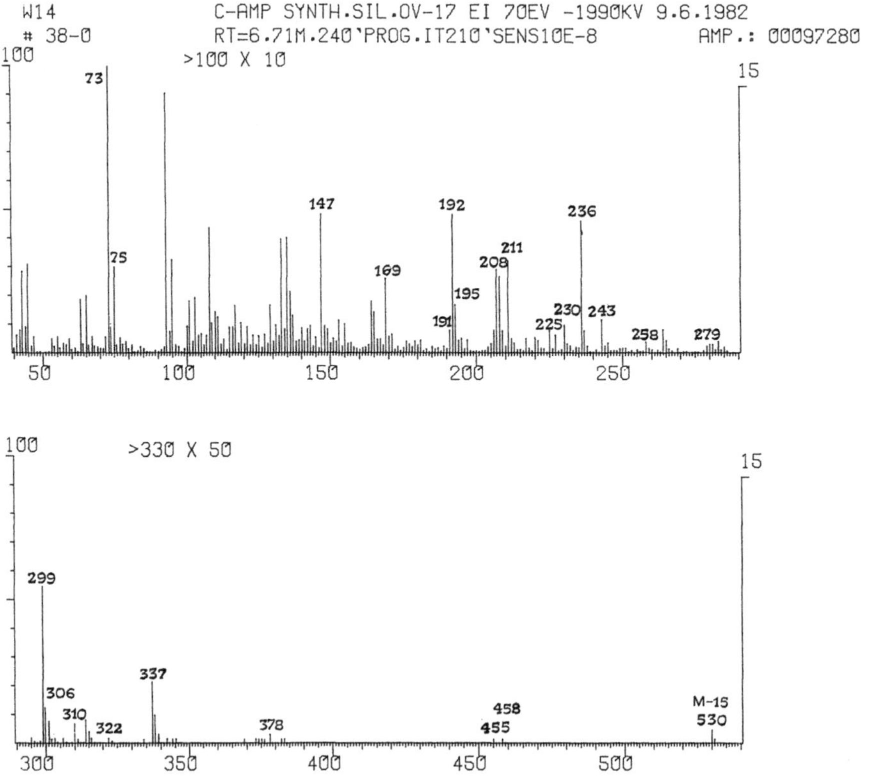

Fig. 2. According to Lawson et al. (1971) the mass spectra of the trisilylated cAMP were recorded on a Finnigan GC (9610)/MS Model 4000 instrument with sample introduction through the gas chromatographic inlet; 4 ft × 2 mm i.d. (glass) 3% OV 17 on Chromosorb W/AW-DMCS; M.P. 80/100 mesh, He-30 ml/min Inj. 240 °; temperature programmed at 15 °/min from 100 ° to 250 °; glas-jet-separator 240 ° and ion source temperatures 220 °; accelerating voltage – 2 kV; ionizing energy 70 eV. The molecular ions m/e 545 and m/e-15 were only observed with sample introduction by direct inlet. The peaks m/e 310, 378, and 392 are unique to the 3′ : 5′-cyclic structure

was used. After detection as described above the cAMP zones were eluted with boiling pyridine. The pyridine was removed under vacuum (23 °C; 0.01 Torr) and residues of 9.042 mg (^{14}C tracer) and 8.985 mg were obtained. The last purification step used ascending paper chromatography (Schleicher & Schüll, Nr. 2043 b. Mgl. 580 by 600 mm) in isobutyric acid-water-ammonia 25% (66 : 33 : 1, v/v, Rf.: 0.56). The chromatograms were dried in a vacuum oven and then scanned for UV-absorption and radioactivity (Metrawatt, LB 280). The paper strips containing cAMP were eluted with methanol. The eluates were evaporated under vacuum and dried over phosphoruspentoxide at 105 °C under vacuum for 12 h. Residues of 1.160 mg (^{14}C tracer) and 1.158 mg were obtained. The residues were stored at −20 °C.

For GC-MS determination sample aliquots were silylated by the following method: cAMP (TMS)$_3$ was formed by dissolving 50 µg of the acid as its diso-

Table 2. cAMP content in maize seedlings

	cAMP-content	
	Minimal amount of cAMP	Maximal amount of cAMP
Per g fresh weight of the maize seedlings	382 pmol	710 pmol
Per g dry weight of the maize seedlings	4763 pmol	8853 pmol

dium or diammonium salt in pyridine-BSTFA-TMCS (9:90:1), followed by heating in a sealed vial at 100 °C for 3 h (Lawson et al. 1971). To remove some turbidity the silylated samples were centrifuged and the supernatants were used for the determination. Synthetic cAMP was also silylated to provide a standard solution.

The silylated genuine cAMP content was quantified by GC detection and referred to silylated cAMP standards. The peaks of the silylated genuine cAMP were examined in a mass spectrometer attached to the gas chromatograph. The cAMP was identified by MS according to Lawson et al. (1971) (Fig. 2).

c-AMP Isolated from Phaseolus vulgaris Resulted in Corresponding MS Data by Direct Inlet (Newton et al. 1980). The cAMP levels found are corrected for recovery of tracer [^{14}C]cAMP which was added to the ethanol/water extract obtained from the freeze-dried seedlings. For loss of silylated cAMP during GC recording a minimum and maximum amount of cAMP was estimated (Table 2). The concentration of cAMP is of the same order of magnitude as that found in animal tissue (10^{-7} M) where cAMP is regulatory (Greengard and Costa 1970).

4 An Isotope Dilution GC-MS Spectrometric Assay for cAMP in Cultured Tobacco Tissue (Johnson et al. 1981 a, b)

By way of this method picomole quantities of cAMP can be determined with high precision and selectivity using gas chromatography and electron impact mass spectrometry with multiple ion detection techniques. Using synthetic [2,8-^{2}H$_2$, 6-^{15}N]-cAMP as the internal standard, suitable specificity was obtained by monitoring the (M-CH$_3$)$^+$ fragment ions of the trimethylsilyl derivatives of cAMP and the internal standard at m/z 530 and m/z 533, respectively. The sensitivity of the assay as judged from the lower limit of detection of the mass spectrometer was 3.0 pmol.

Synthesis, Characterization and Standardization of [2.8-^{2}H$_2$,6-^{15}N]cAMP. 6-Chloropurine riboside cyclic 3′,5′-monophosphate (15 mg), ^{15}NH$_4$Cl (20 mg), dry methanol (250 µg) and triethylamine (50 µg) were heated in a sealed glass ampoule for 12 h at 110 °C. Examination of the product by TLC indicated very little starting material and a major product which co-chromatographed with authentic cAMP [silica gel; 2-propanol:NH_3:H_2O (7:1:3), R_fcAMP $\hat{=}$ 0.6]. The solvent in the reaction mixture was removed under a stream of dry nitrogen and ^2H_2O (1.5 ml, 99% $^2H_2$) containing Raney Ni (20 mg) was added to the residue. This solution was heated in a sealed glass ampule for 24 h at 100 °C. A small aliquot was removed, dried and derivatized (TMS) for GC-MS-analysis. The analysis indicated a mixture of partially labeled cAMP products with the composition; dideuterated, [6-^{15}N]cAMP 60%; monodeuterated, [6-^{15}N]cAMP 30%; [6-^{15}N]cAMP and nonlabeled cAMP 10%. The ^2H_2O in the above reaction mixture was removed and fresh ^2H_2O (1.5 ml) containing Raney Ni (20 mg) was added. This solution was heated for a further 24 h at 100 °C. The catalyst was separated by centrifugation and the final product again analyzed by GC-MS. Analysis of the derivatized (TMS) product indicated the labeled cAMP composition to be [2,8-$^2H_2$,6-^{15}N]cAMP 91%; monodeuterated, [6-^{15}N]-cAMP 8%; [^{15}N]cAMP and nonlabeled cAMP approximately 1%. The final reaction mixture was purified by tlc (silica gel; 20-cm × 20-cm × 0.5-mm plates). The intensive UV-positive band cochromatographing with authentic cAMP was collected and extracted with 50% EtOH:^2H_2O. The solution containing the labeled cAMP was standardized using UV-spectroscopy (λ_{max} = 258 m) and restandardized and checked for purity using GC-MS. The yield of the purified labeled cAMP was 14%. The final standard solution of [2,8-$^2H_2$,6-^{15}N]cAMP (1 μg μl^{-1}) in 50% EtOH:^2H_2O was restored at -20 °C until required, and did not appear to undergo any decomposition under these conditions.

Gas Chromatography/Mass Spectrometry Multiple Ion Detection. The instrumentation consisted of a Varian 1440 series GC connected, via an all-glass single-stage jet separator with solvent dump valve, to a VG-Micromass 7070F mass spectrometer equipped with multiple ion detection facilities. The GC columns used were 4 ft × 1/8 in. o.d., 1/16 in. i.d., silanized glass, packed with 1% OV-17 on (100–120) mesh gas Chrom Q support (Applied Sciences). Column temperature was programmed from 240 °C at 6 °C min^{-1} with the injector temperature maintained at 260 °C. The jet separator line and source re-entrant region were kept at 290 °C. High-purity helium was used as the carrier gas with a flow rate of 25 ml min^{-1}. The mass spectrometer source was operated in the electron impact mode at 70 eV, a filament emission current of 200 μA, and the source block temperature at 200 °C. Under these conditions cAMP has a GC retention time of 3.5 min. An eight-channel MID-unit was used in the alternating accelerating voltage mode using the m/z 207 column bleed peak as the lock-channel reference mass. The detection system consisted of a continuous ynode electron multiplier operated at -3 kV with further amplification of each selected ion channel ($\times$ 500). The mass spectrometer collector slit was adjusted to provide flat top peaks with a resolution (10% valley) of 1000. Selected ion currents as GC elution profiles were recorded on a six-pen Rika Denki chart recorder.

Quantitation was based on peak-height ratios obtained by dividing the respective m/z 530 peak-height response of the (M-CH$_3$)$^+$ fragment ion for the TMS derivative of endogenous cAMP (1b) in the unknown, by the m/z 533 peak-height response of the (M-CH$_3$)$^+$ fragment ion of the added derivatized (TMS) internal standard (1d). Unless there is good reason for concluding that the peak width or

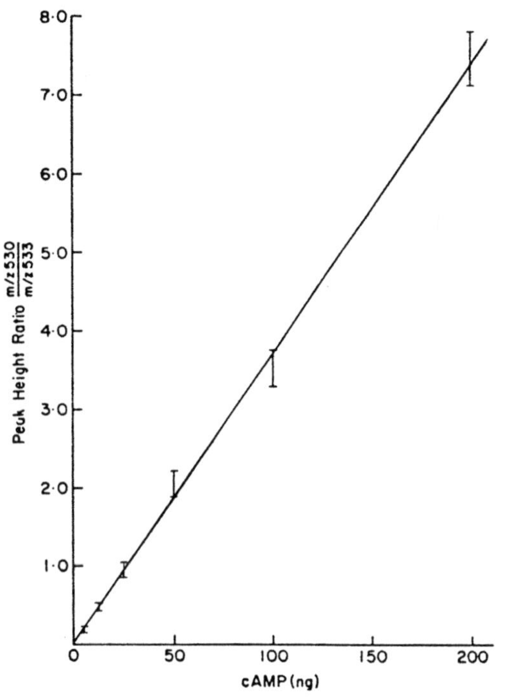

Fig. 3. Standard calibration curve for determining cAMP using MID

shape is changing over successive runs, peak height measurements, as opposed to peak area measurements, are considered to be adequate.

Standard solutions of $[2,8-^{2}H_{2},6-^{15}N]$-cAMP and cAMP were prepared in 50% EtOH:H_2O and appropriate microliter volumes were measured into 1-ml Reacti-Vials to provide GC-MS standards with a range of nonlabeled cAMP concentrations from 0 to 200 ng μl^{-1}, each of which contained a fixed quantity of labeled cAMP, namely 25 ng μl^{-1}. The solutions were dried over a stream of nitrogen, lyophilized overnight, and derivatized as described previously. Injections (1 μl on column) were performed in triplicate for each standard concentration and the peak-height ratios of the m/z 530 and 533 ion responses were plotted as a function of nonlabeled cAMP concentration in the usual manner as shown in Fig. 3. The data were subjected to a linear regression analysis and a line of best fit was determined. The error bars in Fig. 3 represent the spread of observed ratios about the line of best fit.

The tri-TMS derivative of cAMP was readily prepared by adding Regisil RC-2 (99 μl BSTFA/and 1 μl TMCS) and 100 μl pyridine to cAMP in a Teflon-capped Reacti-Vial (Pierce) to provide a sample concentration of approximately 1 μg μl^{-1}. The solution was heated at 70 °C for 10 min and then left overnight at room temperature. The derivatized cAMP remains stable in this solution for many weeks.

cAMP in Cultured Tobacco Tissue (Johnson et al. 1981 a, b). Small pieces of tissue were dropped into methanol-chloroform-formic acid-water (12:5:1:2, by vol.;

5 ml g^{-1} fresh wt tissue) which had been chilled to $-40\,°C$. The following internal standards were added: $c[8\text{-}^3H]AMP$ (25 Ci $mmol^{-1}$; 0.1 pmol g^{-1} tissue); $c[2,8\text{-}^2H_2,6\text{-}^{15}N]AMP$ (4 µg 100 g tissue). After 48 h at $-20\,°C$ to inhibit phosphatase, the mixture was homogenized, stirred at $4\,°C$ for 24 h and finally filtered at $23\,°C$. The plant residue was stirred with methanol-water-formic acid (20:80:1, by vol; 5 ml g^{-1} tissue) at $23\,°C$ for several hours and the mixture was then filtered. The two extracts were evaporated to dryness in vacuo, and aqueous suspensions of the residues were combined.

Aqueous solutions of the extracts were neutralized with 10 M KOH, centrifuged, concentrated and applied to a column (2×10 cm) of neutral alumina (Merck), from which cAMP was eluted with distilled water. The eluate fractions containing $c[^3H]AMP$ were pooled, concentrated and applied to a Dowex-50 ion-exchange column (2×10 cm; 200–400 mesh; H^+ form). Cyclic AMP was eluted with distilled water and the combined ^3H-containing fractions were further purified by TLC on Merck pre-coated silica gel $60F_{254}$ plates using 2-propanol:26% (w/w) aqueous NH_4OH (7:3) as solvent ($R_f cAMP$, $\hat{=}0.6$). Purified methyl blue chromatographed just below $c[^3H]AMP$ and was used to locate cAMP on the TLC plate. A portion of the $c[^3H]AMP$ zone ($\hat{=}25\%$) was discarded to avoid an overlapping fluorescent compound which interfered with the GC-MS-MID assay. The remainder of the TLC zone was eluted with 50% ethanol. Part of the eluate was subjected to GC-MS-MID, while the remainder was further purified by HPLC as outlined below.

An aqueous solution of the evaporated TLC eluate was centrifuged (10,000 g, 10 min) and purified isocratically using a DuPont Zorbax C-8 column (10 µm packing; 4.6 mm $\times$ 25 cm). The mobile phase consisted of 0.16 M acetic acid containing 4% (v/v) methanol and the flow rate was 1.0 ml min^{-1}. In this system the k'-values for the cAMP and its isomer adenosine 2':3'-cyclic monophosphate (2':3'-cAMP) were 1.1. and 0.40 respectively. A peak which eluted at the retention time of authentic cAMP and which exhibited a A_{254}/A_{280} ratio similar to that of cAMP, was collected. This peak, which was homogeneous by analytical HPLC, was further examined by GC-MS-MID. The equipment and experimental details of the GC-MS-MID step in the assay for cAMP are described 4 in this section.

Due to the low cAMP level in plant tissue and the interference from other endogenous compounds, several additional purification steps were necessary for cAMP quantitation in tobacco callus tissue. After inactivation of phosphatase at $-20\,°C$, the tissue was extracted with solvent containing $c[8\text{-}^3H]AMP$ and $c[2,8\text{-}^2H_2,6\text{-}^{15}N]\text{-}AMP$. The former enabled cAMP-containing fractions to be detected by radiotracer monitoring during extract purification and also provided a basis for calculation of cAMP recoveries; the latter served as the internal standard for GC-MS-MID quantitation. The tobacco tissue extract was purified by sequential chromatography on columns of alumina and Dowex-50, and then by TLC on silica gel. The overall recovery of $c[8\text{-}^3H]AMP$ was $\hat{=}20\%$.

The eluate of the $c[8\text{-}^3H]AMP$-containing zone was trimethylsilylated and the trimethylsilyl (TMS) derivatives were analyzed by GC-MS-MID. The monitored fragment ions derived from the tri-TMS cAMP and their identity are as follows:

m/z 533 (M^+-CH_3 of the $^2H_2$, ^{15}N-labeled standard);

m/z 530 (M^+-CH_3 of endogenous cAMP); and

m/z 310 (a ribose phosphate fragment ion, common to both endogenous and isotopically labeled cAMP).

In the GC-system used, the retention times of c[2,8-$^2H_2$,6-^{15}N]-AMP and unlabeled cAMP are 5.8 and 5.9 min respectively, i.e., there is a slight isotopic fractionation. The ioncurrent for m/z 533, 530, and 310 for the TLC-purified extract exhibited peaks at the exact retention times of authentic labeled and unlabeled cAMP, but in addition there were several other prominent peaks at differing retention times. Hence, the purification procedure was not sufficient to remove from the plant extract all compounds which were capable of giving rise to ions at the same m/z values as those used to monitor cAMP and its labeled analog. Although these compounds did not appear to interfere in the quantitation of cAMP, the TLC eluate was further purified by HPLC on a Zorbax C-8 column which effectively separates cAMP from its 2′:3′-isomer. The fraction containing c[^3H]AMP was collected and subjected to GC-MS-MID. The ion-current traces for m/z 533 and m/z 530 each showed only one prominent peak corresponding to the retention time of labeled and unlabeled cAMP, respectively. The ratio of the heights of the 533, 530, and 310 ion-current peaks at the retention time of cAMP was the same as that measured before HPLC purification. The TMS derivative of the compound exhibited the same GC retention time as tri-TMS-cAMP and fragmented to yield high-mass ions characteristic of the TMS derivative of cAMP during electron-impact mass spectrometry.

5 Stability of Cyclic Purine Nucleotides in the Presence of Hydrochloric Acid During Extraction

The following method is useful when it is only cAMP or cGMP that needs to be isolated from plant material (Janistyn and Drumm 1975). Each sample of plant material (1 g) was homogenized in 4 ml 0.1 M HCl at room temperature with an Ultraturrax (20,000 rpm; 1 min). The homogenate was boiled for 10 min in order to hydrolyze all the nucleotides except the cGMP (Sutherland and Rall 1958) or cAMP (own observations).

After centrifugation at 45,000 rpm (15 min) the supernatant fluid was transferred to a slightly basic ion exchange column (MN-2100; ECTEOLA-Cellulose, Machery, Nagel & Co., D-5160 Düren; 8 mm by 160 mm). The final height of the column, which was filled the evening before use, was 130 mm. Elution was carried out with distilled water for cAMP and with 0.1 M aqueous ammonia for cGMP. For example, in the case of *Sinapis alba* (Janistyn and Drumm 1975) the first 190 ml of the eluate were discarded, and the next 80 ml of water containing the cAMP were collected and dried in a rotation evaporator. cGMP was observed when the eluate became basic (pH = 8).

6 Guanosine-3′:5′-Monophosphate (cGMP) in Maize Seedlings (*Zea mays*)

The occurrence of cGMP as cAMP is well known in mammals, but its biological functions are still relatively obscure (Greengard and Costa 1970). Earlier attempts to identify cGMP with certainty in higher plants failed for similar reasons as those with the above-mentioned cAMP. From the same material described for the isolation of cAMP (3), cGMP was isolated by the following procedure: When the pooled and freeze-dried fractions from the anion-exchange column were submitted to preparative thin-layer chromatography (TLC) on pre-coated silica gel plates (F-254, E. Merck Darmstadt, FRG) with the solvent system n-butonol:methanol:ethyl acetate: 25% (w/w) aqueous ammonia, 7:3:4:4 (by vol) the cAMP zone was visible in UV light (254 nm). No zone was detected corresponding to the reference spot of authentic cGMP (R_f-values: cAMP, 0.29; cGMP, 0.15). Because preliminary experiments showed only a little mobilization of authentic cGMP on ECTEOLA-Cellulose (MN 300) TLC plates with water as solvent, the elution of the anion-exchange column was continued with 0.1 M of aqueous ammonia. Fractions of 5 ml were collected. When the eluate became basic (pH 8), a significant absorbance at 252 nm was observed in the next 50 fractions (UV-spectralphotometer, DB 1200, Beckman Munich, FRG). These 50 fractions were pooled and freeze-dried. A light brown residue of 110 mg containing some column material was obtained. This residue was submitted to preparative TLC. A weak blue fluorescent zone was detected under UV light (254 nm) corresponding to a reference spot of authentic cGMP with the R_f 0.15. The R_f 0.15-zone was scraped off and eluted with ethanol:water 8:2 (v/v). The eluate was concentrated on a vacuum rotary evaporator, filtered and freeze-dried. Twenty five mg of substance were obtained in all. The silica containing sample was applied to ascending paper chromatography (Nr. 2043 b Mgl.; 580 × 600 mm; Schleicher and Schüll, Dassel, FRG) in isobutyric acid:water: 25% (w/w) aqueous ammonia, 66:33:1 (by vol). The chromatograms were dried in a vacuum oven at 50 °C and scanned for UV absorption.

The zones with an R_f of 0.28, corresponding to a reference spot of cGMP, were cut out and eluted with methanol. The eluates were evaporated under vacuum and dried over phosphorus pentoxide at 80 °C under vacuum for 12 h. A sample of 1.05 mg was obtained and stored at −20 °C over blue silica gel.

Since at every purification step (preliminary experiments showed the same behavior of cAMP and cGMP for extraction and charcoal absorption) the cAMP/cGMP ratio was found to be 10:1 and the recovery of cAMP had been calculated to be 45%, it was assumed that the recovery of cGMP was also about 45%.

For the GC-MS determination, sample aliquots of 50–80 µg were silylated with 100 µl of a mixture of 50 µl absolute pyridine and 50 µl of N,O-bis(trimethylsilyl)trifluoroacetamide (BSTFA) containing 1% trimethylchlorosilane (TMCS) (Pierce Chemical Co., Rockford, Ill., USA, Nr. 38831) in a Teflon-capped reacti-vial (Pierce) at 70 °C for 30 min. To remove some turbidity, the silylated samples were centrifuged and the supernatants were used for the GC and GC-MS determinations.

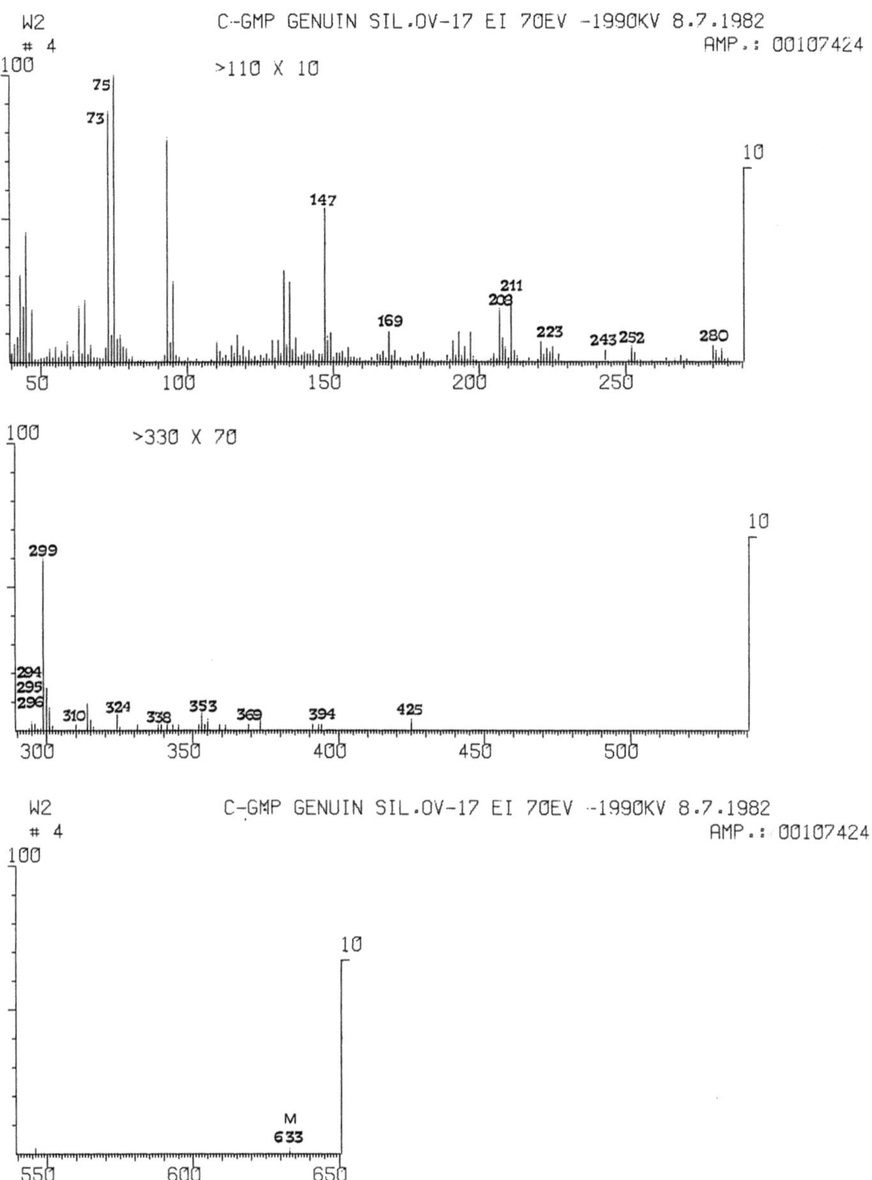

Fig. 4. Mass spectrum of the isolated and silylated cGMP. *Ordinate* relative abundance (%); *abscissa* m/c. Conditions: the mass spectra were recorded on a Finnigan GC (9610)-MS Model 4000 instrument/Finnigan Instruments, Sunnyvale Cal., U.S.A.) with sample introduction through the gas chromatograph inlet; 1% OV-17 on Gas Chrom Q, 100/120 mesh; 120 mm long, 2 mm i.d. glass column, presilylated with SILYL-8 (Pierce); He, 60 ml min^{-1}, 260 °C, column temperature 220 °C at 4 °C min^{-1} to 250 °C (flame ionization detector 200 °C); glass-jet separator 250 °C; ion-source temperature 250 °C; accelerating voltage 2 kV and I.E. 70 eV. The peak of the isolated and silylated cGMP was observed at the retention time of 9 min and derivated authentic cGMP co-chromatographed at the same retention time. Under the given conditions, silylated cGMP and cAMP (6 min 48 s) could be separated clearly

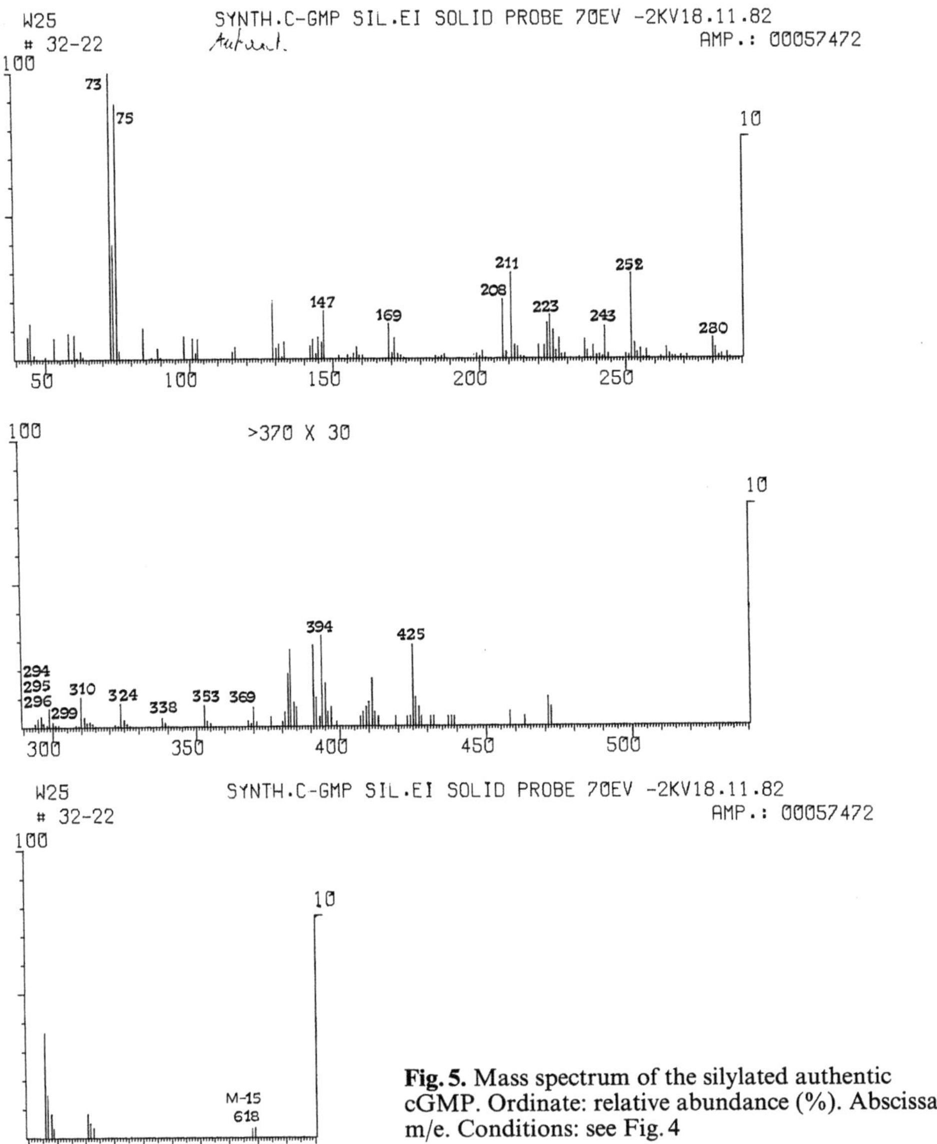

Fig. 5. Mass spectrum of the silylated authentic cGMP. Ordinate: relative abundance (%). Abscissa: m/e. Conditions: see Fig. 4

Table 3. Gas-chromatographic determination of minimum and maximum amounts of cGMP extracted from maize seedlings

	cGMP content (pmol g^{-1} fr. wt. or dry wt.	
	Minimum	Maximum
Fr. wt.	35	72
Dry wt.	436	989

The content of cGMP in the isolated sample, determined as the tetrasilyl derivative using GC detection, was quantified with reference to a standard of silylated authentic cGMP.

As shown in Figs. 4 and 5, the mass spectrum of the isolated and silylated sample agreed with the corresponding spectra of the silylated authentic cGMP. Furthermore, no difference was found between the GC peaks when authentic silylated cGMP was added to the isolated silylated sample (see legend to Fig. 4).

The mass of the molecular ion and all significant fragment ions of the derivated cGMP were derived from the GC-MS investigations of parts of the whole cGMP molecule by Lawson et al. (1971). Also, the molecular ion of m/e 633 or the primer fragment ion of 618 (loss of only one methyl group) was found for the tetrasilyl cGMP.

Referring to a standard of silylated authentic cGMP, the content of the sample of isolated cGMP was measured. Because of the loss of silylated cGMP during the recording of GC, a minimum and maximum amount of cGMP were estimated as shown in Table 3. The data represent the maximal fluctuations observed in all our GC-determinations. The data show that the concentration of cGMP found in maize seedlings is of the same order of magnitude as found in animal tissue (10^{-8} M) (Greengard and Costa 1970).

With the unambiguous identification of cGMP in maize seedlings using GC-MS the occurrence of another biologically important molecule which is known to be regulatory in animals (Greengard and Costa 1970) has now been established in higher plants.

7 GC-Separation of Synthetic cAMP and cGMP in a Mixture

Under the following given conditions the silylderivatives of cAMP and cGMP were satisfactorily separated (Janistyn, unpublished) (Fig. 6):

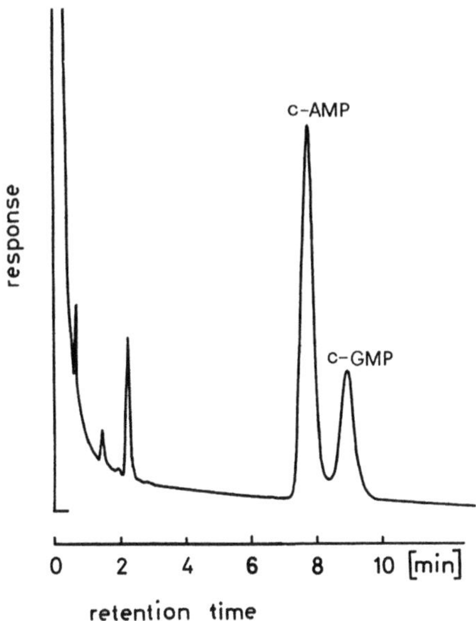

Fig. 6. Gas chromatogram of a mixture of the silylated authentic cAMP (50 µg) and cGMP (20 µg). Conditions: Finnigan GC (9510); 1% OV-17 on Gas Chrom Q, 100/120 mesh; 120 mm–2 mm i.d., glass column, presilylated with SILYL-8 (Pierce); N_2, 60 ml min^{-1}, 260 °C, column temperature 220 °C, 4 °C min^{-1} to 250 °C, FID 200 °C. The peak with a retention time of 7 min and 47 s corresponds to the silylated cAMP and the peak with the retention time of 9 min to the silylated cGMP

8 Cyclic Pyrimidine Nucleotides in Plants?

(Cytidin-3′ : 5′-monophosphate, thymidine-3′ : 5′-monophosphate, and uracil-3′ : 5′-monophosphate)

So far no cyclic pyrimidine nucleotides have been detected in plant species. One of the reasons may be their increased chemical instability compared to cyclic purine nucleotides.

9 Free High Unsaturated Fatty Acids in Oils

The occurrence of linol and linolenic acids in free and many derivative forms has been observed in the oils of different plants.

Arachidonic acid ([all-Z]-5,8,11,14-eicosatetraenoic acid) has hitherto not been confirmed with certainty in plants or in any of their products (Catalano 1967; Laskowski and Kulinowska 1967; Dziedzianowicz-Wierzbicka and Krauze 1970; Hitchcock and Nichols 1971; Kunau 1976). The following method describes the identification and quantification of arachidonic acid in the acid fraction of virgin wheat-germ oil.

Taking into account the quantitative aspect of the problem of detecting such a labile lipophilic compound as arachidonic acid, the investigation was started with 10 kg of virgin wheat-germ oil. Because saponification of the oil by preliminary experiments had raised some doubts whether it was possible to detect arachidonic acid using this method, the free acidic compounds of the oil were extracted using an aqueous sodium bicarbonate solution.

In 1-kg portions, 10 kg cold-pressed (virgin) wheat-germ oil (Keimdiät GmbH, Augsburg, F.R.G.) was added at 5 °C to 2 l ethanol (99%). The mixture was layered on 3 l of a saturated aqueous sodium chloride solution. Under constant but not too vigorous stirring a solution of saturated aqueous sodium bicarbonate was added until a pH of 8.0 was reached. The stirring was continued for 12 h at the same temperature. Thereafter, the phases were allowed to separate in a separating funnel at room temperature. The separated aqueous phase was extracted twice with 1 l diethylether. The organic phases were discarded, the aqueous phase was acidified carefully by adding solid tartaric acid until no further turbidity occurred. Three extractions, each with 1 l diethylether, followed. After separation, the organic phases were dried over sodium sulfate, filtered and pre-evaporated on a rotary vacuum vaporizer. The residual diethylether was removed under high vacuum to yield a light brown residue of 50.5 g (0.5% weight of the starting material) which was kept under nitrogen. As the residue was dissolved in 250 ml hot absol. ethanol and allowed to stand at −20 °C for 48 h, a dense flocculation appeared which was removed by filtration through a glass suction frit. The filtrate was concentrated under a stream of nitrogen at room temperature.

When the volume of the solution was reduced respectively to 150 ml and further to 100 ml filtration and cooling were repeated. The concentration was continued until a highly viscous oil was obtained. Residual ethanol was removed completely under high vacuum; yield 25.4 g (0.25%). The residue was dissolved in 400 ml absol. methanol and after 100 mg of p-toluenesulfuric acid had been added, the solution was boiled by reflux under nitrogen for 8 h. The methyl esters obtained in this way were distilled under nitrogen at 1.3 Pa. The middle fraction with the boiling points 130 °–150 °C (6 g; 0.06%) was redistilled [b.p. 120 °–132 °C; 0.13 Pa; 3 g (0.03%)] and used for the search for methyl arachidonate by gas chromatography and gas chromatography-mass spectrometry (GC-MS).

Small amounts of acid(s) may be better esterified by the following method:

18.6 mg of the residue were dissolved in 95 µl N,N-dimethyl-acetamide. Then, 95 µl of tetramethylammoniumhydroxide and 20 µl of methyliodide were added at room temperature. After the precipitate of the tetramethylammonium-iodide had formed (10 min), the supernatant, including the methylester, were taken directly for GC-MS-analysis.

As shown in Figs. 7 and 8, a peak occurred at the same retention time of 26 min in the gas chromatograms of the methyl ester from the wheat-germ oil and

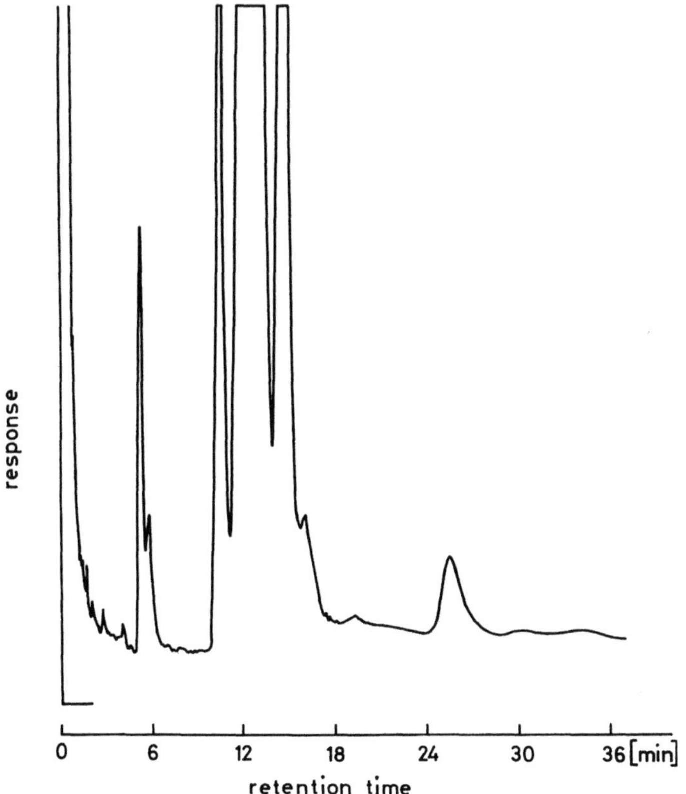

Fig. 7. Gas chromatogram of the methyl ester fraction (120 °–132 °C; 0.13 Pa) isolated from wheat-germ oil. Conditions: Finnigan GC (9510); 10% FFAP on Gas Chrom Q, 100/120 mesh; 2 m–2 mm i.d., glass column; N_2, 50 ml min^{-1}, 240 °C, column temp. 220 °C, FID 200 °C. The peak with a retention time of 26 min corresponds to methyl arachidonate (see Fig. 8)

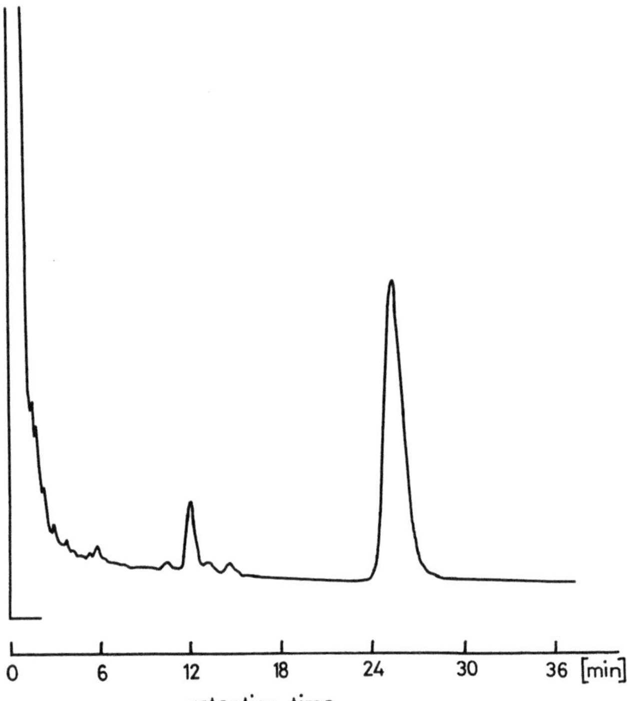

Fig. 8. Gas chromatogram of authentic methyl arachidonate (MA). Conditions: see legend of Fig. 7

from authentic methyl arachidonate. Furthermore, no difference could be discerned between the peaks, when the authentic methyl arachidonate was added to the methyl ester mixture. Gas chromatography-mass spectroscopy analysis of the substance with the retention time of 26 min showed the characteristic fragment ions [m/e 318 (M), 150, 119, 105, 91, 79, 67, 55, 41, and 29] of methyl arachidonate (Myher et al. 1974) (Figs. 9 and 10). This result demonstrates the presence of arachidonic acid in virgin wheat-germ oil.

The content of arachidonic acid in the wheat-germ oil, determined as its methyl ester using gas chromatographic detection, was quantified with reference to a standard of methyl arachidonate. Based on a content of 12% oil in the wheat-germ and the amount of arachidonic acid found in this oil, the content of arachidonic acid was calculated (Table 4).

The mild production method for the oil, as well as the gentle isolation steps used, strongly indicate that the wheat germ itself contains arachidonic acid. The calculated content does not support the idea that this fatty acid may be a reserve compound of the wheat germ. It is well established that arachidonic acid is the predominant physiological precursor of the prostaglandins E_2 and $F_{2\alpha}$ in the animal kingdom (Ramwell 1973). Some questions about the occurrence and the possible role of prostaglandins and their precursors in plants have been reviewed (Saniewski 1979). If the presence of arachidonic acid could be demonstrated in

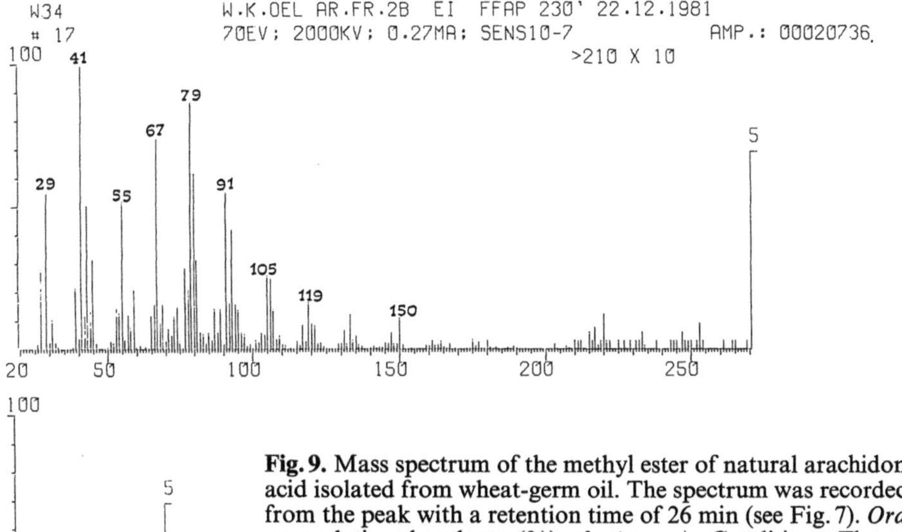

Fig. 9. Mass spectrum of the methyl ester of natural arachidonic acid isolated from wheat-germ oil. The spectrum was recorded from the peak with a retention time of 26 min (see Fig. 7). *Ordinate* relative abundance (%); *abscissa* m/e. Conditions: The mass spectra were recorded on a Finnigan GC (9610)/MS-Model 4000 instrument with sample introduction through the gas chromatograph inlet. 10% FFAP on Gas Chrom Q, 100/120 mesh; 2 m– 2 mm i.d., glass column; He, 50 ml min^{-1}, 250 °C, column temp. 230 °C; glass-jet-separator 260 °C, ion source temp. 270 °C; accelerating voltage – 2 kV and I.E. 70 eV

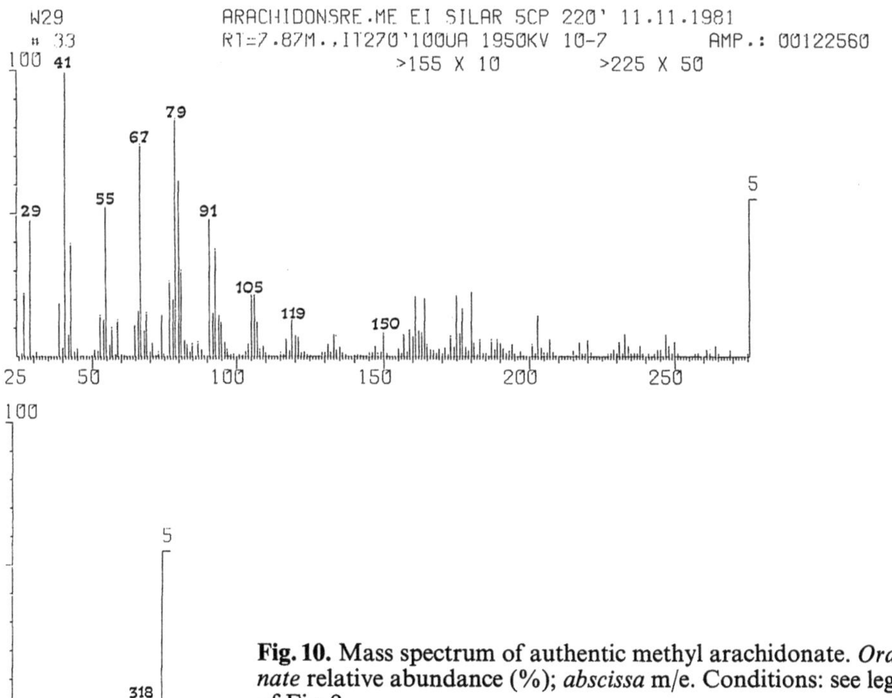

Fig. 10. Mass spectrum of authentic methyl arachidonate. *Ordinate* relative abundance (%); *abscissa* m/e. Conditions: see legend of Fig. 9

Table 4. Arachidonic acid in wheat-germ oil

Arachidonic acid – content of	Amount (pmol g^{-1})
Wheat-germ oil	12,560
Wheat-germ[a] (fresh weight)	1,507
Wheat-germ[a] (dry weight)	1,718

[a] The calculation of the arachidonic acid content of the fresh and dry weight of the wheat-germ is based on 12% total oil content of the wheat-germ and 14% water content data from Keimdiät GmbH, Augsburg, FRG.

higher plants in general, a biosynthesis of the prostaglandins E_2 and $F_{2\alpha}$ via arachidonic acid could be postulated. At this time arachidonic acid has been found only in certain algae, mosses, ferns, and in *Ginkgo biloba* (Hitchcock and Nichols 1971; Kunau 1976). In the red alga *Gracilaria lichenoides*, the prostaglandins E_2 and $F_{2\alpha}$ were found for the first time in the plant kingdom (Gregson et al. 1979). Recently, the prostaglandin $F_{2\alpha}$ was detected in flowering *Kalanchoe blossfeldiana* (Janistyn 1982 b). In this connection it worth noting that the lipoxygenase-2 enzyme from soybeans can transform arachidonic acid into prostaglandin $F_{2\alpha}$ (Bild et al. 1978).

10 Conclusions

The introduction mentioned the problems arising from the isolation of primary vegetable substances (e.g., phythohormones) which, compared to the secondary vegetable substances, occur in very small amounts.

The previous sections have discussed the isolation methods for the substances adenosine-3′:5′-monophosphate (cAMP), guanosine-3′:3′-monophosphate (cGMP) from sterile propagated maize seedlings, and for arachidonic acid from wheat-germ oil. The characterization of these substances was in each case effected by means of gas chromatography-mass spectrometry (GC-MS).

The application of a mass spectrometer as a selective and sensitive detector in gas chromatography has certainly revolutionized the identification and quantification of organic compounds, especially when applied to the physiologically active compounds present in very low concentrations (pmol g^{-1} fresh weight).

Furthermore, we can expect that GC-MS selected ion monitoring methods will gain in importance in the future. Problems arise occasionally in the case of less volatile compounds, since they first have to be changed into more volatile derivatives. The offer of various different means of derivatization (Pierce) has, however, largely solved these problems.

While little is known so far about the potentially plant physiologically active effects of cGMP (Janistyn 1983), arachidonic acid (Janistyn 1982 a) and the prostaglandins connected to this precursor (Janistyn 1982 b); the presence of cAMP-dependent protein kinases in *Lemna paucicostata* has lately been postulated with some probability (Kato et al. 1983, 1984).

Recently now effects had been found of cAMP and cGMP on the autophosphorylation of the elongation factor 1 from wheat embryos, indicating an important role in the translation control of protein biosynthesis at the elongation step (Shin-ichiro Ejiri and Honda 1985). Further, effects of cAMP on the activity of soluble protein kinases in maize coleoptile homogenates could be demonstrated (Janistyn 1986).

References

Berlin J, Barz W, Harms H, Haider K (1971) Degradation of phenolic compounds in plant cell cultures. FEBS-Lett 16:141–146

Bild GS, Bhat SG, Ramadoss CS, Axelrod B (1978) Biosynthesis of a prostaglandin by a plant enzyme. J Biol Chem 253:21–23

Black M, Bewley JD, Fountain D (1974) Lettuce seed germination and cytokinins: Their entry and formation. Planta 117:145–152
The dosage response curve for abscisic acid has also been published in this paper

Brown EG, Newton RP (1981) Cyclic AMP and higher plants (review). Phytochemistry 20:2453–2463

Budde WL, Eichelberger JW (1979) Organic analysis using gas chromatography-mass spectrometry. Ann Arbor Science Publishers, P.O. Box 1425, Mich. 48106

Catalano N (1967) Glyceride composition of olive oil. I. Oils from various areas. Indian Agric 5:529–540

Cleland R (1972) The dosage-response curve for auxin-induced cell elongation: a reevaluation. Planta 104:1–9

Dziedzianowicz-Wierzbicka W, Krauze St (1970) Determination of the composition of fatty acids in beech (*Fagus silvatica*) seed oil. Rocz Panstw Zakl Hig 21:653–664

Galston AW (1961) The life of green plants. Engleword Cliffs: Prentice-Hall

Goldberg ND, Haddox MK (1977) Cyclic GMP metabolism and involvement in biological regulation. Annu Rev Biochem 46:823–896

Greengard P, Costa E (1970) Role of cyclic AMP in cell function. Raven, New York

Gregson RP, Marwood JF, Quinn RJ (1979) The occurrence of prostaglandins PGE_2 and $PGF_{2\alpha}$ in a plant – the red alga *Gracilaria lichenoides*. Tetrahedron Lett 46:4505–4506

Grove MD, Spencer GF, Rohwedder WK, Mandava N, Worley JF, Warthen JD Jr, Steffens GL, Flippen-Anderson JL, Cook JC Jr (1979) Brassinolide, a plant-growth promoting steroid isolated from *Brassica napus* pollen. Nature 281:216–217

Hefendehl FW (1962) Zusammensetzung des ätherischen Öls von *Mentha piperita* im Verlauf der Ontogenese und Versuche zur Beeinflussung der Ölkomposition. Planta Med 241–266

Hitchcock C, Nichols BW (1971) Plant lipids biochemistry. Academic, New York

Janistyn B (1982a) Gas chromatographie-mass spectrometric identification and quantification of arachidonic acid in wheat-germ oil. Planta 155:342–344

Janistyn B (1982b) Gas chromatographic-mass spectroscopic identification of prostaglandin $F_{2\alpha}$ in flowering *Kalanchoe blossfeldiana*. Planta 154:485–487

Janistyn B (1983) Gas chromatographic-mass spectroscopic identification and quantification of cyclic guanosine-3′:5′-monophosphate in maize seedlings (*Zea mays*). Planta 159:382–385

Janistyn B, Drumm H (1975) Phytochrome-mediated rapid changes of cyclic AMP in mustard seedlings (*Sinopsis alba L.*) Planta 125:81–85

Janistyn B (1986) Effects of adenosine-3′:5′-monophosphate (cAMP) on the activity of soluble protein kinases in maize (*Zea mays*) coleoptile homogenates. Z Naturforsch (in press)

Johnson LP, MacLeod JK, Summons RE, Hunt N (1980) Design of a stable isotope dilution gas chromatography-mass spectrometric assay for cAMP: Comparison with standard protein-binding and radioimmuno-assay methods. Anal Biochem 106:285–290

Johnson LP, MacLeod JK, Parker CW, Letham DS (1981 a) The quantitation of adenosine-3′ : 5′-cyclic monophosphate in cultured tobacco tissue by mass spectrometry. FEBS Lett 124:119–121

Johnson LP, MacLeod JK, Parker CW, Letham DS, Hunt NH (1981 b) Identification and quantitation of adenosine-3′ : 5′-cyclic monophosphate in plants using gas chromatography-mass spectrometry and high-performance liquid chromatography. Planta 152:195–201

Jost JP, Rickenberg HV (1971) Cyclic AMP. Annu Rev Biochem 40:741–774

Kato R, Uno I, Ishikawa T, Fujii T (1983) Effects of cyclic AMP on the activity of soluble protein kinases in *Lemna paucicostata*. Plant Cell Physiol 24:841–848

Kato R, Uno I, Ishikawa T, Fujii T (1984) Some characteristics of protein kinases in *Lemna paucicostata*. Planta Cell Physiol 25:691–696

Kunau WH (1976) Chemie und Biochemie ungesättigter Fettsäuren. Angew Chem 88:97–111

Laskowski K, Kulikowska A (1967) Physicochemical properties of walnut oil. Roczpr Panstw Zakl Hig 18:483–486

Lawson AM, Stillwell RN, Tacker MM, Tsuboyama K, McClosky JA (1971) Mass spectrometry of nucleic acid components. Trimethylsilyl derivatives of nucleotides. Am Chem Soc 93:1014–1023

Mohr H, Schopfer P (1978) Lehrbuch der Pflanzenphysiologie. Physiologie der Hormonwirkungen. Springer, Berlin Heidelberg New York, pp 368–390

Myher JJ, Marai L, Kuksis A (1974) Identification of fatty acids by GC-MS using polar siloxance liquid phases. Anal Biochem 62:188–203

Neumüller O-A, Römpps (1985) Chemie Lexikon. Massenspektroskopie, Bd 4, 8. Aufl. 2500. Franckh'sche Verlagshandlung, W. Keller u. Co. Stuttgart

Newton RP, Gibbs N, Moyse CD, Wiebers LJ, Brown EG (1980) Mass spectrometric identification of adenosine 3′ : 5′-cyclic monophosphate isolated from a higher plant tissue. Phytochemistry 19:1909–1911

Pastan JH, Johnson GS, Anderson WB (1975) Role of cyclic nucleotides in growth control. Annu Rev Biochem 44:491–522

Ramwell WP (1973) The prostaglandins, vol 1. Plenum, New York

Rasmussen H (1970) Cell communication, calcium ion and cyclic adenosine monophosphate. Science 170:404–412

Robison GA, Butcher RW, Sutherland EW (1968) Cyclic AMP. Annu Rev Biochem 37:149–174

Saniewski M (1979) Questions about occurrence and possible roles of prostaglandins in the plant kingdom. Acta Hortic 91:73–81

Shin-ichiro Ejiri, Honda H (1985) Effec of cyclic AMP and cyclic GMP on the autophosphorylation of elongation factor 1 from wheat embryos. Biochem Biophys Res Commun 128(1):53–60

Spiteller G (1970) Massenspektrometrische Strukturanalyse organischer Verbindungen. – Eine Einführung. Akad Verlagsgesellschaft, Frankfurt a.M.

Sutherland EW, RAll TW (1958) Fractionation and characterization of a cyclic adenine ribonucleotide formed by tissue particles. Biol Chem 232:1077–1091

Waller GR, Denver OC (1980) Biochemical applications of mass spectrometry. John Wiley, New York

Wellmann E (1983) Encyclopedia of plant physiology, new series, vol 16 B:745–754

GC-MS Methods for Lower Plant Glycolipid Fatty Acids

H. NYBERG

1 Introduction

The aim of this article is to give a description of the GC-MS methods for lower plant glycolipid fatty acids. Although the text deals mainly with examples of algae and mosses, it should be kept in mind that the methods presented are frequently also suitable for higher plants. However, when studying new or not well-known plant material, the suitability of methods should always be tested prior to serious experimental work. It can be stated that, in a sense, higher plants are easier material for glycolipid studies, as they are much more homogeneous with regard to the lipids, compared, for example, with the algae. The algae are also very varied in cell structure, which means that their lipid extractability can at times be difficult, depending on the species studied.

The main subject of this paper is the GC-MS methods, but as the work usually begins from cultured or collected plant material, it was considered appropriate to include chapters also on the extraction and purification of glycolipids from plant material. Perhaps in this way the text will best fill its function as a practical guide to the study of plant glycolipids.

The text is mainly intended to serve those who do not have very much experience on plant lipid studies, and it is hoped that the prospective researcher can start work with the help of this description. However, it is impossible to give a comprehensive account, especially on the GC-MS methodology and result handling here; these aspects usually require some practical experience before operation skill is acquired.

The choice of references and presented objects of lipid studies also represents to some extent the author's own interests, and they can by no means be considered a complete list. The published literature on plant glycolipids and fatty acids is indeed vast, as the glycolipids in plants form a very important component of the photosynthetic membranes, and as such they have attracted a lot of scientific interest.

2 Extraction of the Plant Material

2.1 Handling and Storage

When using lower plants as objects of study for glycolipid fatty acid studies, it is necessary to prevent the degradation of lipids in the plant material. If the material cannot be extracted and processed immediately after harvesting or collecting, it should be stored in deep freeze (at least -18 °C) and preferably lyophilized before use. Lyophilization increases the available storage time and at the same time makes dry weight determinations convenient. Especially in quantitative studies, it is very necessary to know the sample's dry weight. The use of fresh weight seldom gives useful results. In deep freeze, also long storage periods up to several years are possible. During this time the lipid and water contents of the samples do not change markedly if the storage is performed at -18 °C or lower (Rouser et al. 1976). Lipid degradation by enzymatic activity is thus also prevented.

Unicellular algae can usually be extracted without further homogenization, but if the algal cell wall is resistant, it might be useful to homogenize the cells before extraction using, e.g., a French press or a Potter-Elvehjem glass homogenizer; this applies especially to moss spores, for example (Karunen 1977). Larger algae and mosses, as well as all higher plants, should be well homogenized prior to extraction (Kates 1970). This homogenization can be done using several homogenizer types, providing only that they are fairly efficient and do not cause heating of the samples. The homogenization is preferably done in the solvent used for extraction. In these processes, the degradation of lipids caused by phosphatidases and lipases should be taken into account (Kates 1972). These enzymes can be very stable, but may be mostly inactivated, e.g., by using (1) alcohol-containing solvents, (2) dipping larger plant parts into boiling water for 1 minute (which, however, may have a harmful effect especially on highly unsaturated lipids) or (3) extracting the material with hot isopropanol for a few minutes (Kates 1972). Excessive heat should always be avoided.

If the study requires the separation of active chloroplasts, or only the chloroplasts are wanted for study, the cells should be fractionated with the usual buffer homogenization and differential centrifugation procedures. However, if the aim of the study is only to study the glycolipids in an extracted form or their fatty acids, this procedure is not necessary, as glycolipids in plant cells are almost exclusively located in the chloroplasts. Moreover, whole chloroplast separation would be extremely difficult in many algal cells.

2.2 Extraction with Organic Solvents

Ideally, a lipid extraction should quantitatively extract cellular lipids in an undegraded state. Membrane-associated lipids, such as glycolipids, require polar solvents (ethanol or methanol) to disrupt the bonds between the lipids and proteins. Usually only fresh or freshly distilled peroxide-free solvents should be used. If the solvents used need additional purification before use, they should be redistilled with a quartz distilling apparatus.

One of the best and most commonly used extraction procedures is based on the methods of Folch et al. (1957) and Bligh and Dyer (1959). In our laboratory, it has been used for unicellular algae and mosses as follows (Rouser et al. 1976; Koskimies and Simola 1980; Simola and Koskimies-Soininen 1980; Nyberg and Koskimies-Soininen 1984a, 1984b): The lyophilized and weighed material (1–2 g) is extracted in Teflon-stoppered tubes (10–50 ml, size according to amount of material) under nitrogen. The preferred solvent is chloroform-methanol (2:1), 5 ml g^{-1} of material. The extraction time is ca. 12 h at 0 °–+4 °C. The extracted plant material can be filtered away after this and washed with 1–2 ml of solvent; the washings are combined with the extract. This procedure is especially suitable for plant tissues containing easily inactivated degradative enzymes (Kates 1972).

Plant material containing stable degradation enzymes should be homogenized in hot isopropanol (3 mg g^{-1}, Kates 1972) and the residue also washed with it and chloroform-isopropanol (1:1), the filtrate concentrated and the extract then washed with water or 1% NaCl.

Extraction with chloroform-methanol (2:1) has been very much in use for algae, e.g., for *Chlorella* (Nichols 1965 a; Kabata et al. 1979), *Euglena gracilis* (Constantopoulos and Bloch 1967; Schantz et al. 1976; Skowronski and Garrigan 1983), *Volvox* (Moseley and Thompson 1980), *Porphyridium* (Nichols and Appleby 1969; Nyberg and Koskimies-Soininen 1984 a, b), *Ochromonas* (Nichols and Appleby 1969), *Fritschiella* (Wettern 1980) and diverse planktonic green algae (Chuecas and Riley 1969; DeMort et al. 1972; Piorreck et al. 1984). For mosses, it has been used, e.g., by Nichols (1965 b) for *Hypnum cupressiforme*, by Gellerman et al. (1975) for *Mnium, Hylocomium*, and *Pleurozium*, by Solberg (1983) for *Mniobryum*, and by Koskimies and Simola (1980) and Simola and Koskimies-Soininen (1980) for *Sphagnum*.

Some algae, such as *Cyanidium caldarium*, are extremely resistant to extraction and here chloroform-methanol (2:1) is not sufficient. Allen et al. (1970) have used a homogenization treatment and extraction with (1) boiling methanol, (2) cool glacial acetic acid and chloroform, (3) chloroform-methanol (2:1), after which very little lipid was left in the cells. Boiling methanol as a first-stage solvent has also been applied by Kabata et al. (1980) for *Chlorella*, boiling ethanol-chloroform (2:1) by Döhler and Datz (1980) for *Anacystis*, and boiling chloroform-isopropanol (2:1) for moss spores by Karunen (1977). Chloroform-methanol with added water has been used for diatom lipid extraction (Lee and Loeblich 1971; Anderson et al. 1978).

The use of Soxhlet extraction is perhaps not advisable for lipids because of the long extraction time needed at high temperatures; however, it has been applied by Kleinschmidt and McMahon (1970) for *Cyanidium* and by Wright et al. (1980) for several unicellular green algae.

If a long storage of the extract is needed, adding some antioxidant such as tocopherol or BHT (2,6-di-t-butyl-4-methylphenol) is recommended, to a concentration of 0.05% (Kates 1972; Diepenbrock 1981). In this case a storage temperature of −40 °C is recommended (Kates 1972).

2.3 Purification of the Extract

Purification is essential when alcohol-containing solvents are used. It should be remembered that the whole extraction and washing procedures should be performed in a nitrogen or argon atmosphere by blowing gas into the tubes each time after they have been opened.

The extract also contains many water-soluble compounds, which must be removed by washing. First, ca. 2 ml of water is added to each sample tube and the upper layer is discarded after gentle shaking. The lower (lipid) layer should be subsequently washed thrice with the s.c. two-phase of Folch (Folch et al. 1957), which contains chloroform-methanol-water (3:48:47). For each washing, 2 ml additions of this mixture is used. The tubes are shaken, first very gently, but vigorously at least in the last washing. First it is advisable only to turn the tube upside down a couple of times. If the tube is shaken vigorously at first, a milky emulsion very hard to remove can be formed; it can sometimes facilitate the clearing to add some chloroform into the tube; in this way at least part of the lipid layer may be saved. At each washing, the upper aqueous layer is removed, e.g., with a Pasteur pipette, and discarded.

The washed extract is then evaporated until dry. It should be kept dry for a minimum of time, as lipid oxidation and peroxidation easily occur in a dry state. The evaporation is performed by blowing nitrogen or argon into the tube and perhaps warming gently on a water bath (not over 35 °C). The residue is weighed if necessary; it contains, besides the lipid material, also chlorophylls and other pigments.

The purified residue is dissolved in chloroform-methanol (2:1), the tubes are filled with nitrogen or argon and stored in deep freeze (-18 °C or lower). In these conditions the lipids stay practically unchanged for at least some weeks. With an antioxidant added, the time can be extended. Some authors also have used also pure chloroform for storage or different chloroform-methanol mixtures besides the usual 2:1 (Bailey and Northcote 1976; Anderson et al. 1978; Hirayama and Morita 1980; Piorreck et al. 1984). Lee and Loeblich (1971) have used n-hexane to store algal lipids.

3 Separation of Glycolipids from the Total Lipid Extract

3.1 Column Chromatography

For a comprehensive treatise on lipid column chromatography, see Rouser et al. (1976).

Although the separation of polar lipids on a silicic acid column has been known for a long time, it still is one of the most used methods for the fractionation of lipid extracts. We have successfully used the following purified silicic acids: (1) Unisil 100–200 mesh (Clarkson Chem. Co), (2) Bio-Sil A 100–200 mesh (Bio-Rad). On the contrary, the silicic acid 100 mesh (Mallinckrodt 2847) has proved to contain too many small particles, and it cannot usually be used as such without sieving. However, after the smallest particles are removed, it is suitable for use.

As columns we have mostly used nonplugged 145-mm Pasteur pipettes, as they are inexpensive and suitable for handling small samples: Several separations can often be done simultaneously. These columns are stoppered at the lower end with glass wool to prevent the outflow of the silicic acid. A disadvantage with these columns is that the eluent flow cannot easily be interrupted, and thus new eluent must be added fairly often to each column in use. It is advisable to construct some rugged support for the columns.

For larger sample volumes, stopcock columns of the volume of, e.g., 10–50 ml must be used, although they increase the costs as the silicic acid is fairly highly priced. With the stopcock, the eluent flow can then be interrupted at will. It must be remembered to use only Teflon-constructed stopcocks, as others do not tolerate the solvents used.

The column is filled with a slurry of silicic acid in chloroform. The result should be a homogeneous column filled up to ca. 80% (a small volume should be left for the sample). After filling the columns it is recommendable to eluate the samples as soon as possible; the silicic acid must not be left to dry up and no air bubbles should be present in the column. The pipettes (Nyberg and Koskimies-Soininen 1984a) can handle samples of 0.5 ml in the usual conditions. Larger columns can take samples of up to, e.g., 10 ml. With unicellular algae, the sample volume is often small as no large amounts of plant material are usually available. With larger algae and mosses, lack of plant material is, as a rule, no problem.

The elution method we have used for glycolipids is based on those presented in Kates (1972) and Rouser et al. (1976), where the isolation of glycolipids is done by acetone elution. The procedure is as follows: (1) a sample (total lipid extract, obtained as above) of 0.5 ml is pipetted into the pipette column. The sample should contain ca. 150–200 mg of total lipids per 5 ml of extract (Kates 1972). (2) Elution with chloroform (at a rate of about 3 ml min^{-1} if stopcock columns are used) until all pigments have flowed out of the column (about 10 column volumes, Kates 1972). This fraction contains the neutral lipids (hydrocarbons, plant pigments, sterols, sterol esters, glycerides, waxes, fatty alcohols and aldehydes, and free fatty acids; (3) elution with acetone, about five times the volume of the eluted chloroform. This fraction contains mono- and digalactosyl diglycerides (MGDG and DGDG), cerebrosides, steryl glycosides and sulfolipid and small amounts of cardiolipin and phosphatidic acid (this is the fraction processed further in glycolipid studies); (4) elution first with chloroform-methanol (1:1, v/v) and then with pure methanol, both with about the volume of the used acetone. The eluents are combined; the fraction contains mostly phospholipids.

The product Florisil (Floridin, Fisher) can also be used in the column chromatographic separation of cerebrosides, especially; it is a Mg-silicate (Rouser et al. 1976). Preparation of an acid-treated column is time-consuming (Kates 1972) but Florisil is also used as such (O'Brien and Benson 1964; Rouser et al. 1976). Water is harmful if present, because phosphatides are then eluted together with glycolipids.

The Florisil column is filled as the Unisil column, and the sample eluted, e.g., as follows (Carroll et al. 1968): (1) chloroform (neutral lipids and pigments), (2) chloroform-acetone 1:1 (MGDG, cerebrosides), (3) acetone (DGDG, cerebrosides, steryl glycosides, sulfatides and sulfolipid, chlorophyll breakdown products), (4) chloroform-methanol 9:1 (traces of glycolipids, phosphatidic acid, cardiolipin), (5) chloroform-methanol 1:1 (phospholipids).

Column chromatography with DEAE-cellulose has sometimes been used for plant glycolipid separation (Kates 1972; Rouser et al. 1976). The column is prepared by suspending the cellulose in glacial acetic acid. The method can be used either for the original fractionation of total lipid samples (elution: chloroform->neutral lipids; chloroform-methanol 9:1 > glycolipids, cerebrosides, PC, lyso-PC, sphingomyelin etc.; or for the separation of different glycolipids after silicic acid column chromatography (elution: chloroform-methanol 98:2 > MGDG; chloroform-methanol 9:1 > DGDG; methanol > salts; chloroform-methanol (4:1, ammonium salts) > sulfolipid.

Sephadex partition column chromatography and reversed-phase partition chromatography are not recommended for galactolipids.

The silicic acid column chromatography has been used for algal glycolipids by numerous workers, e.g., by Allen et al. (1970) and Kleinschmidt and McMahon (1970) for *Cyanidium*, by Constantopoulos and Bloch (1967) and by Schantz et al. (1976) for *Euglena*, by Anderson et al. (1978) for *Nitzschia alba*, by Moseley and Thompson (1980) for *Volvox*, by Tornabene et al. (1980) for *Dunaliella*, by Nyberg and Koskimies-Soininen (1984, b) for *Porphyridium*, to mention only a few. A DEAE-cellulose column has been employed by, e.g., Opute (1974) for diatom lipid fractionation. For mosses, one of the best descriptions of the use of column chromatography for lipid fractionation is in Gellerman et al. (1975). The elutions are as follows: (1) diethyl ether-hexane 1:99 > wax and steryl esters; (2) diethyl ether-hexane 4:96 > unidentified; (3) larger volume of diethyl ether-hexane 4:96 > triacylglycerols; (4) chloroform > sterols and chlorophylls; (5) acetone-chloroform 1:1 > MGDG and pigments; (6) acetone > DGDG. Silicic acid has been also used for *Sphagnum* by Koskimies and Simola (1980) and Simola and Koskimies-Soininen (1980).

Articles on the glycolipid separation procedures for higher plants are very numerous and outside the scope of this article: however, the important work of Allen et al. (1966) on spinach chloroplast lipids, where the properties and advantages of different methods are compared, could deserve attention.

3.2 Thin-Layer Chromatography (TLC)

Thin-layer chromatography can be used either as the first separation stage to fractionate total lipid extracts or to obtain an overall picture of the lipid composition of the sample under study, or as a device to separate different glycolipids already separated from the other lipids with column chromatography (Renkonen and Luukkonen 1976; Rouser et al. 1976). Today usually commercial silica gel plates (e.g., Merck G60, which contains calcium sulfate as a binder) are used. The procedures for handmade plates are presented, e.g., in Renkonen and Luukkonen (1976).

Acetone is a very useful solvent for glycolipid separations on TLC. Skipski et al. (1967) have developed three solvent mixtures run after each other using the same elution direction. Solvent I is acetone-pyridine-chloroform-water (40:60:5:4) and it moves the neutral and glycolipids along the plate, but the phospholipids are not affected. After the first run the plates are vacuumdried, and solvent II consists of diethyl ether-pyridine-ethanol-2 M ammonia (65:30:8:2). It is allowed to rise almost to the top of the plate (twice that of solvent I). This separates the glycolipids well from each other, but washes away the neutral lipids.

Solvent III, in turn, is used to remove the free fatty acids; it contains diethyl ether-acetic acid (100:3) and does not affect glycolipids.

Clayton et al. (1970) have presented a useful one-dimensional system for the separation of complex mixture of polar plant lipids from wheat. On Silica Gel G, using a solvent chloroform-methanol-30% (w/w) ammonia-water (60:35:5:2.5) all polar lipids were separated as clear spots. MGDG, monoglucosyl ceramide and DGDG were well separated, and also, e.g., other monoglycerides were found in this way. The method presented is useful if no preceding column separation has been made and it is necessary to know the total polar lipid profile of the sample. It will also efficiently separate the glycolipids from each other if this fraction has been already subject to column chromatography.

One-dimensional TLC has been successfully used by Pohl et al. (1970) for *Euglena gracilis* polar lipids. The plates used were Silica Gel H (no binder) and the solvent acetone-benzene-water (91:30:8). MGDG and DGDG are directly separated by this method. However, the use of benzene is not recommended if there is any possibility of inhaling the vapors, as benzene is considered carcinogenic. This system has also been employed, e.g., by Wettern (1980) for *Fritschiella* and by Skowronski and Garrigan (1983) for *Euglena*.

Two-dimensional TLC is usually much more difficult to perform and it is, e.g., often dependent on weather conditions (Renkonen and Luukkonen 1976). Therefore one-dimensional procedures are usually preferred for glycolipids. However, some authors have used 2D-TLC successfully. Kabata et al. (1979) have applied it for *Chlorella ellipsoidea* on Silica Gel H plates. The first solvent is chloroform-methanol-acetic acid-water (70:20:2:2) and the second hexane-diethyl ether-acetic acid (90:10:1). 2D-TLC systems on higher plant chloroplast membrane lipids may be of value also for algae; e.g., the methods of Allen et al. (1966) (Solvent I: chloroform-methanol-water 65:25:4, solvent II: diisobutyl ketone-acetic acid-water 8:5:1), Douce et al. (1973) (Solvent I: chloroform-methanol-water 65:25:4, solvent II: chloroform-acetone-methanol-acetic acid water 100:40:20:20:10) and Wintermans et al. (1981) (Solvent I: acetone-benzene-water 91:30:8, solvent II: chloroform-methanol-7 M ammonia 65:25:4, solvent III for neutral lipid separation: petroleum ether (40–60)-diethyl ether-formic acid 60:40:1.5] may be mentioned.

3.3 Other Applications

Kates and Volcani (1966) have chromatographed diatom lipids on silicic acid-impregnated Whatman 3MM paper with the solvent diisobutyl ketone-acetic acid-water (40:25:5). In TLC, ammonium sulphate-impregnated silica gel G plates have been found to be of use (Khan and Williams 1977, Lem et al. 1980) with the solvent acetone-benzene-water (91:30:8). Rouser et al. (1976) have found silica gel H mixed with 10% (w/w) magnesium silicate very recommendable for polar lipids in general.

Polar lipids can also be fractionated with HPLC; Lynch et al. (1983) have separated galactolipid molecular species with HPLC after having separated MGDG and DGDG with more conventional methods. Recently TLC has been successfully combined with the flame ionization detector (TLC-FID) to obtain a rapid quantitation of polar lipid mixtures (Herslöf 1979; Hirayama and Morita 1980; Tanaka et al. 1984).

4 Isolation of the Glycolipids

4.1 Thin-Layer Chromatography

This method has partly already been presented in Section 3, as the student of glycolipids has, in fact, two alternatives: glycolipids may be isolated as MGDG, DGDG etc. immediately from the total lipid extract, or, as perhaps is more usual and recommendable, they may be isolated and further purified after the glycolipid fraction has first been separated from the total lipid extract by the above-described procedures. This is a common practice, as the glyco- and phospholipids often occupy the same area on the TLC plate (Renkonen and Luukkonen 1976).

The most important method for further fractionation of the glycolipid fraction is TLC, and it has been used in many lower plant studies. In our laboratory, we usually have used Silica gel G plates (20 × 20 cm Merck, with or without a concentrating zone). The preferred solvent is chloroform-methanol-7 M ammonia (65:20:4), which separates the MGDG and DGDG well from each other (Luukkonen et al. 1976). This method has been used for *Porphyridium* (Nyberg and Koskimies-Soininen 1984a) and also for *Sphagnum* (Koskimies and Simola 1980). However, perhaps an even better separation can be obtained by combining the above mentioned solvent with chloroform-acetic acid-methanol-acetone-water (50:10:10:20:5, Rouser et al. 1976) in a two-dimensional run (Simola and Koskimies-Soininen 1984). These methods are recommended for MGDG, DGDG, and cerebrosides in silicic acid-column fractioned glycolipid mixtures, but they are not suitable for total lipid fractionation.

Other solvents for glycolipid TLC are described, e.g., in Kates (1972), Renkonen and Luukkonen (1976) and Rouser et al. (1976). Important solvents used for algae are, e.g., those of Gardner (1968) for *Chlorella* (acetone-acetic acid-water 100:2:1, where phospholipids do not migrate at all), and Pohl et al. (1970) (acetone-benzene-water 91:30:8, used for many different organisms, e.g., for *Euglena* (Schantz et al. 1976; Skowronski and Garrigan 1983), and for *Anacystis* (Döhler and Datz 1980). It was also employed for moss glycolipids by Karunen (1977).

Argentation TLC has been used to separate different molecular species of glycolipids (Nichols and Moorhouse 1969; Siebertz et al. 1980). The argentation plates are prepared by impregnating Silica gel G plates with silver nitrate (spraying with a 10% solution in acetonitrile, 10 ml per 20 × 20 cm plate). The mentioned articles should be consulted if it is necessary to separate possible molecular glycolipid species; the most modern approach to this type of study seems to be the HPLC-based application of Lynch et al. (1983).

For sample application, microsyringes of 10–50 µl (Hamilton, SGE, Terumo, Dynatech etc.) are very useful especially in strictly quantitative work. In qualitative or preparative work, also, e.g., 25–50 µl disposable glass capillary micropipettes may be used (Vitrex or similar), especially for dilute samples, where larger volumes must be applied to the TLC plate. The sample is preferably applied as a row of slightly overlapping very small spots, the row length being about 1 cm and its location 1.5–2 cm from the lower edge of the plate (Renkonen and Luukkonen 1976). This is today made easier by the introduction of TLC plates with a concentrating zone, which improves the results of even less successful plate pipettings. The small s.c. HPTLC-plates may also be of use (Wettern 1980).

4.2 Localization of the Glycolipids on TLC

If the glycolipid fatty acids are to be studied further, only nondestructive stains can be used. For surveying destructive localization methods are also suitable; many such methods are presented, e.g., in Kates (1972) and Renkonen and Luukkonen (1976).

Perhaps the best nondestructive stain for glycolipid localization on TLC is Rhodamine 6G (Kates 1972): a stock solution of 0.12% (w/v) is prepared. This solution is stable in the dark. Before use, 5 ml of the stock solution is diluted to 500 ml with water, to make a 0.0012% solution. The plate is sprayed and viewed while wet in 366 nm UV. Lipid spots appear yellow or blue. The stain is nonspecific for lipids. Another nonspecific stain is iodine vapor, which is very useful for locating the marker spots, as all lipids are temporarily stained by it. However, as iodine reacts with unsaturated fatty acid chains, contamination of the spots to be further analyzed should be avoided (Renkonen and Luukkonen 1976). As MGDG and DGDG usually move well away from each other on the plates, the marker spots beside the sample MGDG and DGDG spots can be stained by iodine and thus locate the sample spots. After they have been scraped off, iodine vapor can be used to ensure that the spots have been removed. The vapor is preferably blown on the plates from a Pasteur pipette containing a few iodine crystals, under a fume hood to avoid breathing it.

If one first wants to make sure that glycolipid spots exist on the plate, specific destructive stains can be used to determine this. The diphenylamine stain (Bailey and Bourne 1960) has been used in our laboratory for the galactolipids and cerebrosides (Simola and Koskimies-Soininen 1984): 4 g diphenylamine, 4 ml aniline and 20 ml of 85% orto-phosphoric acid are dissolved in acetone to make 200 ml. The plates are sprayed well and heated in an oven at 140 °C for 15 min. Glycolipids stain blue, phospholipids do not react. It should be noted that this stain is very poisonous.

The α-naphthol stain has also been used for glycolipids (Kates 1972; Aro and Karunen 1979): The reagent consists of a 2% solution of α-naphthol in ethanol. The plates are first sprayed with this and then with concentrated sulphuric acid, followed by heating at 100 °C for 10 min. Glycolipids stain red-purple, phospho- and neutral lipids stain brown. This stain is very sensitive (Renkonen and Luukkonen 1976). A third much used reagent for glycolipids is the orcinol stain (Renkonen and Luukkonen 1976): 200 mg of orcinol is dissolved in 100 ml sulfuric acid-water (3:1) and stored in the dark at 5 °C. The solution is stable for a week. The plates are sprayed thoroughly and heated at 100 °C for 15 min. The glycolipids stain violet or blue (Skipski et al. 1967).

4.3 Removing the Spots from the TLC Plates

The located spots can be marked, e.g., with a pencil (not touching the spot itself), wetted with 1–2 drops of distilled water to prevent the dusting off of the silica gel, and then scraped off from the plate with a spatula or some other suitable instrument. Great care must be taken not to lose silica gel dust or pieces; perhaps a sucking device can be used to ensure this. The spots are transferred to Teflon-stoppered tubes (e.g., 10 ml) and mixed with, e.g., 5 ml of chloroform-methanol (2:1). Also pure chloroform or n-hexane have been used. After this the tubes are filled with nitrogen or argon and, after the lipids have dissolved (e.g., after 1 h), the silica gel is centrifuged to the bottom and the supernatant removed in a similar tube. Now the lipid samples are ready for derivatization for GLC.

If the individual lipids in the tubes are suspected of being contaminated, it may be necessary to purify them further with silica gel TLC, e.g., after the procedure of Tancrede et al. (1981): the plates are washed by eluting twice with methanol-water (1:1) and the samples run with chloroform-methanol-water (90:25:2). For further details consult Tancrede et al. (1981). Separation of molecular species of glycolipids has already been mentioned (Sect. 4.1). TLC procedures for this are also treated in Renkonen and Luukkonen (1976). On the resolution of lipid molecular species, as well as on many other aspects of lipid chromatography, a valuable source of information is also the review of Kuksis (1983).

5 Derivatization of the Glycolipid Fatty Acids for GC-MS

5.1 General Features

Analytical derivatization is employed for two reasons: (1) to permit analysis of compounds not directly amenable to analysis owing to, e.g., inadequate volatility or stability, or (2) to improve analysis by improving chromatographic behaviour of detectability (Knapp 1979). The classical derivatives of fatty acids have been the methyl esters. Butyl and other alkyl esters have also been used, and silylated derivatives can also be useful. Trimethylsilyl derivatives can readily be formed, but they are somewhat unstable. The t-butyl-dimethylsilyl derivatives are more stable to hydrolysis and have a more favorable MS fragmentation.

5.2 Formation of Methyl Esters

Methyl esters can be prepared directly from the glycolipid fatty acids by esterification in methanol with a suitable catalyst. In our laboratory, usually the procedure of Klopfenstein (1971) is used; it is based on methanolysis of the lipids and subsequent esterification of the fatty acids. Procedure: (1) The purified lipid sample (which is in, e.g., chloroform-methanol 2:1) is evaporated to dryness by blowing nitrogen or argon and perhaps heating very gently (not over 35 °C!); (2) After evaporation, 1 ml of analytical grade boron trichloride (14% in methanol) is added (it is recommendable to use a reagent packed in ampules, e.g., that of Allied Science. An ampule should not be left to stand opened) under a fume hood. Boron trichloride, like many other derivatization reagents, is very poisonous; (3) The Teflon-stoppered tube is then filled anew with nitrogen and heated, well sealed, at 90 °C for 30 min. Klopfenstein (1971) gives 120 °C for 90 min, but we found (Nyberg and Koskimies-Soininen 1984b) our values to be sufficient; (4) After heating, 4 ml of $CHCl_3$ and 1 ml of water are added to the tube and after vigorous shaking the upper layer is discarded. The organic phase is then washed three times with 2 ml of the Folch 2-phase (Folch et al. 1957) and the upper layer always discarded. (5) The lower layer containing the methyl esters is evaporated to dryness with nitrogen and the methyl esters at once dissolved into 30 µl of n-hexane. After this the sample is ready for GLC analysis.

Many authors also use boron trifluoride-methanol (10 or 14%) in the methylation (Morrison and Smith 1964), but it has been shown to produce artifacts and cause a larger loss of especially highly unsaturated acids (Klopfenstein 1971; Knapp 1979). However, the procedure is still in fairly wide use: (1) To 4–16 mg of lipid, a mixture of 14% boron trifluoride-methanol-benzene-methanol (5:4:11) is added under nitrogen; (2) Heating at 100 °C for 30 min (depends on lipid types); (3) After cooling, 2 ml of n-pentane and 1 ml of water is added and the tubes shaken. The organic layer is separated and analyzed with GLC.

A somewhat less used method fo the methylation of lipid fatty acids is that described by Hirvisalo and Renkonen (1970) and used by Nyberg and Koskimies-Soininen (1984a) for algal glycolipids: (1) the purified samples in Teflon-sealed tubes are evaporated to dryness and 1 ml of a mixture of concentrated sulfuric acid-methanol (6:94) is added; (2) the sealed tubes (under nitrogen) are heated at 70 °C for 6 h; (3) the methyl esters are washed as in the boron trichloride method, the organic phase evaporated and the esters dissolved in, e.g., n-hexane.

The plant glycolipid fatty acids are generally long-chain acids, so butyl esters are not essential as derivatives as excessive volatility of the esters is no problem (Knapp 1979).

5.3 Silylation and Other Methods

Trimethylsilyl (TMS) esters can very well be used as fatty acid derivatives. For TMS reactions, preferred solvents are pyridine, dimethylformamide, tetrahydrofuran and acetonitrile. Presence of water should be avoided, as it decomposes both TMS reagents and derivatives. It is also necessary to remember that only all-glass injector ports are to be used with TMS derivatives, as stainless steel is corroded by them. However, stainless steel columns can be used.

One of the most common silylation procedures is that based on the method of Sweeley et al. (1963): (1) Introduce 5 or 10 ml of sample and 1.0 ml of the reagent TRI-SIL (Pierce) into a small screw cap vial; (2) Shake vigorously to dissolve. The reaction may be warmed to 75–85 °C; (3) Allow to stand for 5 min or until derivatization is complete. The mixture is then ready for injection into the chromatograph. Organic acids etc. are completely derivatized within 5 min.

A typical procedure is also that of Kuksis et al. (1969): The fatty acid or lipid sample is dissolved in a solution of hexamethyl-disilazane (HMDS)-trimethyl-chlorosilane (TMCS)-pyridine (2:1:10) and bis(trimethylsilyl)acetamide (BSA)-pyridine (2.5 mEq ml^{-1}). A mixture of BSA-pyridine (1:5) can be also directly used for silylation (0.2–0.3 ml/2–3 g of sample; Drozd 1981). The mixture is shaken for 1 min and then analyzed directly. This seems to be a more suitable procedure than the use of an HMDS-TMCS-mixture.

To form t-butyldimethylsilyl (TBDMS) esters of the fatty acids, the sample is treated with 2 molar equivalents of N-TBDMS-N-methyltrifluoroacetamide in dimethylformamide. The reaction is complete at once and the esters can be analyzed by GLC (Knapp 1979).

Fatty acid pyrrolidines have been formed by Řezanka et al. (1983) from fatty acids of green algae: (1) The samples are first converted to fatty acid methyl esters (FAME) with boron-trifluoride-methanol as in Morrison and Smith (1964); (2) 10 mg of FAME's are boiled with 1 ml of pyrrolidine and 0.1 ml of acetic acid for 1 h. The pyrrolidines are purified by TLC (light petroleum-diethyl ether 1:1) on silica gel G.

6 GLC Instrumentation for Fatty Acid Analysis

6.1 General Features

The following account follows in the main the descriptions of Schill (1977) and Willard et al. (1981).

GLC accomplishes a separation of fatty acids or their derivatives by partitioning solutes between a mobile gas phase and a stationary liquid phase held on a solid support. A sample containing the solutes is injected into an injector, where it is vaporized and carried into the column inlet. The solutes are adsorbed at the head of the column by the stationary phase and then desorbed by fresh carrier gas. Each solute will travel at its own rate through the column, and then they are eluted sequentially in the increasing order of their partition ratios and enter a detector attached to the column exit. The signals appear as peaks on the plot. The time of emergence (retention time) is characteristic for each component. The peak area is proportional to the concentration of the component in the mixture.

Basically, a GC consists of six parts: (1) a supply of carrier gas in a high pressure cylinder with pressure regulators and flow meters, and a valve to introduce extra make-up gas to some detectors: (2) a sample injection system, (3) the separation column in an oven to regulate temperature; (4) the detector; (5) an electrometer and a recorder (printer/plotter), today usually also an integrator; (6) injector and detector thermostated parts to regulate their temperature. The scope of this description does not permit a comprehensive account of the many different GC instruments in use today, so the prospective worker should consult, e.g., the handbooks of the different manufacturers and the detailed descriptions in the works mentioned in the beginning of this chapter.

6.2 Carrier Gas System

The purpose of the carrier gas is to transport the sample through the column to the detector. The selection of carrier gas is an important decision and usually dictated by the type of detector used. As the flame ionization detector (FID) is almost exclusively in use with the fatty acids, suitable gases are argon, nitrogen, helium, and hydrogen; the relative response of the FID for the gases decreases in the order stated above (Schill 1977). If short retention times are desired, hydrogen is to be recommended as the carrier gas, but the potential dangers associated with the use of hydrogen (flammability) should be considered. In our laboratory, we have used nitrogen and helium; helium can be recommended for most GLC work.

The flow rate is dependent on the column used and it should be decided and determined specifically for each system. For a glass capillary column, we have used, e.g., the rate of 30 ml min^{-1}; flow rates of 20–40 ml min^{-1} seem to be common. It is very essential that the flow rate is controlled and stable for many reasons, e.g., retention time reproducibility, baseline stability and difficulties in quantitation.

Automatic flow controllers are a useful part of any GC instrument. A soap-bubble flowmeter is cheap and easy in use, and it should be found in every GC laboratory. Today, thermal mass flowmeters are also available. In some cases, the column temperature affects the flow rate.

It is very important that the carrier gas used is extremely pure and dry, as contamination causes baseline drift and is harmful to the detector function. Drying filters or cartridges are commercially available (e.g., Chrompack) and can be used to keep the inflowing gases dry.

6.3 Injection and Injectors

The sample must be introduced as a vapor in the smallest possible volume and in a minimum of time, without decomposition or fractionation occurring. Both quantity of sample introduced and the manner of introduction must be reproducible with a high degree of precision. The sample can be injected either manually or using an autosampler. The autosampler is very useful, as it releases the operator to other duties and the injection is performed reproducibly and more precisely than manually. The autosampler can be connected to most modern GC instruments.

Liquid samples (volume usually 1–10 µl) are injected by a microsyringe (Hamilton, Dynatech, SGE, etc.) through a self-sealing silicone rubber septum. The septum easily begins to bleed as it is repeatedly punctured and therefore it should be replaced as soon as it shows signs of damage. The sample must be vaporized instantaneously, and thus the injection zone temperature generally must exceed the boiling points of all sample components. For fatty acid methyl esters, 200 °–210 °C is usually enough (Nyberg and Koskimies-Soininen 1984a). The sample should not contain much over 1% of the individual solutes; more concentrated samples should be diluted accordingly. Generally a smaller sample produces a better peak form. Too large a sample overloads the column and the result is an asymmetric peak with a trailing front. When very dilute samples are to be analysed, the use of a concentration precolumn, such as 2,6-di-phenyl-p-phenylene oxide porous polymer (Willard et al. 1981), allows the quantitative transfer of up to 20 µl of sample.

Packed metal columns (length, e.g., 2.5 m) can be used for fatty acid analysis, especially when rapid elution is desirable and, e.g., if only some important acids are to be investigated. They have been used for algal fatty acids (Matucha et al. 1972). However, for the higher separation capacity, capillary columns are recommended. The advantages of different column types are discussed later in this paper (Sect. 7.1).

The packed columns require a simpler injection system than the capillary columns (usually a splitless flash vaporizer injection port, Willard et al. 1981), or the sample is injected directly onto the end of the column (on-column injection). In general, on-column injection gives better precision for narrow-boiling-range mixtures, but flash vaporization injection is more suitable for wide-boiling-range mixtures (Schill 1977).

The capillary columns need a small sample volume (1–2 µl) and even this is reduced by the use of an injector splitter, which permits only a small part of the sample to enter the capillary column, the rest being vented through the split. The split ratio can vary between 1:10 and 1:1000. The injector body contains a glass insert that has an annular splitter at the end; at this point the vapors are split into two streams (Fig. 1). For good mixing of the sample before the split takes place, the insert is packed with a bed of silanized glass beads or a series of baffles are used (Willard et al. 1981). The split region must be in a heated zone of constant temperature.

All-glass splitter systems have received attention, as the use of glass capillary columns is common today. These are available from several manufacturers. A solvent-less all-glass injector system is described by Sisfontes et al. (1981). It may be used for fatty acid mixtures, when high boiling components are to be separated.

Splitless systems can also be used for capillary columns; the entire sample is then injected into the column (Kozuharov 1980).

With the new "cold" sampling systems, the liquid sample enters at temperatures which are not higher than the column itself, or are even lower (Schomburg et al. 1983a). The complete vaporization of the liquid sample is not initiated before the temperature of the injector or of the column inlet is elevated by slow or fast temperature programming. The system allows some advantages, especially

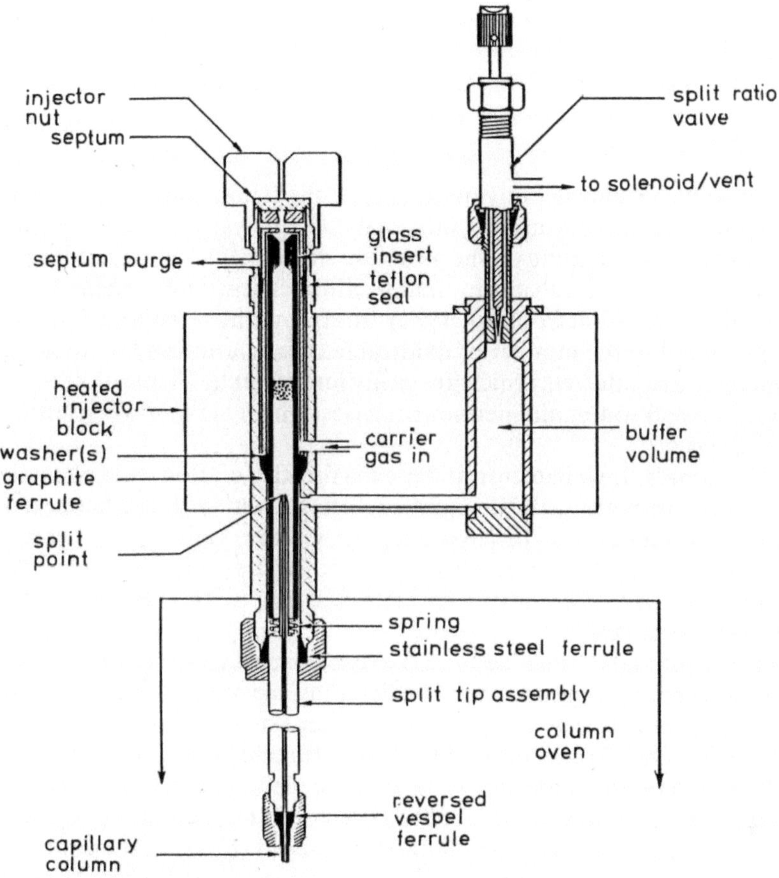

Fig. 1. Cross-sectional view of capillary splitter/injector with splitter type insert installed. (Courtesy of Varian Associates, Inc.)

the repeatability of the quantitative results is improved with capillary columns (Schomburg et al. 1983b).

In splitless injection involatile by-products have a significant effect on quantitative analysis; the transport of medium and high boiling solutes from the vaporizing chamber into the column is hindered by the retention power of the "dirt" (Grob and Bossard 1984). The extent of material transferred is not equal for clean calibration mixtures and dirty samples. Clean mixtures give data which are not applicable to the sample (area/concentration for the external standard method, or response factors for the internal standard method). Systematic errors can be excluded and dirty samples analyzed by the splitless method with a quantitation procedure by calibrating with additions of standards to the sample (Grob and Bossard 1984). Also a fused silica capillary precolumn (length several metres), where dirty sections can be cut away after every few injections, can be used (Grob 1984). Quantitation in splitless injections for involatile material containing samples can be improved by the use of a light packing of glass wool in the injector insert (Grob and Neukom 1984). The standard deviation of the results is reduced in this way and the solute transfer onto the column is better.

6.4 The Detector

This description will concentrate only on the use of the flame ionization detector (FID), as it is almost exclusively used in fatty acid GC analysis. It is sensitive, has a wide range, and a great reliability. The structure of the FID is shown in Fig. 2.

The FID responds only to substances that produce charged ions when burned in a hydrogen/air flame (Willard et al. 1981). In an organic compound the response is proportional to the number of oxidizable carbon atoms. Only those inorganic compounds are detected, which are easily ionized in the flame at 2100 °C. The FID is insensitive to water and permanent gases, which makes it very suitable for fatty acid analysis.

As the FID is a mass flow detector, it depends directly on flow-rate of carrier gas. It also varies in response with the applied voltage, and with the flame temperature, which is a function of the hydrogen/air mixture ratio. The operating temperature range for the FID itself is 100 °–420 °C, and programmed temperature applications are easy. For fatty acid methyl esters, a FID temperature of 200 °–210 °C is quite suitable.

In a FID flame, not all of the air enters the reactive part of the flame. Usually five to ten times more air is needed than predicted by the reaction stoichiometry (Sullivan 1977); this means that an eye should be kept on the air supply gas cylinder, as it is the first one to run empty. If the flow is directed along the jet through a narrow collector, as in some designs, less air is needed. The FID also requires that the carrier flow from the column to the detector is temperature controlled, with no cold spots. The hydrogen flow should be turned off when the column is removed from the GC oven, or when the GC is not in use to avoid risk of explosion. All unused detector fittings should be capped tightly to ensure that the hydrogen does not flow into the oven.

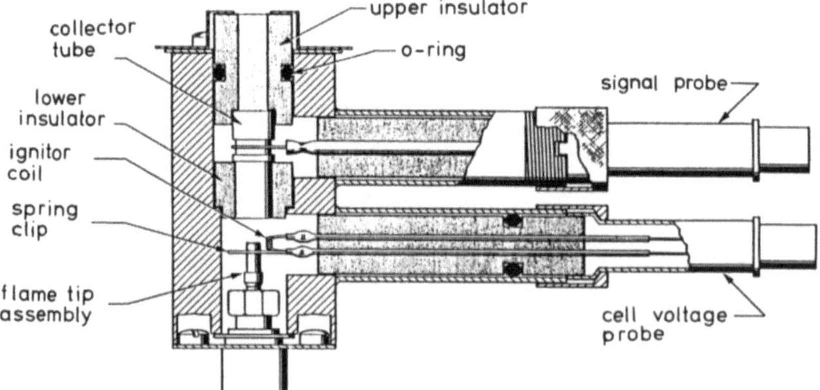

Fig. 2. Flame ionization detector cross-sectional view. (Courtesy of Varian Associates, Inc.)

The operation principle of the FID is as follows (Willard et al. 1981): Column effluent enters the burner base through a filter, is mixed with hydrogen, and the mixture burned at the tip of the jet with air (preferably a mixture of 20% oxygen – 80% nitrogen). Ions and free electrons are formed in the flame. These enter the gap between two electrodes, the flame jet and the collector. Across the two electrodes is imposed an applied potential of about 400 V. This lowers the resistance across the gap and causes a current to flow. Normally an external bucking voltage is provided to balance the potential generated by the ions and free electrons formed in a pure hydrogen/air flame. This ensures that a net current flows only when ionized material enters the gap, thus enhancing the differential sensitivity of the detector. This current flow across an external resistor is sensed as a voltage drop, is amplified, and displayed on a recording system. When CH_2-groups are entered into the flame, a complex process takes place in which positively charged carbon species and electrons are formed, and the current is greatly increased.

The optimum hydrogen flow rate for maximum FID sensitivity varies according to type and manufacturer of the instrument, and the instrument manual should be consulted. It is commonly in the range of 30–40 ml min^{-1}, and to some extent also dependent on the compounds analyzed with the FID. By increasing the hydrogen flow rate above that required to give maximum sensitivity, the linear range of the FID is increased; improved quantitative results are obtained by optimizing for linearity, rather than for sensitivity (Albertyn et al. 1982).

7 Column Selection for Fatty Acid GLC Analysis

7.1 Column Types

Both packed and open tubular (capillary) columns can be used. Packed columns are stainless steel, copper or glass tubing, commonly 1.6–3.2 mm bore and <3 m in length, filled with a narrowly sieved, inert support that has been coated for a liquid phase. It is possible to pack both packed or some capillary columns in your own laboratory, but this requires practice and suitable equipment, and is therefore not described here. A general account is found, e.g., in Kates (1972). Examples of commonly used packed columns are presented in Table 1.

The wall-coated open tubular (WCOT) capillary columns are long narrowbore (ID 0.25–0.45 mm) tubings of glass; their length can be 20–150 m. Silane-treated Pyrex glass the most desirable features but also soda glass columns are used (Kozuharov 1980). WCOT columns possess limited sample capacity, which is partly overcome with the support-coated open tubular (SCOT) capillary columns, in which a porous layer is formed on the inside wall of the tubing. The SCOT columns can be used much shorter than the WCOT columns (about 16 m). Recently, the fused silica capillary columns (FSCC) have gained an ever greater popularity. Examples of common capillary columns are presented in Table 2.

The obvious advantages of fused silica capillary columns in GC detection are a high resolution converting to higher sensitivity and higher precision, more rapid

Table 1. Examples of packed columns used for fatty acid ester GLC

Column type and ID	Length	Support (treatment)	Stationary phase	Temperature	Ref.
Steel-packed 4 mm	2 m	Chromaton N-AW-HDMS 60/80	DEGS	190 °C	a
Steel-packed 3 mm	1.5 m	Chromaton N-AW-NMDS 60/80	Apiezon L	230 °C	a
Steel-packed 3 mm	2 m	Chromosorb G-AW-DMSC 80/100	XE-60	160° > 200 °C	b
Steel-packed 3 mm	1 m	Not stated	EGS	180° > 230 °C	c
Steel-packed 3.2 mm	2.5 m	Not stated	SP 2340	170° > 200 °C	d
Steel-packed 4 mm	1.8 m	Gas Chrom W	BDS	180 °C	e
Steel-packed 4 mm	1.2 m	Cromosorb Q Cromosorb P	EGSS-X Silar 10C	Not stated Not stated	f
Steel-packed 3.2 mm	2.6 m	Cromosorb W 80–100	SE 30	90° > 270 °C	g
Steel-packed		Gas Chrom Q 100–120	EGSS-X SE 30	185 °C	h
Glass-packed 4 mm	2 m	Supelcoport 80–100	DEGS-PS	165 °C	i
Glass-packed 4 m	1.8 m	Gas Chrom Q 100–120	Silar 10C	165 °C	j
Glass-packed 2.6 mm	2.5 m	Diatomite CAW 80–90	SE 30	220° > 280 °C	k

References: a) Matucha et al. (1972), b) Podojil et al. (1978), c) Kabata et al. (1980), d) Clarkson et al. (1980), e) Anderson et al. (1978), f) Lem et al. (1980), g) Johns and Perry (1977), h) Simola and Koskimies-Soininen (1980), i) Albertyn et al. (1982), j) Williams and MacGee (1983), k) Barta and Kömives (1984)

analysis and a high degree of inertness; in MS detection a greater sensitivity and selectivity are obtained, together with improved peak resolution by multi-channel detection (Settlage and Jaeger 1984).

Relatively short capillary columns have been used for the GLC separation of fatty acid esters to reduce the time of instrumental analysis while maintaining the degree of separation typical of packed columns, and to obtain better results more quickly as a result of faster elution (Lercker 1983). Excellent results have been obtained with 10–15 m glass or fused silica capillary columns.

7.2 Supports, Liquid Phases, and Their Characteristics

The support provides an inert surface onto which the stationary liquid phase can be placed in a packed column. Only the diatomaceous earths and Teflon are used to any extent. The diatomaceous materials can be either firebrick-derived, as Chromosorb P and Gas Chrom R, or materials derived from filter aids, which include Chromosorb W, Anakrom ABS, Gas Chrom Q etc. (Willard et al. 1981). Deactivation of all diatomaceous supports is necessary for most applications.

Table 2. Examples of capillary columns used for fatty acid ester GLC

Column type and ID	Length	Column treatment	Stationary phase	Temperature	Ref.
Steel WCOT 0.25 mm	45.7 m 30.5 mm	Not stated	DEGS, BDS Apiezon L	175 °C	a
Steel SCOT 0.5 mm	45.7 m 30.5 m	Not stated	DEGS, BDS Apiezon L	175 °C	a
Glass WCOT 0.3 mm	35 m	HCl etching	FFAP	220 °C	b
Glass WCOT 0.25 mm, 0.3 mm	Not st.	HCl etching (BaCO$_3$-dep.)	SP-1000 Silar 10C Silar 5CP	150° > 230 °C 140° > 220 °C 100° > 220 °C	c
Glass WCOT 0.7 mm	40 m	HCl etching	DEGS	30° > 180 °C (non-vap. inj.)	d
Glass WCOT Fused silica	20–25 m 15 m	Not stated	Carbowax 20M OV-17 OV-1	200 °C 160 °C 150° > 260 °C	e
Glass WCOT 0.25 mm	40 m	Superox-4	OV-25 OV-225 SP-2300	200 °C 210 °C 200 °C	f
Glass WCOT 0.3 mm	30 m	Not stated	Silar 10C	120° > 185 °C	g
Glass SCOT 0.3 mm	60 m 40 m 30 m	Not stated	SS-4, SP-2340 Silar 10C Silar 9CP, OV-275	175 °C 190 °C	h
Glass SCOT 0.45 mm	100 m	Silica T40	Silar 10C Alltech CS-10	180 °C	i
Glass SCOT 0.5 mm	77 m	Not stated	SE-30	100° > 290 °C	j
Fused silica 0.25 mm	25 m	Superox-4	Silar 10C	150° > 240 °C	k

References: a) Nelson (1974), b) Karunen (1977), c) Sisfontes et al. (1981), d) Badings and de Jong (1983), e) Lercker (1983), f) Golovnya et al. (1984), g) Nyberg and Koskimies-Soininen (1984a), h) Kobayashi (1980), i) Kozuharov (1980), j) Řezanka and Podojil (1984), k) Arrendale et al. (1983).

The stationary liquid phase provides separation of the sample. It should have a suitable selectivity for fatty acids and their derivatives, and a reasonable thermal and chemical stability. It has an upper temperature limit, which must not be transgressed; preferably the highest operation temperature should be at least 10 °–15 °C lower than the limit to prevent detector fouling and to prolong column life. Column bleed would be very harmful in a GC-MS system. Table 3 presents the most popular and widely used stationary phases of today.

The Kovats retention indices (R.I.) indicate where compounds will appear on a chromatogram with respect to n-alkanes injected with the sample. By definition, the R.I. for pentane is 500, for hexane 600, for heptane 700, etc. regardless of the column used or the operating conditions, although the exact conditions and column must be specified. It is only necessary to run a standard set of alkanes on a particular column to determine retention times of compounds for which the R.I. are known for that column and under the particular operating conditions.

Table 3. The most commonly used stationary phases for fatty acid methyl esters

Phase	Chemical type	Maximum °C	Manufacturer
Apiezon L	Saturated hydrocarbon lubricant	250	Apiezon Ltd
Carbowax 20M	Polyethylene glycol	220	Union Carbide
DEGA	Diethylene glycol adipate	200	Several
DEGS	Diethylene glycol succinate	200	Several
EGSS-X	Ethylene glycolsuccinate silicone	225	Applied Science
FFAP	Nitroterephthalic ester of Carbowax 20M	250	Several
OV-17	Polyphenyl methyl siloxane	325	Ohio Valley Co.
OV-101	Dimethyl polysiloxane	350	Ohio Valley Co.
OV-225	Cyanopropyl (phenyl) methyl polysiloxane	265	Ohio Valley Co.
OV-275	Diacyanoallyl-polysiloxane	250	Ohio Valley Co.
SE-30	Dimethyl polysiloxane	300	General Electric
Silar 5CP	γ-Cyanopropylphenyl siloxane	275	Silar Lab. Inc. (Applied Science)
Silar 10C	Di-γ-cyanopropyl siloxane	275	Silar Lab. Inc. (Applied Science)
SP-2300	Cyanopropyl polysiloxane	275	Supelco Inc.
SP-2340	Di-γ-cyanopropyl polysiloxane	275	Supelco Inc.
XE-60	Cyanopropyl (ethyl) polysiloxane	265	General Electric

Data collected from Haken (1984); Kates (1972); Supina (1977); Willard et al. (1981); and the catalogs of several leading distributors.

Table 4. McReynolds constants of the stationary phases in Table 3. (After Supina 1977; Willard et al. 1981; Haken 1984; and data from distributors' catalogs)

Phase	x'	y'	z'	u'	s'
Apiezon L	32	22	15	32	42
Carbowax 20M	322	536	368	572	510
DEGA	378	603	460	665	658
DEGS	499	751	593	840	860
EGSS-X	484	710	585	831	778
FFAP	340	580	397	602	627
OV-17	119	158	162	243	202
OV-101	17	57	45	67	43
OV-225	228	369	338	492	386
OV-275	781	1006	885	1177	1089
SE-30	16	55	44	65	42
Silar 5CP	319	495	446	637	530
Silar 10C	523	775	659	942	801
SP-2300	316	495	446	637	530
SP-2340	520	757	659	942	800
XE-60	204	371	339	492	365

$x' = \Delta I$ for benzene; $y' = \Delta I$ for n-butanol; $z' = \Delta I$ for 2-pentanone; $u' = \Delta I$ for nitropropane; $s' = \Delta I$ for pyridine. Increasing values indicate increasing polarity for each type of compound. The retention indices are compared for each compound to those obtained for the same compounds on a squalane column (Supina 1977).

Classification of stationary phases by their ability to retard compounds (the s.c. Rohrschneider constants or the later developed McReynold's constants, Table 4) involves measurement of the R.I. for index compounds on a given column, as compared to those same compounds on a squalane column to determine the degree to which each is retarded (Willard et al. 1981). The difference, ΔI, gives a measure of solute-solvent interaction.

As Tables 3 and 4 show, many popular stationary phases today are very polar. Those with lower polarity, as Apiezon L, SE-30 and OV-101 are also suitable to a wide range of other compounds than the fatty acid methyl esters. Many phases are also marketed under different trade marks in spite of their essential uniformity (e.g., Silar 5CP – SP-2300 and Silar 10C – SP-2340).

Today the polysiloxane phases are the dominant group used in all gas chromatography and very numerous different polysiloxane phases are available. Perhaps the cyanoalkyl polysiloxanes are the most efficient modern phases for fatty acid and lipid separations. An excellent review on the development and uses of polysiloxane phases has been published (Haken 1984).

The phases can be in principle used with any column type, and most combinations are commercially available. An ever-extending range of phases in fused silica WCOT columns is today also on the market, and their use is to be recommended in many cases; these columns are very flexible and not as easily broken as glass columns. The material is very inert and the columns efficient and easily installed. Fused silica WCOT columns with chemically bonded stationary phases are to be in many cases preferred instead of ordinary WCOT fused silica columns; these columns have a much longer column life.

8 Interpretation of GC Data and Calculation of Results

8.1 Identification of Peaks Using Standard Compounds

The FID detector in fatty acid analysis generates an analog signal, which is recorded in the form of a series of peaks. Each fatty acid ester produces its own peak when eluted from the column, and different acids can be identified by their retention times by comparing these with the retention times of known standard compounds run with the same GC instrument under similar conditions. Modern instruments have reliable printer/plotters as a rule, older models may have only a graphical plotter.

When running the standard fatty acid esters on GC, it has to be remembered that the column and its temperature programming, and the injector and detector temperatures should be exactly the same as the values used in the analysis itself. The standard compound amounts should be approximately on the same level with the sample, as larger samples often have a slightly longer retention time than small samples. This is especially true in sensitive capillary columns. High-quality commercial standards are available from many suppliers (e.g., Applied Science Labs.; Nu-Check-Prep, Inc.; Supelco Inc.; and Serva GmbH, to mention only a few).

A corresponding commercial standard can be found for most fatty acids occurring in lower plant glycolipids, but often they are not available for the more rarely found acids. It should be kept in mind that the identification by commercial standards is always only tentative, and for more exact information on the identity of the observed peaks, other methods should be used, such as the s.c. equivalent chain length (ECL) determinations and mass spectrometry. The standards do not tell very much about the double bond positions in unsaturated acids, either. In routine analyses, when the composition of the sample is essentially known they can, however, in many cases be relied on and also used to follow-up the gradual changes in retention times as the column grows older.

8.2 Quantitation of Results

In FID detectors, the peak area is directly proportional to the total mass and there is no dependency on the flowrate of the mobile phase (gas), as the FID responds to mass flowrate and not to the concentration of solute. The area under the peaks can be integrated in several ways and so make quantitative determinations possible.

The manual methods of calculating chromatograms include (1) multiplying the actual peak height times the width at half-height (probably the most widely used manual method); (2) measurement of the peak height (correct drawing of the peak base line is essential in both of these); (3) planimetry (area measurement); (4) weighing the peak cut from the paper (paper used must be very high-quality and homogeneous).

Today, however, by far the most accurate measurements are made with a computing integrator, which provides complete automation to the peak measurements and a printout of the results in table form. The analog chromatographic signal is digitized by a hardware ADC (analog-to-digital converter). The software program can then detect peaks, correct the baseline and the possible base line drift, calculate areas and retention times, determine concentrations of components using stored calibration factors and make a complete report (Willard et al. 1981).

Among the many convenient features of a chromatography data system are, e.g., the following: (1) automatic baseline drawing to the plot; (2) correct peak integration also for very long peaks (the peak does not have to fit on the paper used as in purely graphical plotters; ADC overange occurs very seldom); (3) programmability to leave out very small (dirt) peaks or any chosen retention time window from the results; e.g., the solvent peak and any clear contamination peaks can be excluded from the results in this way; (4) calculation and detection of tangent peaks; (5) fused peak area allocation; (5) possibility of changing the calculation program to calculate some results anew, if, e.g., unexpected contamination appears; and (6) possibility of storing the plot and/or the results in memory or on discs for an indefinite time. Space does not permit a comprehensive account of chromatography data systems, but the reader is referred to, e.g., the systems and operator manuals of Varian Associates, Inc. and Hewlett-Packard Co., to mention a couple of the leading manufacturers.

The three principal evaluation methods are calibration by standards, area normalization and internal standard. In calibration by standards only the peaks of interest need to be measured, but the same injection volume is required each time. The necessary calibration standards should be run under the same instrument operating conditions as the sample. The percent concentrations are obtained by rationing the volume of each component of interest to the sample size. Relative response factors must be taken into account and they are best obtained by analyzing standard samples.

Area normalization may be used for evaluation when the chromatogram represents the entire sample and all peaks have been resolved. The area of each peak is measured, then divided by its response factor to give the peak's calculated area. Adding these together provides the total calculated area. The % by volume for individual components is obtained by multiplying the individual calculated area by 100, and then dividing by the total calculated area.

All these calculations can be done with a chromatography data system automatically, if required.

8.3 External and Internal Standardization

External standardization means the preparation of standards at the same levels of concentration as the unknown compounds in the same matrix as the unknowns (Debbrecht 1977). These standards are then run chromatographically under identical conditions as the sample. A direct relationship can be established between peak size and composition of one or more components. The unknowns are then compared to the standards for analysis.

Internal standardization permits the operating conditions to vary, and the injection volume does not necessarily have to be identical every time. The internal standard is a known component added to the sample in a known quantity, which is completely eluted and resolved in the column. It is necessary that it is not confused with any natural component in the sample. In fatty acid analysis, the internal standard is preferably a long-chain fatty acid not present in nature or present only in minute amounts; as such, e.g., heptadecanoic acid has been used (Karunen 1977).

To summarize the differences between internal normalization, external standardization and internal standardization, the following points have been put forward by Debbrecht (1977): (1) In internal normalization the analyzed peaks of both the standard and the sample total 100%; (2) In external standardization the amounts actually injected of the standard and unknown sample are accurately known; (3) in internal standardization the amount of different material added to an accurately known amount of the standard and unknown is exactly known.

9 Fatty Acid Ester Structure Determination by GC-MS

9.1 Equivalent Chain Lengths (ECL)

In this chapter, only straight-chain saturated and unsaturated fatty acids will be considered: information on the identification of, e.g., branched-chain, hydroxy, methoxy, cyclic, and substituted fatty acids can be found, e.g., in the reviews of Jamieson (1970), Stein et al. (1976) and Lie Ken Jie (1980). However, in plant glycolipids these acids occur only rarely.

The ECL values have long been used to express the elution sequence of fatty acid (methyl) esters from a GC column. They are determined from a reference curve obtained by plotting the logarithms of the retention times of two or more known, normal, saturated monocarboxylic methyl esters against the number of carbon atoms in the acid. ECL values of other esters, chromatographed under identical operation conditions, are then read from the reference curve using observed retention times. The ECL values may be calculated using an equation similar to the determination of the Kovats R.I.'s (Jamieson 1970):

$$ECL = 2 \left[\frac{\log Rx - \log Rn}{\log Rn + 2 - \log Rn} \right] + n .$$

In this equation, Rx is the retention time of the unknown ester and Rn and Rn + 2 are the retention times of saturated esters of chain lengths n and n + 2.

It is sometimes more convenient to use as the reference curve the semilog plot of retention times of homologous mono-olefinic esters against chain lengths; values read from this curve are designated as "modified ECL" or MECL values (Ackman 1963).

ECL values of olefinic esters increase as the polarity of the stationary phase increases, owing to the increased interaction of their double bonds with the stationary phase. In strongly polar phases, eicosanoic acid (20:0) is eluted before α-linolenic acid (18:3ω3): the Silar- and SP-phases mentioned in Tables 3 and 4 belong to this category. ECL values are dependent of the concentration of the stationary phase: if its concentration is decreased and all other factors remain con-

Table 5. Fatty acids found in lower plant glycolipids (data compiled from Gellerman et al. (1975), Karunen (1977), Anderson et al. (1978), Aro and Karunen (1979), Karunen and Aro (1979), Moseley and Thompson (1980), Nyberg and Koskimies-Soininen (1984 a, b). Extensive lists of lower plant total fatty acids are to be found, e.g., in Matucha et al. (1972), Podojil et al. (1978), Johns et al. (1979), Řezanka et al. (1983), and Řezanka and Podojil (1984)

12:0	trans 16:1ω13	18:1ω9	20:1ω9	22:0
14:0	16:2ω6	18:1ω7	20:2ω9	22:1
14:1	16:3ω6	18:2ω6	20:2ω6	22:5ω3
14:3	16:3ω3	18:3ω6	20:3ω6	22:6
15:0	16:4ω?	18:3ω3	20:4ω6	24:0
16:0	17:0	18:4ω3	20:4ω3	24:1
16:1ω9	17:1ω8	19:0	20:5ω6	
16:1ω7	18:0	20:0	20:5ω3	

stant, also the ECL values decrease, or the apparent polarity of a polyester phase will become less as its concentration is decreased (e.g., data on EGSS-X, Jamieson 1970). Column age also affects the ECL values and it is sometimes useful to compare the results obtained with a new and an aged column for identification of the fatty acid esters.

There is available a large amount of published retention data for methylene-interrupted polyolefinic methyl esters and much of this data has been collected and tabulated in ECL form (Jamieson 1970).

Table 5 presents a list of the fatty acids found in lower plant glycolipids. Using mainly the ECL values and semilog correlations, many isomers differing in the double bond positions have been successfully determined.

9.2 Semilogarithmic Correlations

When the logarithms of the retention times of fatty acid esters are plotted against the number of carbon atoms, curves that approach straight lines are obtained for members of different homologous series (Jamieson 1970, 1975). This relationship is linear over a wide range of chain lengths but departure from lincarity has been observed at shorter chain lengths. Semilog plots are of value in predicting retention times of compounds not available for comparison and also for discovering if compounds belong to the same homologous series.

The relationship is based on chain length, the number of olefinic bonds, and the position of unsaturation relative to the terminal methyl group (carbon end-chain, Jamieson 1975). Mono-olefinic acids with the same carbon end-chain give a straight line plot of log retention time vs. carbon number, those with the shorter end-chain generally have longer retention times, and acids with different end-chains give a series of parallel straight lines. It is advantageous to use the points for $\omega 9$ mono-olefinic acids as reference points rather than the points for saturated acids, since the $\omega 9$ acids occur widely in plant lipids.

Jamieson (1975) has published a large collection of calculated ECL values for some frequently encountered C18, C20, and C22 polyunsaturated fatty acids, based on their relationship to the ECL values of $18:3\omega 3$ for the same column and analytical conditions.

9.3 Mass Spectrometers and Their Function Principles

It is obvious that no universally valid description of fatty acid structure determination using mass spectrometry (MS) can be given in an article like this, and for deeper studies the reader is referred to the papers mentioned in the following chapters. The mass spectrometer is a complicated instrument and its use needs some skill, which can only be obtained by actually operating it.

A MS system can be divided in the following main parts: (1) sample inlet systems (most often the MS is connected with a GC and the sample is, in fact, first subjected to GC fractionation); (2) ion source; (3) ion analyzer system; (4) ion detector; (5) spectrum recording system; (6) vacuum chamber and pumping system; (7) electronic power and control system (Willard et al. 1981). Usually the data is in modern instruments collected on a data handling basis (on a tape) which allows a rapid accumulation of the large amount of data generated and a rapid reconstruction of the mass spectra or the chromatograms (in GC-MS).

Two types of MS instrument are used for GC-MS work: magnetic sector MS and quadrupole mass filters. There seems to be no clear preference for one or the other type of instrument. GC-MS seems to be applicable to virtually all samples that can be separated by GC. An overview of the GC-MS coupled system possibilities is presented by Smith (1984).

According to McCloskey (1970) the production of a mass spectrum involves the following sequential steps: (1) introduction and vaporization of the sample into the high vacuum of the MS; (2) ionization of some of the sample molecules through bombardment by a monoenergetic beam of electrons; (3) rapid decomposition of most of the primary ions along a number of energetically favored pathways; (4) continuous acceleration of the positive ions thus produced, by a negative potential of several thousand volts; (5) separation of the ions according to their mass to charge ratio (m/e) by passage through a strong magnetic field and (6) collection of the ions and recording of their relative abundances. The masses and abundances of the ions produced in this manner are then related to the structure of the original molecule undergoing ionization.

9.4 Interpretation of Mass Spectra of Fatty Acid Esters

The mass spectrum of a compound contains the masses of the ion fragments and the relative abundance of these ions plus often the parent ion. The dissociation fragments will always occur in the same relative abundance for a particular compound. Thus the mass spectrum becomes a "fingerprint" for each compound, as no two molecules will be fragmented and ionized in exactly the same manner on electron bombardment (Willard et al. 1981). The size and structure of the molecule can often be reconstructed from the fragment ions in the spectrum of a pure compound. Often the most intense peak (base peak) is normalized to a value of 100 and the other peaks are reported as percentages of the base peak. The mass spectrum is tabulated as a series of peaks in a coordinate system (x = m/e; y = relative abundance). Figures 3 and 4 give the mass spectra of palmitic (16:0) and eicosapentaenoic (20:5ω3) acid methyl esters as examples of a saturated and an unsaturated acid.

The mass spectra of all normal chain saturated esters are very similar and simple. The molecular ions (M) are well defined. The relative abundance of M increases with methyl pentanoate upwards (McCloskey 1970). Identification of M can be verified by the acylium ion (M-31), due to loss of the saturated ester. The base peak of the spectrum is m/e 74, the ion produced by the gamma hydrogen migration to a double bond followed by beta cleavage (McLafferty rearrangement). Ions of the series $CH_3OCO(CH_2)+n$ are arithmetically found at m/e (59+14n), i.e., m/e 87, 101, 115, 129, 143, 157 etc. The lowest potential member is m/e 87, and it derives its stability from the enol form.

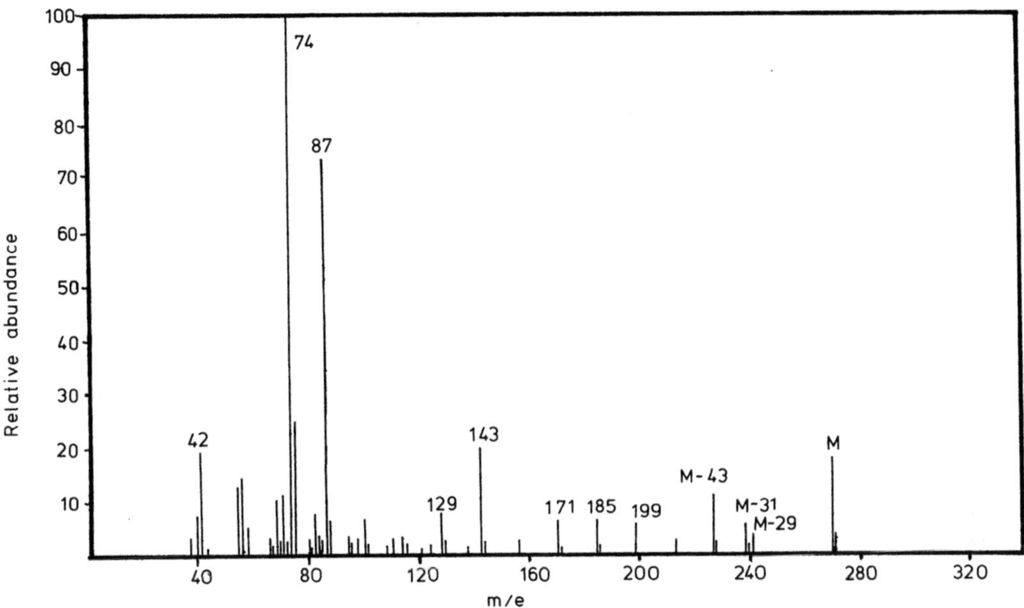

Fig. 3. Mass spectrum of methyl palmitate from *Sphagnum magellanicum* glycolipids

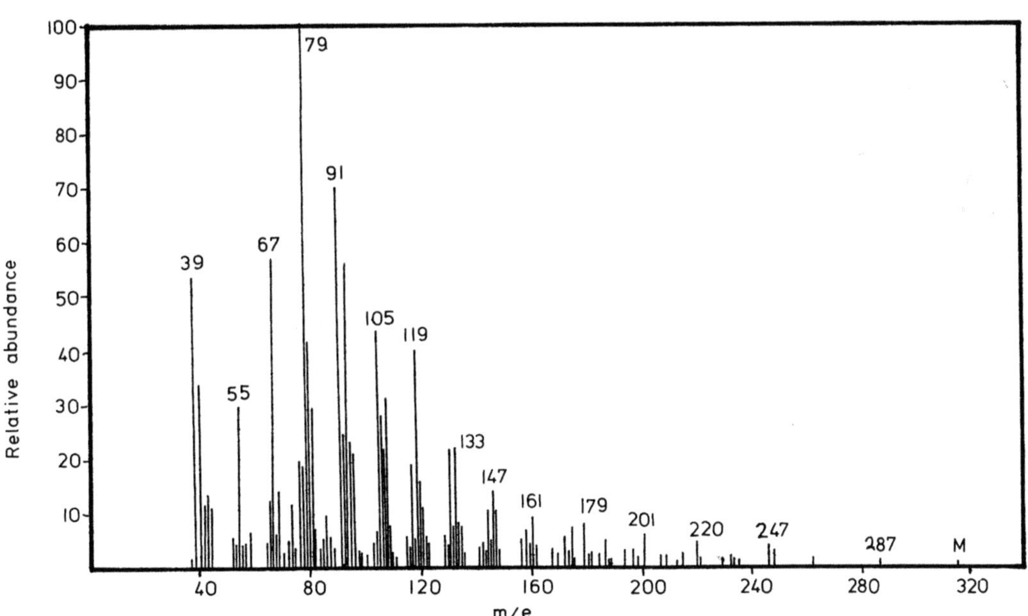

Fig. 4. Mass spectrum of methyl eicosapentaenoate from *Porphyridium purpureum* glycolipids

Both simple cleavage and rearrangement processes contribute to the formation of hydrocarbon ions, the most prominent of which are m/e 69, 83, and 97. They are relatively low in abundance. In short, a saturated fatty acid methyl ester can always be known by the prominence of m/e 74 and 87 and the molecular ion peak (M); the other peaks are quite inconspicuous.

The mass spectra of unsaturated fatty acid esters are very different from those of the saturated ones, and they also differ from each other somewhat according to the degree of unsaturation. The molecular ions are abundant, and are confirmed in each case by a loss of a methoxyl radical (M-31) and by elimination of methanol. The presence of M-74 and M-56 are usually characteristic and therefore useful. The lower mass regions are heavily populated with hydrocarbon ions which are of little use for structural determinations.

The location and stereochemistry of the double bond cannot usually be determined directly by mass spectrometry. An account of suitable methods for the more accurate characterization of unsaturated fatty acid esters is published in McCloskey (1970). The mass spectra of the methyl esters of 18:1, 18:2, and 18:3 have been investigated by Hallgren et al. (1959). The MS of perdeuterated fatty acid esters from the green alga *Scenedesmus* has been studied by Wendt and McCloskey (1970).

10 Abbreviations

ADC	analog-to-digital converter	m/e	mass to charge ratio
BSA	bis(trimethylsilyl)acetamide	MECL	modified ECL
DGDG	digalactosyl diacylglyceride	MGDG	monogalactosyl diacylglyceride
ECL	equivalent chain length	MS	mass spectrometry
FAME	fatty acid methyl ester	R.I.	Kovats retention index
FID	flame ionization detector	SCOT	support-coated open tubular
FSCC	fused silica capillary column		column
GC	gas chromatography	TBDMS	t-butyldimethylsilyl
GLC	gas liquid chromatography	TLC	thin layer chromatography
HMDS	hexamethyldisilazane	2D-TLC	two-dimensional TLC
HPLC	high performance liquid	TMCS	trimethylchlorosilane
	chromatography	WCOT	wall-coated open tubular column
HPTLC	high performance TLC	w/v	weight/volume
M	molecular ion	v/v	volume/volume

The solvent mixture compositions are always given as v/v, if not otherwise stated.

Acknowledgments. The author is very grateful to Prof. Liisa Simola for allowing me good working facilities, and to Ms. Kristiina Koskimies-Soininen, M. Sc., for useful and friendly discussions. The English language was checked by Ms. Patricia Poussa, M.A.

References

Ackman RG (1963) Structural of unsaturated fatty acid esters through graphical comparison of gas-liquid chromatographic retention times on a polyester substrate. J Am Oil Chem Soc 40:558–564

Albertyn DE, Bannon CD, Craske JD, Hai NT, O'Rourke KL, Szonyi C (1982) Analysis of fatty acid methyl esters with high accuracy and reliability I. Optimization of flame-ionization detectors with respect to linearity. J Chromatogr 247:47–61

Allen CF, Good P, Davis HF, Chisum P, Fowler SD (1966) Methodology for the separation of plant lipids and application to spinach leaf and chloroplast lamellae. J Am Oil Chem Soc 43:223–231

Allen CF, Good P, Holton RW (1970) Lipid composition of Cyanidium. Plant Physiol 46:748–751

Anderson R, Livermore BP, Kates M, Volcani BE (1978) The lipid composition of the non-photosynthetic diatom Nitzschia alba. Biochim Biophys Acta 528:77–88

Aro E-M, Karunen P (1979) Effect of changed environmental conditions on glycolipids of the mosses Pleurozium schreberi and Ceratodon purpureus. Physiol Plant 45:201–206

Arrendale RF, Chapman GW, Chortyk OT (1983) Gas chromatographic analyses of fatty acids on laboratory-prepared fused silica Silar 10C capillar columns. J Agric Food Chem 31:1334–1338

Badings HT, de Jong C (1983) Glass capillary gas chromatography of fatty acid methyl esters. A study of conditions for the quantitative analysis of short- and long-chain fatty acids in lipids. J Chromatogr 279:493–506

Bailey RW, Bourne EJ (1960) Colour reactions given by sugar and diphenylamine-aniline spray reagents on paper chromatograms. J Chromatogr 4:206–213

Bailey DS, Northcote DH (1976) Phospholipid composition of the plasma membrane of the green alga, Hydrodictyon africanum. Biochem J 156:295–300

Barta IC, Kömives T (1984) Gas-liquid chromatographic method for the rapid analysis of the epicuticular wax compositions of plants. J Chromatogr 287:438–441

Bligh EG, Dyer WJ (1959) A rapid method of total lipid extraction and purification. Can J Biochem Physiol 37:911–917

Carroll KK, Cutts JH, Murray GD (1968) The lipids of Listeria monocytogenes. Can J Biochem 46:899–904

Chuecas L, Riley JP (1969) Component fatty acids of the total lipids of some marine phytoplankton. J Mar Biol Assoc UK 49:97–116

Clarkson DT, Hall KC, Roberts JKM (1980) Phospholipid composition and fatty acid desaturation in the roots of rye during acclimatization of low temperature. Planta 149:464–471

Clayton TA, MacMurray TA, Morrison WR (1970) Identification of wheat flour lipids by thin-layer chromatography. J Chromatogr 47:277–281

Constantopoulos G, Bloch K (1967) Effect of light intensity on the lipid composition of Euglena gracilis. J Biol Chem 242:3538–3542

Debbrecht FJ (1977) Qualitative and quantitative analysis by gas chromatography II. Quantitative analysis. In: Grob RL (ed) Modern practice of gas chromatography. John Wiley, New York, p 166

DeMort CL, Lowry R, Tinsley I, Phinney HK (1972) The biochemical analysis of some estuarine phytoplankton species I. Fatty acid composition. J Phycol 8:211–216

Diepenbrock W (1981) Zur Umweltvariabilität der Fettsäure-Zusammensetzung von Galaktolipiden in Rapsblättern. Fette Seifen Anstrichm 83:297–302

Döhler G, Datz G (1980) Effect of light on lipid and fatty acid composition of cyanobacteria, Anacystis nidulans (Synechococcus). Z Pflanzenphysiol 100:427–435

Douce R, Holtz RB, Benson AA (1973) Isolation and properties of the envelope of spinach chloroplasts. J Biol Chem 248:7215–7222

Drozd J (1981) Chemical derivatization in gas chromatography. Elsevier, Amsterdam

Folch J, Lees M, Sloane-Stanley (1957) A simple method for the isolation and purification of total lipids from animal tissues. J Biol Chem 226:497–509

Gardner HW (1968) Preparative isolation of monogalactosyl and digalactosyl diglycerides by thin layer chromatography. J Lipid Res 9:139–141

Gellerman JL, Anderson WH, Richardson DG, Schlenk H (1975) Distribution of arachidonic and eicosapentaenoic acids in the lipids of mosses. Biochim Biophys Acta 388:277–290

Golovnya RV, Kuz'menko TE, Vasil'ev AV (1984) Stable and reproducible selective glass capillary columns with polysiloxane stationary phases for the analysis of fatty acid methyl esters. J Chromatogr 292:49–55

Grob K (1984) Effect of "dirt" injected on-column in capillary gas chromatography; analysis of the sterol fraction of oils as an example. J Chromatogr 287:1–14

Grob K, Bossard M (1984) Effect of dirt on quantitative analyses by capillary gas chromatography with splitless injection. J Chromatogr 294:65–75

Grob K, Neukom HP (1984) Glass wool in the injector insert for quantitative analysis in splitless injection. Chromatographia 18:517–519

Haken JK (1984) Developments in polysiloxane stationary phases in gas chromatography. J Chromatogr 198:153–155

Hallgren B, Ryhage R, Stenhagen E (1959) The mass spectra of methyl oleate, methyl linoleate, and methyl linolenate. Acta Chem Scand 13:845–847

Herslöf B (1979) Application of TLC/FID in lipid analysis. In: Appelqvist L-Å, Liljenberg C (eds) Advances in the biochemistry and physiology of plant lipids. Elsevier/North-Holland, Amsterdam, p 301

Hirayama O, Morita K (1980) A simple and sensitive method for the quantitative analysis of chloroplast lipids by use of thin layer chromatography and flame ionization detector. Agric Biol Chem 44:2217–2219

Hirvisalo EL, Renkonen O (1970) Composition of human serum sphingomyelins. J Lipid Res 11:54–59

Jamieson GR (1970) Structure determination of fatty esters by gas liquid chromatography. In: Gunstone FD (ed) Topics in lipid chemistry, vol 1. Logos, London, p 107

Jamieson GR (1975) GLC identification techniques for long-chain unsaturated fatty acids. J Chromatogr Sci 13:491–497

Johns RB, Perry GJ (1977) Lipids of the marine bacterium Flexibacter polymorphus. Arch Mikrobiol 114:267–271

Johns RB, Nichols PD, Perry GJ (1979) Fatty acid composition of ten marine algae from Australian waters. Phytochemistry 18:799–802

Kabata K, Sadakane H, Miyachi M, Nagata K, Hatano S, Watanabe T (1979) Studies on lipid changes during the development of frost hardiness in Chlorella ellipsoidea. J Fac Agric Kyushu Univ 23:155–161

Kabata K, Sadakane H, Kurose M, Kobayakawa A, Watanabe T, Hatano S (1980) Changes in fatty acid composition of membrane fractions during hardening of Chlorella ellipsoidea. J Fac Agr Kyushu Univ 25:91–97

Karunen P (1977) Determination of fatty acid composition of spore lipids of the moss Polytrichum commune by glass capillary column gas chromatography. Physiol Plant 40:239–243

Karunen P, Aro E-M (1979) Fatty acid composition of polar lipids in Ceratodon purpureus and Pleurozium schreberi. Physiol Plant 45:265–269

Kates M (1970) Plant phospholipids and glycolipids. Adv Lipid Res 8:225–265

Kates M (1972) Techniques of lipidology: isolation, analysis and identification of lipids. North-Holland/American Elsevier, Amsterdam

Kates M, Volcani BE (1966) Lipid components of diatoms. Biochim Biophys Acta 116:264–278

Khan M-U, Williams JP (1977) Improved thin-layer chromatographic method for the separation of major phospholipids and glycolipids from plant lipid extracts and phosphatidyl glycerol and bis(monoacylglyceryl)phosphate from animal lipid extracts. J Chromatogr 140:179–185

Kleinschmidt MG, McMahon VA (1970) Effect of growth temperature on the lipid composition of Cyanidium caldarium II. Glycolipid and phospholipid components. Plant Physiol 46:290–293

Klopfenstein WE (1971) On methylation of unsaturated acids using boron trihalide-methanol reagents. J Lipid Res 12:773–776

Knapp DR (1979) Handbook of analytical derivatization reagents. John Wiley, New York

Kobayashi T (1980) Gas-liquid chromatographic separation of geometric isomers of unsaturated fatty acid methyl esters using a glass capillary column. J Chromatogr 194:404–409

Koskimies K, Simola LK (1980) The fatty acid composition of some Sphagnum species. Can J Bot 58:259–263

Kozuharov S (1980) Coating support-coated open tubular columns with Silar 10C or Alltech CS-10 and Silica T40 for separation of isomers of fatty acid methyl esters. J Chromatogr 198:153–155

Kuksis A (1983) Lipids. In: Heftmann E (ed) Chromatography: Fundamentals and applications of chromatographic and electrophoretic methods part B: Applications. Elsevier, Amsterdam, p 75

Kuksis A, Stachnyk O, Holub BJ (1969) Improved quantitation of plasma lipids by direct gas-liquid chromatography. J Lipid Res 10:660–667

Lee RF, Loeblich AR (1971) Distribution of 21 : 6 hydrocarbon and its relationship to 22 : 6 fatty acid in algae. Phytochemistry 10:593–602

Lem NW, Khan M, Watson GR, Williams JP (1980) The effect of light intensity, day length, and temperature on fatty acid synthesis and desaturation in Vicia faba L. J Exp Bot 31:289–298

Lercker G (1983) Short capillary columns in the analysis of lipids. J Chromatogr 279:543–548

Lie Ken Jie MSF (1980) The characterization of long-chain fatty acids and their derivatives by chromatography. In: Giddings JC, Gruschka E, Cazes J, Brown PR (eds) Advances in chromatography, vol 18. Marcel Dekker, New York, p 1

Luukkonen A, Kääriäinen L, Renkonen O (1976) Phospholipids of Semliki forest virus grown in cultured mosquito cells. Biochim Biophys Acta 450:109–120

Lynch DV, Gundersen RE, Thompson GA (1983) Separation of galactolipid species by high performance liquid chromatography. Plant Physiol 72:903–905

Matucha M, Žilka L, Švihel K (1972) Gas chromatographic analysis of the higher fatty acids of the alga Chlorella vulgaris (pyrenoidosa). J Chromatogr 65:371–376

McCloskey JA (1970) Mass spectrometry of fatty acid derivatives. In: Gunstone FD (ed) Topics in lipid chemistry, vol 1. Logos, London, p 369

Morrison WR, Smith LM (1964) Preparation of fatty acid methyl esters and dimethylacetals from lipids with boron fluoride-methanol. J Lipid Res 5:600–608

Moseley KR, Thompson GA (1980) Lipid composition and metabolism of Volvox carteri. Plant Physiol 65:260–265

Nelson GL (1974) Elution characteristics of fatty acid methyl esters on capillary columns. Lipids 9:254–263

Nichols BW (1965a) Light induced changes in the lipids of Chlorella vulgaris. Biochim Biophys Acta 106:274–279

Nichols BW (1965b) The lipids of a moss (Hypnum cupressiforme) and of the leaves of green holly (Ilex aquifolium). Phytochemistry 4:769–772

Nichols BW, Appleby RS (1969) The distribution and biosynthesis of arachidonic acid in algae. Phytochemistry 8:1907–1915

Nichols BW, Moorhouse R (1969) The separation, structure and metabolism of monogalactosyl diglyceride species in Chlorella vulgaris. Lipids 4:311–316

Nyberg H, Koskimies-Soininen K (1984a) The glycolipid fatty acids of Porphyridium purpureum cultured in the presence of detergents. Phytochemistry 23:751–757

Nyberg H, Koskimies-Soininen K (1984b) The phospholipid fatty acids of Porphyridium purpureum cultured in the presence of Triton X-100 and sodium desoxycholate. Phytochemistry 23:2489–2495

O'Brien JS, Benson AA (1964) Isolation and fatty acid composition of the plant sulfolipid and galactolipids. J Lipid Res 5:432–436

Opute FI (1974) Lipid and fatty acid compositions of diatoms. J Exp Bot 25:823–835

Piorreck M, Baasch K-H, Pohl P (1984) Biomass production, total protein, chlorophylls, lipids and fatty acids of freshwater green and blue-green algae under different nitrogen regimes. Phytochemistry 23:207–216

Podojil M, Lívanský K, Prokeš B, Wurst M (1978) Fatty acids in green algae cultivated on a pilot-plant scale. Folia Microbiol 23:444–447

Pohl P, Glasl H, Wagner H (1970) Zur Analytik pflanzlicher Glyko- und Phospholipide und ihrer Fettsäuren I. Eine neue dünnschichtchromatographische Methode zur Trennung pflanzlicher Lipoide und quantitative Bestimmung ihrer Fettsäure-Zusammensetzung. J Chromatogr 49:488–492

Renkonen O, Luukkonen A (1976) Thin-layer chromatography of phospholipids and glycolipids. In: Marinetti GV (ed) Lipid chromatographic analysis, vol 1, 2nd edn. Marcel Dekker, New York, p 1

Řezanka T, Podojil M (1984) The very long chain fatty acids of the green alga. Chlorella kessleri. Lipids 19:472–473

Řezanka T, Vokoun J, Slavíček J, Podojil M (1983) Determination of fatty acids in algae by capillary gas chromatography-mass spectrometry. J Chromatogr 268:71–78

Rouser G, Kritchevsky G, Yamamoto A (1976) Column chromatographic and associated procedures for separation and determination of phosphatides and glycolipids. In: Marinetti GV (ed) Lipid chromatographic analysis, vol 3, 2nd edn. Marcel Dekker, New York, p 713

Schantz R, Blee E, Duranton H (1976) Métabolisme des galactolipides chez Euglena gracilis. Physiol Vég 14:141–157

Schill R (1977) Instrumentation. In: Grob RL (ed) Modern practice of gas chromatography. John Wiley, New York, p 289

Schomburg G, Husmann H, Behlau H, Schulz F (1983 a) Cold sample injection with either the split or splitless mode of temperature-programmed sample transfer. Design and testing of a new, electrically heated construction for universal application of different modes of sampling. J Chromatogr 279:251–258

Schomburg G, Husmann H, Schulz F, Teller G, Bender M (1983 b) Cold sample injection with either the split or splitless mode of temperature-programmed sample transfer. Comparison to cold on-column injection with a commercial device. J Chromatogr 279:259–267

Settlage J, Jaeger H (1984) Advantages of fused silica capillary gas chromatography for GC-MS applications. J Chromatogr Sci 22:192–197

Siebertz HP, Heinz E, Joyard J, Douce R (1980) Labelling in vivo and in vitro molecular species of lipids from chloroplast envelopes and thylakoids. Eur J Biochem 108:177–185

Simola LK, Koskimies-Soininen K (1980) The effect of fluoride on the growth and fatty acid composition of Sphagnum fimbriatum at two temperatures. Physiol Plant 50:74–77

Simola LK, Koskimies-Soininen K (1984) Comparison of glycolipids and plastids in callus cells and leaves of Alnus and Betula. Plant Cell Physiol 25:1329–1340

Sisfontes L, Nyborg G, Svensson L, Blomstrand R (1981) Separation of complex long-chain fatty acid mixtures by high-performance glass capillary gas chromatography. J Chromatogr 216:115–125

Skipski VP, Smolowe AF, Barclay M (1967) Separation of neutral glycosphingolipids and sulfatides by thin-layer chromatography. J Lipid Res 8:295–299

Skowronski G, Garrigan OW (1983) Glycolipids and thylakoid proteins in chloroplasts and streptomycin-bleached lamellae of Euglena. Lipids 18:539–544

Smith SL (1984) Coupled systems: Capillary GC-MS and capillary GC-FTIR. J Chromatogr Sci 22:143–148

Solberg Y (1983) Lipid constituents of the moss Mniobryum wahlenbergii var. glaciale. Cryptogam Bryol Lichenol 4:129–143

Stein RA, Slawson W, Mead JF (1976) Gas-liquid chromatography of fatty acids and derivatives. In: Marinetti GV (ed) Lipid chromatographic analysis, vol 3, 2nd edn. Marcel Dekker, New York, p 857

Sullivan JJ (1977) Detectors I. In: Grob RL (ed) Modern practice of gas chromatography. John Wiley, New York, p 213
Supina WR (1977) Columns and column selection in gas chromatography. In: Grob RL (ed) Modern practice of gas chromatography. John Wiley, New York, p 113
Sweeley CC, Bentley R, Makita M, Wells WW (1963) Gas-liquid chromatography of trimethylsilyl derivatives of sugars and related substances. J Am Chem Soc 85:2497–2507
Tanaka M, Takase K, Ishii J, Itoh T, Kaneko H (1984) Application of a thin-layer chromatography-flame ionization detection system for the determination of complex lipid constituents. J Chromatogr 284:433–440
Tancrede P, Chauvette G, Leblanc RM (1981) General methods for the purification of lipids for surface pressure studies. J Chromatogr 207:387–393
Tornabene TG, Holzer G, Peterson SL (1980) Lipid profile of the halophilic alga, Dunaliella saline. Biochem Biophys Res Commun 96:1349–1356
Wendt G, McCloskey JA (1970) Mass spectrometry of perdeuterated molecules of biological origin. Fatty acid esters from Scenedesmus obliquus. Biochemistry 9:4854–4866
Wettern M (1980) Lipid variation of the green alga Fritschiella tuberosa during growth in axenic batch culture. Phytochemistry 19:513–517
Willard HH, Merritt LL, Dean JA, Settle FA (1981) Instrumental methods of analysis, 6th edn. Wadsworth, Belmont
Williams MG, MacGee J (1983) Rapid determination of free fatty acids in vegetable oils by gas liquid chromatography. J Am Oil Chem Soc 60:1507–1509
Wintermans JFGM, van Besouw A, Bögemann G (1981) Galactolipid formation in chloroplast envelopes II. Isolation-induced changes in galactolipid composition of envelopes. Biochim Biophys Acta 663:99–107
Wright DC, Berg LR, Patterson GW (1980) Effect of cultural conditions on the sterols and fatty acids of green algae. Phytochemistry 19:783–785

Analysis of Phospholipid Molecular Species
by Gas Chromatography
and Coupled Gas Chromatography-Mass Spectrometry

D. V. LYNCH and G. A. THOMPSON, JR.

1 Introduction

Every plant membrane contains a complex, heterogeneous population of lipids, the physicochemical properties of which influence the physical state and (therefore) the physiological function of the membrane. The diverse lipid mixture present in each membrane seems to make possible a broader latitude for maintaining functional physical properties in the event of sudden environmental change. Phospholipids are the principal structural elements of most membrane bilayers. The physical properties of a phospholipid are significantly influenced by the polar head group, the nature of the two fatty acyl groups, and even the arrangement of the acyl groups on the glycerol backbone.

Increasing awareness of these complex physical interactions has prompted many recent efforts to analyze phospholipid composition in greater detail. It is no longer considered adequate to simply quantify the different lipid classes (e.g., phosphatidylcholine, phosphatidylglycerol, etc.) from a given membrane and analyze the bulk fatty acid composition of each. Any comprehensive analysis must include detailed information on the molecular species composition of each phospholipid class, i.e., a quantitative accounting of which two fatty acids are bound together on each phospholipid molecule, and where possible, the precise distribution of the acyl chains on the *sn*-1 position and the *sn*-2 position. Recent investigations have suggested that even subtle changes in phospholipid molecular species composition of membranes can be of critical importance to organisms acclimating to environmental stress (Lynch and Thompson 1984c). Such studies have also provided new insights into aspects of plant membrane lipid metabolism.

Until recently, there were no convenient ways to analyze phospholipid molecular species in a quantitative way. The preferred method was to purify each class of phospholipid and then further resolve this mixture of molecular species into groups by preparative thin-layer chromatography on silver nitrate-impregnated silica gel plates (Van Golde et al. 1968). This process of argentation TLC separated the lipids according to their degree of fatty acid unsaturation but not according to chain length. Therefore, each band on the plate would usually contain several molecular species. With luck, the fatty acyl pairings present in each band could be deduced from gas chromatography of fatty acid methyl esters prepared from the mixed species eluted from the TLC plate. Because of the need to manipulate separately a number of different subsets of each phospholipid class, this approach was not only time consuming but under most conditions only semi-quantitative.

In this chapter we describe a much more satisfactory procedure for determining phospolipid molecular species composition by gas chromatography (GC) or coupled gas chromatography-mass spectrometry (GC-MS). The development of these GC-based procedures was facilitated by advances in capillary GC and particularly by the formulation of new stationary phases stable at temperatures up to 250 °C or higher. Some of the early work with these systems was done by Kuksis and associates (Breckenridge et al. 1976; Holub and Kuksis 1978). The principle is to convert the heterogeneous mixture of molecular species comprising a given phospholipid class to a form which is volatile enough for gas chromatography and retains the natural pairings of fatty acids. The method of choice, as described below, is to remove the phospholipid polar head group by incubation with the enzyme phospholipase C, and then convert the resulting mixture of diacylglycerols to trimethylsilyl or other types of volatile derivatives for analysis by GC or GC-MS. In addition to detailing the procedures for derivatization and for analysis by gas chromatography, related techniques for the extraction and purification of individual phospholipids are also presented.

Before proceeding to a detailed description of the necessary experimental protocols, we should offer this cautionary reminder. The molecular species composition of any particular phospholipid class can differ markedly from one membrane type to another, even within the same cell (Lynch and Thompson 1984 a, b; Dickens and Thompson 1982; Ramesha et al. 1982), thus one must be careful to undertake such analyses using the most homogeneous membrane preparation possible (see Vol. 1). Otherwise, the most detailed analysis will have little biological relevance.

2 Lipid Preparation

2.1 Lipid Extraction

One of the most effective procedures used for the quantitative extraction of lipids from membranes is that described by Bligh and Dyer (1959) using chloroform and methanol as extracting solvents. The procedure has the advantages of being simple, rapid, and efficient. In some plant tissues it may be necessary to employ more rigorous treatments/procedures [e.g., dipping tissue (leaves) in boiling isopropanol] in order to inactivate endogenous lipases, thereby preventing lipid degradation during extraction (Kates 1972; Quinn and Williams 1978).

We have not found such procedures to be needed when extracting lipids from plant cell fractions that we have used. The safest policy with any unfamiliar plant tissue is to compare by thin layer chromatography (TLC) and gas chromatography (GC) of fatty acids using the lipids extracted using hot isopropanol (Kates and Eberhardt 1957) with those extracted by the more convenient Bligh Dyer method. If there are no detectable differences, such as increased levels of phosphatidic acid or losses of unsaturated fatty acids, the Bligh Dyer procedure, slightly modified as outlined below, should prove satisfactory. Note that the following protocol is designed for use with isolated cell organelles.

Procedure. To 0.8 volume of aqueous membrane suspension is added three volumes chloroform: methanol (1 : 2, v/v). This is mixed and allowed to stand for 10 min. Then one additional volume of chloroform is added, and, after mixing, one volume of distilled water is also added and the sample is mixed again. Two liquid phases should now separate. To remove protein (which forms a flocculent white precipitate at the interface between the upper and lower phases), the extract may be filtered using Whatman no. 1 filter paper, or, preferably, diatomaceous earth [Hyflo-Super Cel: Johns Manville, prewashed immediately before use with chloroform: methanol (1 : 2, v/v) tamped into a 3–5-mm deep pad in a scintered glass funnel. In both cases, suction should be employed to facilitate filtration. The residue remaining on the filter may be scraped off and extracted again with a small volume of chloroform, filtered, and combined with the total extract. The extract is poured into a separatory funnel. After the two phases have cleared, the lower (chloroform) phase containing the lipids may be drawn off, taken to dryness by rotary evaporation and resuspended in a small volume of chloroform for column chromatography. One or 2 ml of ethanol may be added prior to rotary evaporation to aid in removing any traces of water by forming an azeotropic solution.

2.2 Purification of Phospholipids

The crude lipid extract obtained above will contain, in addition to phospholipids, all the other membrane lipids, including sterols, nonsterol isoprenoids, and, in the case of chloroplasts, copious amounts of glycolipids. Before the phospholipids can be subjected to GC analysis, they must be (1) separated from the other types of lipid and (2) resolved into classes defined by the polar head group involved. The most satisfactory purification strategy is to first separate the phospholipids in bulk from the other lipids by column chromatography and then resolve the phospholipids into their individual classes by preparative TLC. Procedures for achieving this are described below.

2.2.1 Column Chromatography

The lipid extract may be conveniently fractionated into three classes, namely, neutral lipids, glycolipids, and phospholipids, using silicic acid column chromatography.

Procedure. Two g silicic acid (Mallinkrodt, 100 mesh) and 1 g Hyflo Supercel are mixed dry and washed sequentially with chloroform: methanol (1 : 1, v/v) and then chloroform in a scintered glass funnel. A slurry of the powder (in chloroform) is poured into a glass column (0.5–1.0 cm diameter) having a Teflon stopcock and a plug of glass wool to retain adsorbent. The adsorbent is packed, first by gravity and then, after adding more chloroform and opening the stopcock, under 0.5 psi of N_2. When the chloroform level has fallen just to the top of the silicic acid, the lipid extract (≤ 50 mg lipid in chloroform) is carefully added taking pains to avoid disturbing the top of the column packing. The lipid is allowed to enter the column, and small volumes of chloroform are added to wash down the sides and keep the adsorbent covered at all times. Following application of the lipids to the column, the neutral lipids are eluted with approx. 30 ml chloroform. This is followed by 80–100 ml acetone to elute the glycolipids and then 30 ml chloroform: methanol (1:1, v/v) to elute the phospholipids. Each solvent should be allowed to descend just to the top of the column packing before carefully adding the next solvent. The lipids should be eluted at a flow rate of 1–3 ml min^{-1}, facilitated by pressurizing the headspace above the column solvent reservoir with nitrogen gas. Each fraction is taken to dryness, resuspended in a small volume of chloroform: methanol (6:1, v/v), and checked by TLC to confirm that lipid separation was complete.

Some plants contain non-phospholipids that elute with the phospholipids in the above protocol. For example, diacylglyceryltrimethylhomoserine is present, sometimes as a major component, in many algae (Eichenberger 1982) and higher plants (Sato and Furuya 1983).

2.2.2 Thin Layer Chromatography

The phospholipid fraction obtained from silicic acid column chromatography must be further separated into phospholipid classes (i.e., PC, PE, PG) by high performance liquid chromatography (Ashworth et al. 1981; Kaduce et al. 1983; Christie 1985) or by TLC. The former method is advantageous in that the dry lipids are not exposed to air thereby diminishing the chance of oxidation. However, it is more complicated and requires greater expenditure of money and equipment. Thin layer chromatography has the advantages of being relatively inexpensive, highly reproducible and immune to "equipment failure". The major drawback, i.e., the possible oxidation of dry lipids exposed to air while on the TLC plate, may be satisfactorily overcome by handling the plates in an inert atmosphere and/or avoiding prolonged exposure of the dry lipids to air.

Procedure. The phospholipid mixture, containing 5–10 mg lipid dissolved in a small volume (≤ 100 µl) chloroform: methanol (6:1, v/v), is applied as a streak across the bottom of a 10 cm wide plate coated with silica gel H (Merck, Darmstadt, W. Germany) and placed in a paper-lined TLC tank containing chloroform: acetic acid: methanol: water (70:25:5:2.2, v/v/v/v)/(Lynch and Thompson 1984a) (Fig. 1). When the solvents have migrated to within 1–2 cm of the top of the plate, it is removed from the tank and immediately placed in a chamber flushed with nitrogen. The plate is flushed for 2–5 min to remove the solvents (although traces of acetic acid will persist) and the lipid bands are detected by iodine vapors. To avoid exposing the entire plate to iodine vapors, it is advisable to employ a Pasteur pipette containing iodine crystals (retained by glass wool) through which nitrogen is passed (Kates 1972). By directing the nitrogen stream containing iodine vapors over one vertical edge of the plate, a small portion of each band will become colored, and lightly straining bands may be selectively exposed for longer times.

Before the iodine color fades, the location of the bands is marked on the plate with a needle, and the bands of silica gel containing lipid are immediately scraped

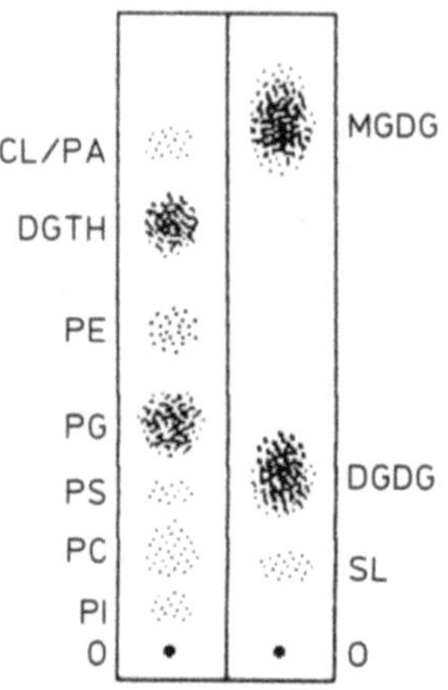

Fig. 1. Separation of common phospholipids on a silica gel H plate developed in the solvent system chloroform:acetic acid:methanol:water (70:25:5:2.2, v/v/v/v). *Left lane* illustrates resolution, starting from origin (*O*), of phosphatidylinositol (*PI*), phosphatidylcholine (*PC*), phosphatidylserine (*PS*), phosphatidylglycerol (*PG*), phosphatidylethanolamine (*PE*), diacylglyceryltrimethylhomoserine (*DGTH*), and phosphatidic acid (*PA*) plus cardiolipin (*CL*). An incomplete removal of glycolipids during the column chromatography step would result in contamination of the phospholipids by one or more of the glycolipids shown resolved by the same solvent system in the right lane: sulfolipid (*SL*), digalactosyldiglyceride (*DGDG*), monogalactosyldiglyceride (*MGDG*)

from the plate into tubes containing 3 ml chloroform:methanol:water (3:5:1, v/v/v). This relatively polar solvent mixture is required to elute the phospholipids from the silica gel. After vortexing, the silica gel is sedimented by centrifugation in a clinical centrifuge, and the supernatant containing each eluted lipid is decanted. The elution step is repeated twice, and the supernatants are pooled. The solvents are dried under a stream of nitrogen (until acetic acid can no longer be detected), and each individual phospholipid class is resuspended in a small volume of chloroform:methanol (6:1, v/v). The molar proportions of the various phospholipid classes are quantified by a phosphorus analysis (Marinetti 1962) of each eluted band. Alternatively, a phosphorus determination can be carried out on the bands scraped from a separate TLC plate run specifically for that purpose (Rouser et al. 1970).

3 Formation of Derivatives for GC or GC-MS

Phospholipids must be converted to volatile, nonpolar derivatives for gas chromatographic-mass spectrometric analysis. This is accomplished by removal of the polar head group by phospholipase C hydrolysis and conversion of the resulting mixture of diacylglycerols to silyl derivatives. During these procedures, opportunities for lipid oxidation and acyl chain migration (forming 1,3-diacylglycerols) must be avoided.

3.1 Phospholipase C Treatment

Phospholipids may be most effectively converted to diacylglycerols by incubating the isolated lipids with phospholipase C. The conditions for the enzymatic cleavage are mild, and lipid degradation/oxidation is not typically a problem. Because this step involves the use of a commercial enzyme preparation, which may vary in activity from batch to batch, care must be taken to ensure complete conversion in order to assure reproducible results. The conversion of phospholipids to diacylglycerols should be monitored when employing this technique for the first time, when analyzing lipids which may contain "unusual" components, (e.g., ether-containing lipids, nonphosphorus-containing lipids, etc.), or when using a new batch of phospholipase C. Enzyme preparations from certain sources are more effective than others at cleaving different phospholipid classes and/or molecular species, especially those containing alkyl side chains (Dickens and Thompson 1982). In such cases, enzymes from a variety of sources (and companies) should be tried. The procedure described below is from Lynch and Thompson (1984a).

Procedure. A chloroform-methanol solution containing 0.2–3 µmol of a single phospholipid class purified by TLC is placed in a 5 ml Reactivial (Pierce Chemical Co., Rockford, IL), concentrated to dryness under a stream of nitrogen, and resuspended in 1 ml peroxide free ethyl ether. Special care must be taken to remove all traces of chloroform, which is known to inactivate phospholipase C. To this is added 30 to 35 µl enzyme suspension (phospholipase C grade II from *Bacillus cereus*, approx. 800 units mg^{-1}, Boehringer Mannheim, Indianapolis IN) and 250 µl buffer 10 mM in Na phosphate (pH 7.4), 250 µM in $CaCl_2$, and 100 µM in $ZnCl_2$. The zinc salt is added to the buffer immediately before use to avoid precipitation during storage. The Reactivial is capped and incubated at 37 °C for 1.5–2 h with moderate shaking.

Following incubation, the diacylglycerols are extracted by decanting the ether phase and subsequent re-extraction of the aqueous phase with 1 ml ethyl ether:petroleum ether (1:1, v/v). The ether fractions are combined and taken to dryness for conversion to silyl derivatives (see Sect. 3.2).

In instances where a complete extraction of all lipid material is required, e.g., when confirming that the conversion of phospholipids to diacylglycerols is quantitative, an alternative procedure should be employed: Following incubation, the Reactivial is uncapped and placed under a stream of nitrogen to evaporate the ether phase, and the remaining aqueous phase is brought to 0.8 ml with water, and subjected to a routine lipid extraction as described above (Sect. 2.1) with the exception that the extract need not be filtered and the chloroform phase (total volume < 3 ml) may be dried under nitrogen following removal from the Reactivial with a Pasteur pipette.

The effectiveness of the enzyme in converting phospholipids to diacylglycerols may be assessed by several methods, most of them involving TLC to separate diacylglycerols from unreacted phospholipids. A suitable solvent system for the separation of these two lipids consists of petroleum ether:ethyl ether:acetic acid (70:30:1, v/v/v). In this system, diacylglycerols have an R_f of approx. 0.3 while phospholipids remain at the origin.

The most convenient method for quantifying conversion entails the use of lipids derived from cells/tissue which have been supplied with a radioactive fatty acid or fatty acid precursor such as acetate or CO_2. The lipids which were prelabeled biosynthetically may be treated with phospholipase C, and the bands from

TLC (corresponding to diacylglycerols and any remaining phospholipids) may be quantified by liquid scintillation counting (Kates 1972).

The possibility of isomerization of 1,2-diacylglycerols to 1,3-diacylglycerols is monitored by TLC using borate-treated thin layer plates (approx. 2% boric acid) developed in the solvent system hexane:ethyl ether 1 : 1 (v/v) (Myher 1978).

3.2 Conversion of Diacylglycerols to Silyl Derivatives

The diacylglycerols obtained by enzyme treatment must be converted to derivatives suitable for separation by gas chromatography and characterization/quantitation by mass spectrometry. The ideal derivative should be (1) volatile and relatively nonpolar, exhibiting the shortest possible retention times on appropriate GC columns, (2) sufficiently stable so as to produce abundant characteristic fragments subsequent to ionization in the mass spectrometer, and (3) chemically stable, allowing storage and/or further processing of the lipid derivatives without

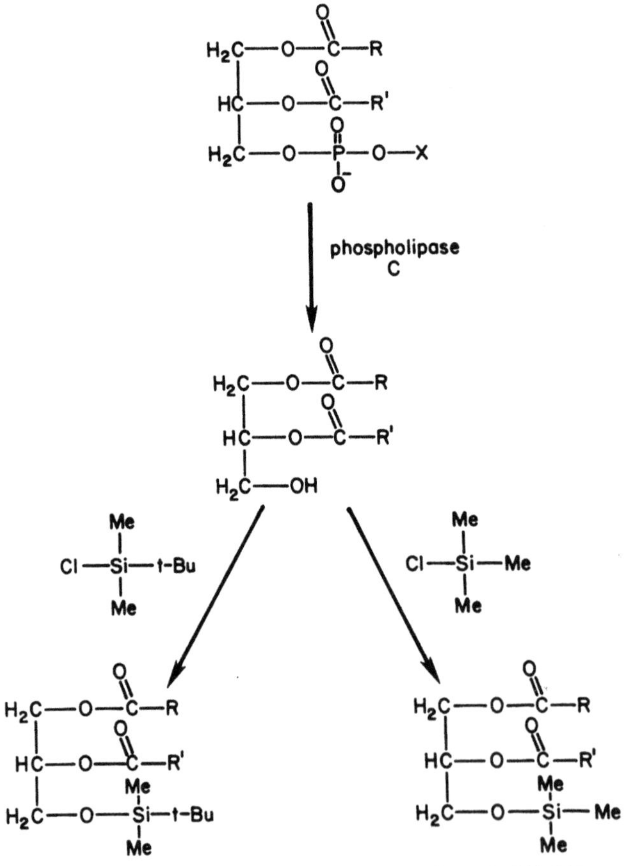

Fig. 2. Steps in the formation of tert-butyldimethylsilyl (*lower left*) and trimethylsilyl (*lower right*) derivatives of diacylglycerols

degradation. Additionally, the derivatization procedure should be mild, thereby minimizing the opportunity for lipid degradation.

At present, two of the most commun modifications employed for gas chromatographic separation of molecular species involve the formation of trimethylsilyl (TMS) or tert-butyldimethylsilyl (t-BDMS) derivatives (Myher 1978) (Fig. 2). Acetate derivatives have also been employed for analysis of lipid molecular species by TLC and GC (Gregor 1977; Nishihara et al. 1980), but because of their relatively long retention times and poor mass spectrometric properties, they are not recommended for these types of analyses.

3.2.1 Formation of Trimethylsilyl Derivatives

The TMS derivatives have relatively short retention times and require very mild conditions for their formation. The derivatives, however, are rather unstable, particularly in the presence of water, and do not produce a high mass fragment in great abundance, although the fragment ions of the acyl chains are abundant. TMS derivatives are recommended for routine gas chromatographic analyses and GC-MS analyses (identification) of molecular species based solely on acyl chain fragments.

Procedure. Diacylglycerols are transferred into a 5 ml Reactivial and taken to complete dryness under nitrogen. Any trace of water must be removed for derivatization to be complete. While maintaining the dry sample under nitrogen, 250–500 µl Sylon HTP (Supelco; pyridine:hexamethyldisilazane:trimethylchlorosilane, 9:3:1) is added. The vial is capped under nitrogen and the mixture is reacted for 30–60 min at 30 °–35 °C. Following this, the solvent is then removed under nitrogen, leaving a white or tan residue. The derivatives are extracted from this using 2 ml petroleum ether, and centrifuging the undissolved material.

The TMS ethers may be stored at -10 °C under nitrogen in dry chloroform for limited periods (2 weeks). Caution should be taken to warm samples removed from the freezer prior to opening vials to prevent condensation of H_2O. The use of TLC to purify TMS derivatives or assess the completion of the reaction is possible, but should be employed judiciously. Because the silating reagent mixture is sensitive to traces of water, the purchase and storage of the mixture in small aliquots (1 ml volumes in sealed ampules) is strongly recommended.

3.2.2 Formation of tert-Butyldimethylsilyl Derivatives

The t-BDMS derivatives of diacylglycerols are chemically stable (up to 10^4 times more stable than TMS derivatives), thus they may be exposed to traces of water (e.g., during chromatography by TLC), and can be stored for relatively long periods (months). They are also more stable during fragmentation/ionization in the mass spectrometer, producing an abundant fragment/ion of high mass, the $[M-57]^+$ ion. The derivatization conditions typically used are somewhat harsher than needed for TMS ether formation, and the relative retention times for t-BDMS derivatives on commonly used capillary columns are slightly longer than those of the corresponding TMS derivatives. The use of t-BDMS derivatives is recommended for GC-MS analysis when a large $[M - 57]^+$ ion is desired, or if further manipulation/processing of the derivatives (e.g., TLC or argentation TLC) is anticipated.

Procedure. Diacylglycerols are placed in a 2-ml glass ampule and taken to dryness under nitrogen. An aliquot (250–500 µl) of tert-butyldimethyl chlorosilane/imidazole reagent (Applied Science Laboratories, Deerfield, IL) is added and the ampule is sealed under nitrogen. The ampule is heated in a sand bath at 80 °C for 20 min. After cooling, the material is transferred to a test tube and mixed with 5 ml petroleum ether and 0.5 ml water. The ether phase is then removed and taken to dryness. To remove traces of water the ether extract may be dried over anhydrous Na_2SO_4, or ethanol (2–3 drops) may be added prior to evaporation under nitrogen.

Because the high temperature required for the formation of t-BDMS derivatives might favor some oxidation of polyunsaturated fatty acids, it is suggested that in early trials, fatty acid methyl ester analyses be made of the phospholipid starting material as well as the t-BDMS derivatives to ensure that the fatty acid pattern is unchanged by the reaction. Recently, Pierce Chemical Co. introduced a new t-BDMS derivatizing reagent, N-Methyl-N(tert-butyldimethylsilyl) trifluoroacetamide (MTBSTFA) for gas chromatographic analysis which has the advantage of being reactive at room temperature. Use of this reagent would diminish the chances of lipid degradation during derivatization.

4 Gas Chromatography

Gas chromatographic separation of lipids, particularly fatty acid methyl esters, has been widely used for many years. The chromatography of larger, complex lipid species, however, is relatively new. The primary impediment had been the lack of liquid phases which are sufficiently polar and are stable at high temperatures. Indeed, until the recent advent of high temperature polar phases and wall coated open tubular columns containing such phases, the use of coupled GC-MS for the analysis of molecular species was a necessity: the mass spectrometer was required to identify multiple components which often eluted together in a single incompletely resolved peak when using nonpolar thermally stable liquid phases (such as OV-1 or SP 2100) and packed columns (Dickens and Thompson 1982).

Coupled GC-MS has the advantage that all peaks may be unambiguously identified with respect to the molecular species contained therein, although quantitation requires a number of corrections not needed when using a standard method of detection (i.e., flame ionization detection) on single-component peaks. In this section, guidelines for the gas chromatographic separation of lipid molecular species (as silyl derivatives of diacylglycerols) will be discussed, and MS procedures will be considered in Sect. 5.

The most important variables relating to the chromatography of silyl derivatives include column type (packed vs. capillary) and liquid phase. Many of the molecular species analyses employing GC-MS have utilized packed columns containing thermally stable nonpolar liquid phases such as 3% OV-1 at temperatures from 240 °–350 °C (Dickens and Thompson 1982; Myher et al. 1978). The chromatographic separation of molecular species was based on the combined total number of acyl chain carbon atoms per molecule.

Thus, a sample might yield five cleanly resolved peaks containing molecular species of carbon number 28, 30, 32, 34, and 36, respectively. However, some of the peaks would be heterogeneous; for example the peak having derivatives with 34 acyl carbon atoms might contain species having C_{18} fatty acids combined with C_{16} fatty acids and also species in which C_{14} fatty acids were associated with C_{20} fatty acids. Varying degrees of unsaturation would be found within the same peak as well. The mass spectrometer was relied upon to identify and quantify these various components within a given peak.

The recent development of polar liquid phases having relatively high thermal stabilities (>270 °C), and the technology to successfully apply these liquid phases as wall coatings in capillary columns have led to great improvements in analyzing lipid molecular species. Increased chromatographic resolution of closely related molecular species also simplified the interpretation of mass spectra from the peaks. To date, the capillary columns used most commonly for lipid molecular species analyses have been made of glass. Although fused silica columns have many advantages (such as ease of manipulation, inherently high inertness) the ability to obtain such a column efficiently coated with a polar liquid phase has been limited.

Fortunately, coating technology and related advances (such as the development of bonded phases) are progressing rapidly, and many companies now offer fused silica columns containing polar liquid phases which appear suitable for these types of analyses. The reader is advised to consult current product catalogues for the most up-to-date information on the availability and suitability of such columns.

To date, good success has been obtained using glass capillary columns coated with SP 2330 (Supelco, Bellefonte, PA), a cyanopropyl/phenylsiloxane polymer which allows the separation of species based on both acyl chain carbon number and degree of unsaturation. Equivalent coatings are available from other companies. Optimal operating conditions and chromatograph settings may depend on several variables. Two different sets of operating parameters are detailed below and may be used as guidelines when initially testing such columns.

Chromatography of the TMS derivatives of diacylglycerols derived from *Dunaliella* phospholipids (Lynch and Thompson 1984 a, b) was performed using a 10 m × 0.25 mm i.d. open tubular glass column coated with SP 2330 (Supelco) (Fig. 3). Column temperature was programmed from an initial temperature of 200 °C (held for 5 min) to 250 °C at 10 °C min^{-1}. Injector and detector temperatures were maintained at 270 °C and 300 °C, respectively. The carrier gas was nitrogen, with head pressure maintained at 0.5 kg cm^{-2}. Split injection (split ratio approx. 20 : 1) was used.

Trimethylsilyl and t-BDMS derivatives of diacylglycerols from several sources were analyzed using a column identical to that above but with different operating conditions (Myher and Kuksis 1982). Column temperature was programmed

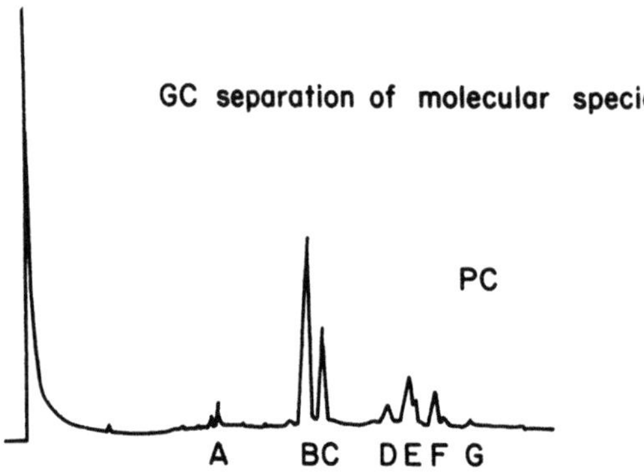

Fig. 3. Gas chromatographic separation of *Dunaliella salina* phosphatidyl-choline-derived diacylglycerol trimethylsilyl ethers. Key to major peaks: *A* 14:2/16:0; *B* 16:0/18:2; *C* 16:0/18:3; *D* 18:1/18:2; *E* 18:2/18:2 and 18:1/18:3; *F* 18:2/18:3; *G* 18:3/18:3

from 190 °C (held for 0.6 min) to 250 °C at 20 °C min^{-1}. Injector and detector temperatures were maintained at 270 °C and 300 °C, respectively. The carrier gas was hydrogen [1], with head pressure maintained at 10 psi. Splitless injection was employed.

One important step in preparing the injector and column for chromatography of silyl derivatives is the deactivation of all surfaces which come in contact with the derivatives. This is most easily accomplished by injecting μl quantities of silating reagent or commercial column conditioner such as Silyl-8 (Pierce Chem. Co.).

In addition to straightening the end of a glass column for insertion into the injection port of the chromatograph, accommodations must also be made to couple a glass capillary column to the inlet of the mass spectrometer. This can be done using a short piece of fused silica tubing which is attached to the end of the capillary column using an appropriate ferrule (Supelco), Teflon tubing or General Electric silicone sealant.

5 Mass Spectrometry

The efficiency of capillary columns in separating molecular species is excellent, and continually improving. One hurdle, however, lies in the identification of molecular species. The various combinations of paired acyl chains are numerous, and the chromatographic elution patterns are often too complex to identify each peak (especially the smaller ones) by comparison to known standards, by inference from the overall acyl chain compositions, by calculation of equivalent chain length values, or by argentation TLC to prefractionate molecular species based on degree of unsaturation. The most effective method for identifying all species is through the use of coupled GC-MS. In many cases, particularly when using

---

[1] Hydrogen as a carrier gas gives shorter retention times and better peak resolution than nitrogen because of its smaller molecular size (Myher and Kuksis 1982). However, care must be taken to avoid fire hazards that could result from gas accumulations.

nonpolar packed columns which separate species by carbon number (but not by double bond number), the use of mass spectrometric data to identify and quantitate the component molecular species in a given peak is a necessity.

5.1 Instrumentation

The availability of GC-MS instruments is steadily becoming more widespread. Indeed, small but powerful mass spectrometers designed for interfacing with standard chromatographs are now available (e.g., from Hewlett Packard, Palo Alto, CA). Although the basic principles of operation are similar for all instruments, the components of each system may be somewhat different. The reader is advised to consult the manual of the specific instrument to be used or discuss the experimental problem with the technician/primary operator of the GC-MS instrument.

Mass spectra consist of characteristic patterns representing molecular fragments with different mass/charge ratios (expressed as m/e or m/z) produced by ionization of the sample. To date, most GC-MS analyses of lipid molecular species (as silyl derivatives) have employed electron impact ionization, although other modes of ionization, especially chemical ionization, have properties which make them potentially very useful for such work. This section will concern itself solely with aspects of mass spectrometry employing a Finnigan Model 4000 quadrupole instrument and electron impact ionization.

Mass spectra of eluted chromatographic peaks are acquired, accumulated and manipulated with the aid of a dedicated computer interfaced to the instrument. This greatly enhances the power of such instruments and allows for different modes of data analysis. Thus, in addition to acquiring information directly from mass spectra (or mass lists), information may also be obtained by employing selected ion monitoring (also called selected ion retrieval). In this mode, the chromatographic elution profiles for selected ions (fragments) may be continuously monitored.

5.2 Operating Conditions

The mass spectrometer must be tuned so that it is capable of detecting ions in the range of 200–800 m/z. Typically, the mass spectrometer is operated at an ionization voltage of 70 eV (electron impact) and an ion source temperature of 250 °–300 °C (Dickens and Thompson 1982; Myher et al. 1978). Optimal conditions will vary slightly with each instrument.

5.3 Identification of Molecular Species

The characteristic fragmentation patterns obtained using silyl derivatives may be used to identify the component(s) of a chromatographic elution peaks. The TMS and t-BDMS derivatives of diacylglycerols produce many of the same characteristic ions, especially those derived from the constituent acyl chains. In addition, t-BDMS derivatives also yield an abundant ion corresponding to the loss of the tert-butyl moiety and designated as $[M - 57]^+$. The $[M - 57]^+$ion is most useful in that it indicates the mass of the intact diacylglycerol, thus allowing the unambiguous determination of acyl carbon number and degree of unsaturation. In some instances, a chromatographic peak may contain two different molecular species, e.g., 14:0/18:0 and 16:0/16:0 that yield the same values for m/z for the $[M - 57]^+$ion. In such a case, the different species may be distinguished by the acyl fragments. Key fragments which are in relative abundance and which are diagnostic of lipid molecular species are given in Table 1. These may be compared to the spectrum shown in Fig. 4. A number of other fragments/ions, some of which are diagnostic, are also produced but usually in low abundance. More complete lists of lipid-derived ions are given elsewhere (Myher 1978; Myher et al. 1978).

Table 1. Key diagnostic fragments/ions of TMS and t-BDMS derivatives and m/z values for selected species

Ion/ fragment type	Structure	Values of m/z for designated species[a]		
		16:0/16:0	16:0/18:0	16:0/18:2
M−156[b]	Molecular ion minus methyl group from TMS moiety	667	695	691
M−57[c]	Molecular ion minus t-butyl group	625	653	649
M−R₁COO[c] M−R₂COO	Molecular ion minus acyl chain moiety from sn-1 or sn-2 position	427	427/455	427/451
R₁CO+74 R₂CO+74	Acyl chain fragment from sn-1 or sn-2 position plus $[O-Si(CH_3)_2]$	313	313/341	313/337

[a] Values for other species may be calculated as follows: Increasing acyl chain length by two carbons results in an increase in m/z by 28. Addition of a double bond results in a decrease in m/z by 2.
[b] TMS derivatives only (not very abundant).
[c] t-BDMS derivatives only.

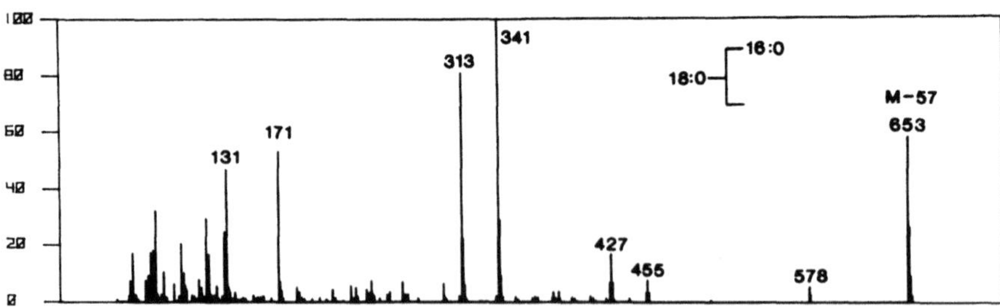

Fig. 4. GC-MS spectrum of the t-BDMS ethers of 1-palmitoyl-2-stearoyl-*rac*-glycerol. Numbers on spectrum represent m/e of major fragment ions. Ordinate is percent relative intensity. (Myher et al. 1978)

Table 2. Values of m/z for $[M-57]^+$ and $[RCO+74]^+$ ions of selected molecular species (t-BDMS derivatives)

Species	$[M-57]^+$	$[RCO+74]^+$
14:0/16:0	597	285, 313
16:0/16:0	625	313
16:0/16:1	623	313, 311
16:0/18:0	653	313, 341
16:0/18:1	651	313, 339
16:0/18:2	649	313, 337
16:0/18:3	647	313, 335
18:0/18:0	681	341
18:1/18:1	677	339
18:1/18:2	675	339, 337
18:1/18:3	673	339, 335
18:2/18:2	673	337
18:2/18:3	671	337, 335
18:3/18:3	669	335

From the mass spectrum of an eluting peak, one can thus determine the identities of the molecular species composing the peak from the $[M-57]^+$ ions and ions of acyl fragments $[RCO+74]^+$. A list of m/z values for $[M-57]^+$ and $[RCO+74]^+$ ions of typical plant phospholipid species is given in Table 2.

Under certain conditions, more detailed information may be gleaned from mass spectra. For example, 1,3 isomers of derivatized diacylglycerols give rise to spectrum peaks representing $[M-CH_2OCOR]^+$ ions (Myher et al. 1978). An indication of the positional distribution of acyl chains at the 1- and 2-positions of the glycerol backbone of diacylglycerols can also be gained by GC-MS of t-BDMS derivatives (Myher et al. 1978). This determination was based on the tendency of acyl chains bound at the sn-2 position to be cleaved more readily than those bound at the sn-1 position, reflected in the relative abundances of $[M-RCOO]^+$ ions produced by reverse isomers of totally saturated species. In biological samples, however, the use of this method to establish the positional distribution of acyl chains is limited; the presence of unsaturated chains, which are prone to enhanced fragmentation, and the potential for a single peak to contain isomers and/or related molecular species complicate such analyses of mass spectra based on relative quantitative differences in ion abundance. Determination of positional distribution of acyl chains by treatment of the isolated phospholipid classes using phospholipase A_2 (see below) is recommended.

5.4 Quantitation of Molecular Species by GC-MS

Mass spectrometry may be employed to quantitate lipid molecular species, although relatively complex calculations using predetermined correction factors are required. Use of an appropriate capillary column which resolves most if not all molecular species based on acyl carbon number and degree of unsaturation diminishes the need to use mass spectrometry for quantitation. In cases where complete separations of species are not achieved (e.g., when using packed columns), the procedures outlined below may be used to estimate the relative proportions of lipid molecular species. This procedure relies on the mass data for the estimation of the relative proportions of molecular species within a given chromatographic peak. Theoretically, the estimates may be calculated from the relative abundances of the respective $[M - 57]^+$ ions. In practice, the low sensitivity of many mass spectrometers in this high mass range and the possibility of different species having the same value of m/z for the $[M - 57]^+$ ion prevent this. The calculations described below are performed using the relative abundances of acyl chain fragments which are of smaller mass and are characteristic of the individual species.

The basic steps in quantifying lipid molecular species by GC-MS are the following: (1) The relative proportions (weight percent) of species differing in acyl carbon number (i.e., separated on a nonpolar packed column) are obtained by GC and flame ionization detection. (2) The weight percent values are then converted to mole percent values using averaged molecular weights for the species constituting the individual elution peaks. (3) Using GC-MS, mass spectra (mass lists) for each chromatographic peak are obtained. (4) From each mass list, the integrator units representing the acyl fragments $[RCO + 74]^+$ of each species are corrected for unequal fragmentation (see below) and summed. In this way a value for the relative proportion of different species constituting each chromatographic peak is obtained. This is then multiplied by the value for peak mole percent (#2 above) to establish the mole percent of each molecular species.

The major drawback to the use of GC-MS as a quantitative tool lies in the unequal fragmentation of the silyl derivatives. Two outstanding problems relating to fragmentation are discussed below.

Enhanced fragmentation of polyunsaturated acyl chains is the most serious disadvantage of GC-MS. Unsaturated acyl chain fragments $[RCO + 74]^+$ have a tendency to undergo further fragmentation, resulting in substantial losses. In these situations, quantitation requires the use of large correction factors determined empirically from standards containing species with acyl chains varying in degree of unsaturation (Kuksis et al. 1978; Dickens and Thompson 1982). For example, it was established that the observed values for acyl ions derived from 18:2 and 18:3 must be multiplied by factors of 1.8 and 6.1, respectively, to correct for unequal fragmentation, thereby obtaining accurate estimates of the acyl ions and corresponding species (Dickens and Thompson 1982). The need for this type of correction can be largely eliminated by reduction of all double bonds (see Sect. 5.5) or by using chemical ionization (Crawford and Plattner 1984), fast atom bombardment (Gross 1984), field desorption (Sugitani et al. 1982) or some other type of soft ionization MS technique.

Unequal cleavage of acyl chains at the 1- and 2-positions can influence the relative ion abundance based on mass spectra (see discussion by Myher et al. 1978). For purposes of quantitation, summing the acyl fragments and careful comparison of the fragment ions representing different species should suffice to overcome this source of error.

5.5 Quantitation of Molecular Species by GC-MS Following Reduction of Double Bonds Using Deuterium

As an alternative to the use of empirical factors to correct for enhanced fragmentation of polyunsaturated acyl chains, a method for specifically reducing the double bonds of acyl chains with deuterium was developed (Dickens et al. 1982). This step eliminates the need for such correction factors inasmuch as all chains are saturated, but at the same time it preserves the mass spectrometric information on acyl chain unsaturation through changes in m/z resulting from the specific addition of deuterium (Fig. 5). Deuteration gives all of the originally unsaturated acyl chains unique masses that are higher by 2 mass units/double bond than are the masses of ordinary saturated fatty acids of equivalent chain length. In addition, deuteration of the acyl chains of phospholipid-derived diacylglycerols improves the chromatographic properties of the corresponding t-BDMS derivatives by improving resolution of peaks on non-polar columns.

Procedure. Diacylglycerols (200–500 µg) obtained from phospholipase C treatment of phospholipids are transferred in benzene to a 5 ml Reactivial (modified as shown in Fig. 6) and the solvent is evaporated under nitrogen. One milliliter

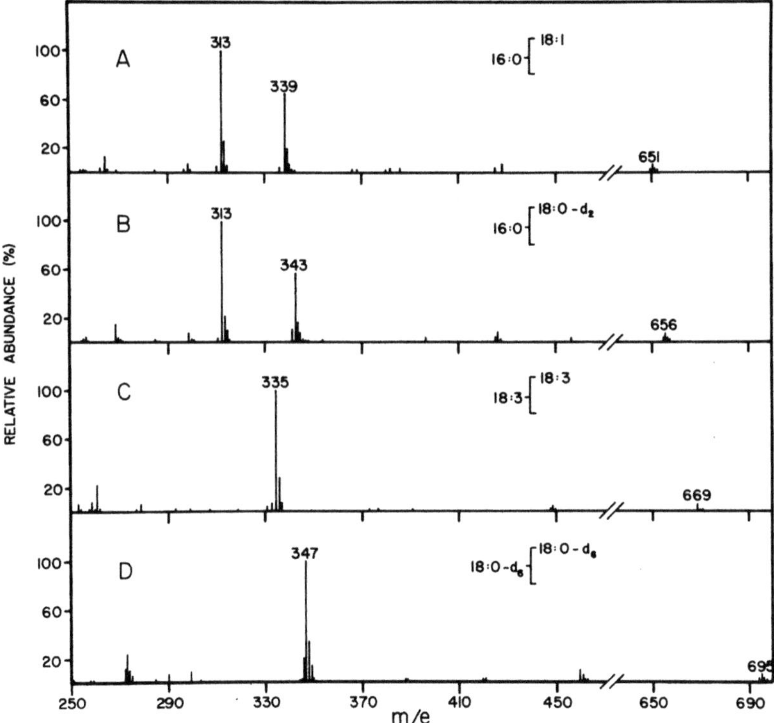

Fig. 5. Mass spectra of t-BDMS derivatives of 1-oleoyl-2-palmitoyl glycerol **A** before and **B** after reduction with deuterium and of 1,2-dilinolenoyl glycerol **C** before and **D** after reduction with deuterium. (Dickens et al. 1982)

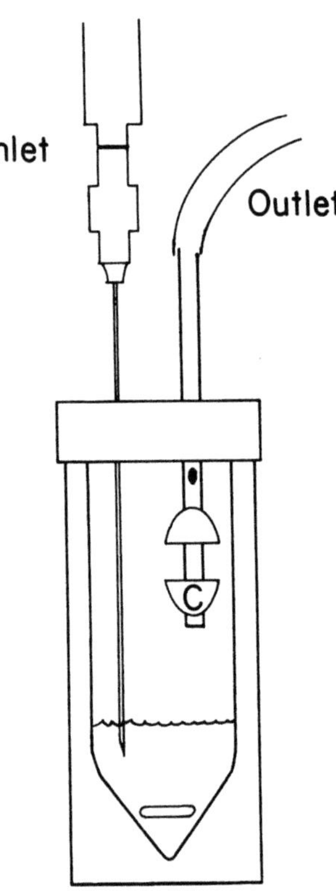

Fig. 6. Apparatus for the catatylic deuteration of diacylglycerols as described in the text. *C* Teflon cup for catalyst (Dickens et al. 1982)

of P-dioxane and 50–100 µl methanol-d_4 (Cambridge Isotope Laboratory, Cambridge, MA) are added to the vial along with a Teflon-coated stirring flea. After capping, the vial is deoxygenated by evacuation and subsequent purging with nitrogen, repeated three times. The sample is then gassed for 3 min at room temperature with deuterium (C.P. grade 99.5% D_2, Union Carbide, Linde Div.) at a rate of 1.6 ml min^{-1}. Next, the vial is placed in a sand bath heated to 50 °C and bubbled for an additional 10 min with D_2 gas. The Teflon cup (see Fig. 6) containing 4 mg of dry tris (triphenylphosphine)rhodium(I) chloride (Wilkinson catalyst: Tridom Chem., Hauppage, NY) is then lowered below the surface of the solvent, wetting the catalyst for the first time. After the catalyst is dissolved (approx. 1 min) the cup is pulled out of the solvent and the flow of D_2 reduced to 0.1 ml min^{-1}. The mixture is stirred slowly for 4 h at 50 °C.

Following reduction, the solvent is removed in vacuo and the diacylglycerols are separated from the catalyst by TLC using petroleum ether:ethyl ether:acetic acid (70:30:1, v/v/v) as the developing solvent system. The diacylglycerols are detected with iodine, eluted from the silica gel using petroleum ether:ethyl ether (1:1, v/v), and then converted to t-BDMS derivatives as described above.

5.6 Direct MS Analysis of Underivatized Phospholipids

Recent advances in mass spectrometry have created the potential for quantifying phospholipid molecular species by direct MS analysis without prior GC separation. This has been achieved using chemical ionization MS (Foltz 1977; Crawford and Plattner 1984) field desorption MS (Wood et al. 1977), and other "soft" ionization techniques. The advantages of this approach are the saving in time of derivatization and the avoidance of potentially selective losses of certain molecular species during sample preparation. We shall briefly describe the isobutane chemical ionization MS procedure as employed by Crawford and Plattner (1984).

Samples of mixed molecular species of PC (0.05–5 µg) are applied in chloroform solution to the probe tip of a Finnigan 4535/TSQ tandem quadrupole mass spectrometer. When the solvent has dried, the probe is inserted into the ion source and heated rapidly at a rate that causes complete lipid vaporization within 5–6 s. Data are collected throughout the total run of 30 s by utilizing either a multiple ion detecting probe or by scanning over the mass range of m/z 500–800 and then summing across the phospholipid peak.

Quantification is based on measurement of the diglyceride $[MH-183]^+$ ions. Because of differing detectability of diglyceride ions of different molecular weights, response factors need to be calculated based on the analysis of lipid standards (Crawford and Plattner 1984). Corrections can also be made to differentiate between molecular species that differ by only 2 mass units. This requires developing a system of linear equations to correct for changes in mass distribution due to hydrogen abstraction or to natural isotopic abundances.

Many classes of phospholipids from natural sources have two or more molecular species with identical molecular weights. For example, mixtures containing 18:1/18:2 PC and 18:0/18:3 PC might be encountered. Although these components cannot be resolved by a single mass spectrometer, the MS/MS capabilities of many modern instruments can quantify the individual lipids. If the diglyceride ions are selectively passed into the collision chamber and scanned with a different quadrupole, the $[RCO]^+$ ion for each fatty acid of the individual molecular species can be observed and quantified.

6 Determination of Positional Distribution of Acyl Chains Using Phospholipase A_2

Determination of acyl chain distribution at the sn-1 and sn-2 positions of phospholipids is needed to gain a complete picture of the lipid molecular species composition. In plants, the positional distribution of fatty acids is suggestive of the metabolic pathway giving rise to the lipid species (Roughan and Slack 1982). To date, positional analyses of plant lipids indicate that individual phospholipids of isolated membrane types have few (if any) reverse isomers. This simplifies the positional analysis and designation of molecular species. The procedure employs phospholipase A_2, which cleaves the acyl chain bound to the sn-2 position. The free fatty acid and lysophospholipid products are then isolated by TLC and analyzed (as fatty acid methyl esters) by GC (Ramesha and Thompson 1983; Watanabe et al. 1981). The method described below is essentially that used by Ramesha and Thompson (1983).

Procedure. Phospholipid (obtained by TLC) is dried under nitrogen in a 5 ml Reactivial and resuspended in 1 ml ethyl ether. Fifty μl of 100 mM sodium borate (pH 7.6) and 0.2 mg phospholipase A_2, 600–1200 units mg^{-1} from *Crotalus adamanteus* venom (Sigma Chem. Co., St. Louis, MO) in 50 μl of 100 mM Tris-HCl (pH 7.6) containing 0.4 mM $CaCl_2$ are added. For larger amounts of lipid (>2 mg), the respective volumes of reagents may be scaled up, and as much as 4 mg enzyme may be added. The mixture is capped tightly and incubated for 2 h at 37 °C with vigorous shaking. The ethyl ether is then evaporated, and the products are extracted (Bligh and Dyer 1959) from the residual buffer after acidifying with a few drops of acetic acid. Separation of the resultant free fatty acids, lysophospholipids and any unreacted phospholipids is achieved by TLC in chloroform:acetic acid:methanol:water (70:25:5:2.2, v/v/v/v). If it is known from previous experience that the enzymatic reaction is complete, free fatty acids may be separated from lysophospholipids by TLC using petroleum ether:ethyl ether:acetic acid (70:30:1, v/v/v).

Following detection of lipid with iodine, the bands of silica gel containing lipids are scraped directly into BF_3/methanol (Sigma Chem. Co.), sealed in glass ampules under nitrogen and heated at 100 °C for 10 min (Morrison and Smith 1964). The resulting methyl esters are extracted with hexane and analyzed by gas chromatography (Lynch and Thompson 1984a).

7 Conclusion

GC and GC-MS have proved to be the first practical techniques for the quantitative analysis of phospholipid molecular species. Through their use we have gained a much more complete picture of plant membrane lipid composition. It has become apparent that subtle but perhaps very important lipid compositional changes triggered by sudden environmental stress can often be detected *only* by a molecular species analysis (Lynch and Thompson 1984c). These findings justify the inclusion of molecular species analysis as a standard part of every phospholipid characterization.

The methodology for phospholipid molecular species quantification has evolved rapidly as a result of many technical advances. Coupled GC-MS was initially essential because chromatography alone could not achieve good peak resolution. Improvements in GC stationary phases and capillary column design have now made it possible to cleanly resolve many phospholipid classes into their component species for quantitative determination without the use of MS, except for the initial molecular species identification.

Not all lipid mixtures are amenable to the type of analysis described here. For GC, it is necessary to prepare volatile derivatives, and this is not easily achieved with glycolipids and certain other lipids of interest. In some cases procedures have been developed that will permit an analysis of these by direct MS of the underivatized lipids. In other cases it has become possible to achieve excellent separations of lipid molecular species by reverse phase high performance liquid chromatography. The methodology utilized in this technique will be described by Smith and Thompson (1986) in another volume of this series. It is fair to say that all classes of plant lipids may now be resolved into their individual molecular species and accurately quantified. The major hurdle that remains is to determined why such complex lipid mixtures exist.

Acknowledgments. Work from the authors' laboratory supported in part by grants from the National Science Foundation (PCM 8200289), the Robert A. Welch Foundation (F-350), and the National Cancer Institute (1 T32 CA09182).

References

Ashworth EN, St John JB, Christiansen MN, Patterson GW (1981) Characterization of the phospholipid composition of wheat roots using high performance liquid chromatography. J Agric Food Chem 29:879–881

Bligh EG, Dyer WJ (1959) A rapid method of total lipid extraction and purification. Can J Biochem Physiol 37:911–917

Breckenridge WC, Yeung SKF, Kuksis A, Myher JJ, Cham M (1976) Biosynthesis of diacylglycerols by rat intestinal mucosa *in vivo*. Can J Biochem 54:137–144

Christie WW (1985) Rapid separation and quantification of lipid classes by high-performance liquid chromatography and mass (light scattering) detection. J Lipid Res 26:507–512

Crawford CG, Plattner RD (1984) Phospholipid molecular species quantitation from mass spectra of underivatized lipids. J Lipid Res 25:518–522

Dickens BF, Thompson GA Jr (1982) Phospholipid molecular species alterations in microsomal membranes as an initial key step during cellular acclimation to low temperature. Biochemistry 21:3604–3611

Dickens BF, Ramesha CS, Thompson GA Jr (1982) Quantification of phospholipid molecular species by coupled gas chromatography-mass spectrometry of deuterated samples. Anal Biochem 127:37–48

Eichenberger W (1982) Distribution of diacylglyceryl-0-4' (N,N,N-trimethyl) homoserine in different algae. Plant Sci Lett 24:91–95

Foltz RL (1977) Chemical ionization mass spectrometry in structural analysis. Lloydia 35:344–353

Gregor HD (1977) Lipid composition of *Daucus carota* roots. Phytochemistry 16:953–955

Gross RW (1984) High plasmalogen and arachidonic acid content of canine myocardial sarcolemma: a fast atom bombardment mass spectroscopic and gas chromatography-mass spectroscopic characterization. Biochemistry 23:158–165

Holub BJ, Kuksis A (1978) Metabolism of molecular species of diacylglycerophospholipids. Adv Lipid Res 16:1–126

Kaduce TL, Norton KC, Spector AA (1983) A rapid isocratic method for phospholipid separation by high-performance liquid chromatography. J Lipid Res 24:1398–1403

Kates M (1972) Techniques of lipidology. In: Work TS, Work E (eds) Laboratory techniques in biochemistry and molecular biology, vol 3, part 2. Elsevier/North-Holland, Amsterdam, pp 265–610

Kates M, Eberhardt FM (1957) Isolation and fractionation of leaf phosphatides. Can J Bot 35:895–905

Kuksis A, Breckenridge WC, Myher JJ, Kakis G (1978) Replacement of endogenous phospholipids in rat plasma lipoproteins during intravenous infusion of an artificial lipid emulsion. Can J Biochem 56:630–639

Lynch DV, Thompson GA Jr (1984a) Microsomal phospholipid molecular species alterations during low temperature acclimation in *Dunaliella*. Plant Physiol 74:193–197

Lynch DV, Thompson GA Jr (1984b) Chloroplast phospholipid molecular species alterations during low temperature acclimation in *Dunaliella*. Plant Physiol 74:198–203

Lynch DV, Thompson GA Jr (1984c) Retailored lipid molecular species: a tactical mechanism for modulating membrane properties. Trends Biochem Sci 9:442–445

Marinetti GV (1962) Chromatographic separation, identification, and analysis of phosphatides. J Lipid Res 3:1–20

Morrison WR, Smith LM (1964) Preparation of fatty acid methyl esters and dimethylacetals from lipids with boron trifluoride methanol. J Lipid Res 5:600–608

Myher JJ (1978) Separation and determination of the structure of acylglycerols and their ether analogues. In: Kuksis A (ed) Handbook of lipid research, vol 1. Fatty acids and glycerides. Plenum, New York

Myher JJ, Kuksis A (1982) Resolution of diacyglycerol moieties of natural glycerophospholipids by gas-liquid chromatography on polar capillary columns. Can J Biochem 60:638–650

Myher JJ, Kuksis A, Marai L, Yeung SKF (1978) Microdetermination of molecular species of oligo- and polyunsaturated diacylglycerols by gas chromatography-mass spectrometry of their *tert*-butyl dimethylsilyl derivatives. Anal Chem 50:557–561

Nishihara M, Yokota K, Kito M (1980) Lipid molecular species composition of thylakoid membranes. Biochim Biophys Acta 617:12–19

Quinn PJ, Williams WP (1978) Plant lipids and their role in membrane function. Prog Biophys Molec Biol 34:109–173

Ramesha CS, Thompson GA Jr (1983) Cold stress induces in situ phospholipid molecular species changes in cell surface membranes. Biochim Biophys Acta 731:251–260

Ramesha CS, Dickens BF, Thompson FA Jr (1982) Phospholipid molecular species alterations in *Tetrahymena* ciliary membranes following low temperature acclimation. Biochemistry 21:3618–3622

Roughan PG, Slack CR (1982) Cellular organization of glycerolipid metabolism. Annu Rev Plant Physiol 33:97–132

Rouser G, Fleischer S, Yamamoto A (1970) Two dimensional thin layer chromatographic separation of polar lipids and determination of phospholipids by phosphorus analysis of spots. Lipids 5:494–498

Sato N, Furuya M (1983) Distribution of diacylglyceryltrimethylhomoserine in green plants. Plant Cell Physiol 24:1113–1116

Sugitani J, Kino M, Saito K, Matsuo T, Matsuda H, Katakuse I (1982) Analysis of molecular species of phospholipids by field desorption mass spectrometry. Biomed Mass Spectrom 9:293–301

Van Golde LMG, Pieterson WA, van Deenen LLM (1968) Alterations in the molecular species of rat liver lecithin by corn oil feeding to essential fatty acid-deficient rats as a function of time. Biochim Biophys Acta 152:84–95

Watanabe T, Fukushima H, Kasai R, Nozawa Y (1981) Studies on thermal adaptation in *Tetrahymena* lipids. Changes in positional distribution of fatty acids in diacyl-phospholipids and alkyl-acyl-phospholipids during temperature acclimation. Biochim Biophys Acta 665:66–73

Wood GW, Lou PY, Morrow G, Rao GHS, Schmidt DE, Tuebner J (1977) Field desorption mass spectrometry of phospholipids. III. Survey of structural types. Chem Phys Lipids 18:316–333

GC-MS of Plant Sterol Analysis

G. Combaut

The structures of the sterol side chains and ring systems discussed in this chapter
are presented in Figs. 1 and 2 respectively.

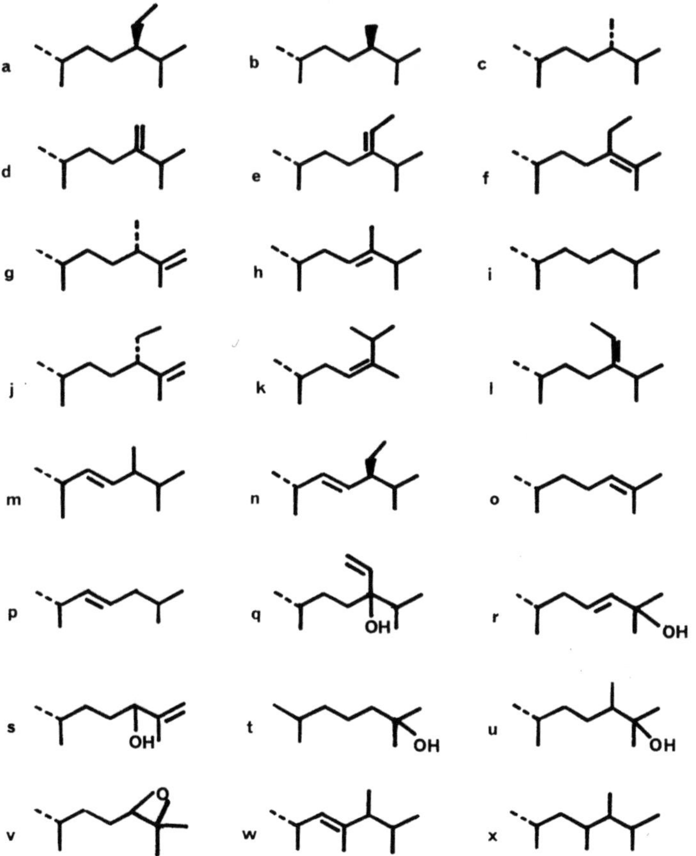

Fig. 1. Structures of sterol side chains

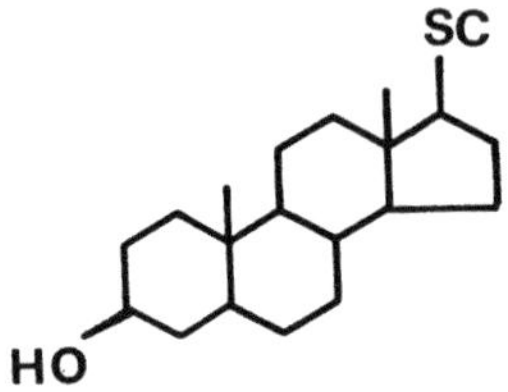

Fig. 2. Structures of sterol ring systems

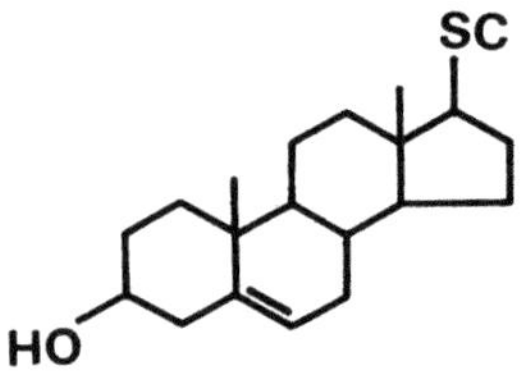

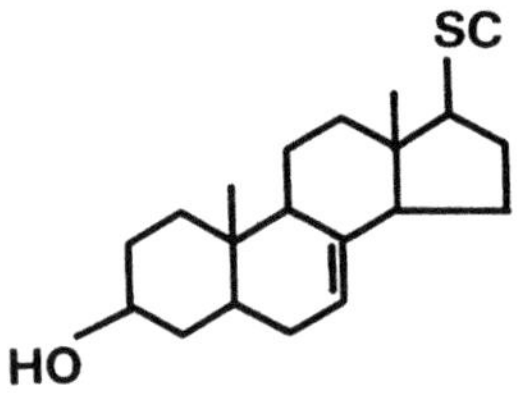

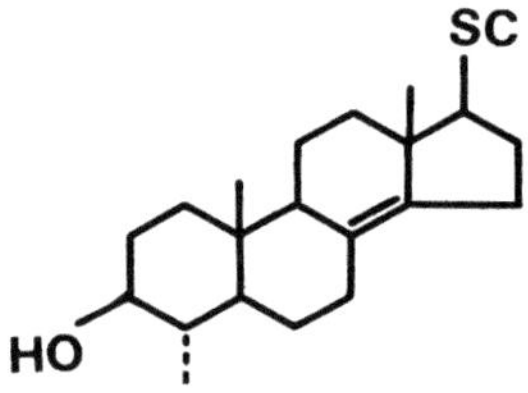

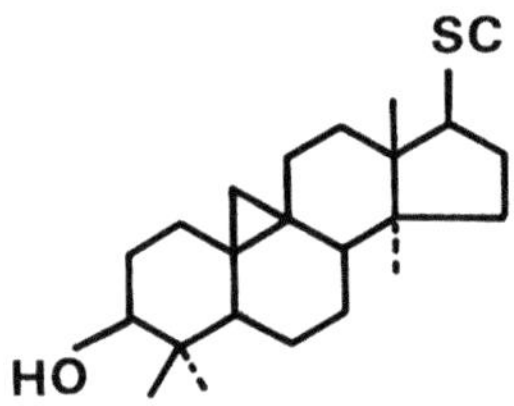

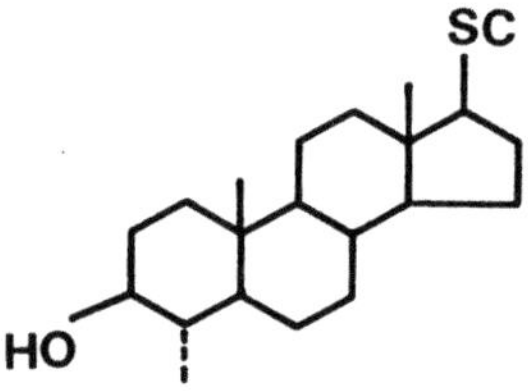

1 Introduction

An example taken from the abundant literature will illustrate the decisive role played by GC-MS in sterol analysis. Matsumoto et al. (1982) reported the first analysis of sterols in Antarctic lake sediments, and observed the surprising result that 24-ethylcholesterol (24-ethylcholest-5-en-3β-ol 2a) is the dominant sterol in all these sediment samples. Thus the C_{29}/C_{27} sterol ratios are high and comparable with those of lake sediments from temperate zones, attributed to the sterol input from vascular plants. However, as pointed out by Matsumoto et al. (1982), no vascular plants are present in the dry valley areas. The author showed that aeolian dust is not an important source of the sterol 2a and that sedimentary rocks are also not the source of this sterol. Moss contains 24-methylcholest-5-en-3β-ol (2b–2c) and the sterol 2a, but the distribution of mosses in the areas studied is restricted. Thus mosses are also unlikely sources of the sterol 2a in the sediments. In fact, the epibenthic algae, mostly the widely distributed blue-green algae especially *Phormidium* spp., are found to be the source of 24-ethylcholesterol 2a in Antarctic dry valley lake sediments.

This work is one of the many studies dedicated to sterol analysis by the GC-MS method. In the last 20 years, the GC-MS technique has become a viable tool for obtaining structural information concerning minor components in complex sterol mixtures. According to Djerassi (Popov et al. 1976) in his fundamental review entitled *Minor and Trace Sterols in Marine Invertebrates*, present knowledge of sterol mass spectrometric behavior has made such progress that often GC-MS data are sufficient to suggest a structure of newly detected sterols. Thus the limitation is obvious: being a pertinent method for studying complex mixtures of already known sterols, GC-MS, can only suggest hypothetic structures for new sterols. In addition, synthesis and semipreparative GC or HPLC are used to yield enriched fractions. Structural features including C-24 configurational assignments can be obtained on purified sterols by high field ^{1}H NMR and X-ray diffraction analysis (see for example Dow et al. 1983; Kokke et al. 1979; Withers et al. 1979 a, b; Swenson et al. 1980).

2 Development of GC-MS Plant Sterol Analysis

Knights (1967) after the pioneering work of Eneroth et al. (1964, 1965) used combined GC-MS, to determine the mass spectra of 15 plant sterols, and closely related compounds. Free sterols, acetate-, trifluoroacetate-, and trimethylsilyl-derivatives were investigated and the principal ions listed for Δ^5 and Δ^7 monounsaturated sterols. In the same way Brooks et al. (1968) presented a list of the methylene unit values, and of the principal mass spectrometric data for trimethylsilyl ethers of 28 sterols including the major natural sterols. The collective book *Method of Steroid Analysis* is a general survey, dealing with various aspects of plant sterol analysis: *Qualitative and Quantitative Analysis of Plant Sterols by Gas-Liquid Chromatography* (Knights 1973), *Some Aspects of Mass Spectrometry in Steroid Analysis* (Brooks and Middleditch 1973), *Derivatization and Gas Chromatography in the Mass Spectrometry of Steroids* (Vandenheuvel et al. 1973). In the *Specialist Periodical Report*, Brooks and Middleditch (1979) reviewed the different applications of GC-MS including sterol analysis in biological material. In this review several books pertaining to GC-MS are listed.

During the 1970's, several studies of plant sterol are published in which GC-MS is largely used. For example the sterols of *Zea mays* (seeds, coleoptiles, or shoots) previously analyzed by GLC (Kemp et al. 1967, 1968 a; Kemp and Mercer 1968; Hartmann et al. 1975) are studied by GC-MS (Knights and Smith 1976; Comita and Klosterman 1976; Suemitsu et al. 1977; Itoh et al. 1981). This period also coincides with the development of reversed phase chromatographic methods for the separation of homologous sterols and the advent of high field NMR for the identification of new sterols (Scheid and Benveniste 1979) and determination of configuration at C-24 (Nes et al. 1976, 1977).

Concerning marine sterols, Ikekawa et al. (1968) used for the first time GC-MS for sterol investigation in some green and brown algae. This is the starting point of several studies using GC-MS for sterol analysis of algae (see, for example, Knights and Brooks 1969; Fattorusso et al. 1975, 1976; Morisaki et al. 1976; Francisco et al. 1977). The method for marine sterol analysis has been systematized by Popov et al. (1976); a first general review was presented by Goad (1978) and using computer-assisted structure manipulation, Varkony et al. (1978) predicted the existence of 1178 3-hydroxy natural sterols. Two recent reviews are related to marine sterols, *Marine Sterols* (Djerassi 1984) and *Dinoflagellate Sterols* (Withers 1983).

3 Operations Before GC-MS Sterol Analysis

3.1 Extraction and Isolation of Plant Sterols

Numerous extraction procedures have been employed for the isolation of plant sterols. Dried tissues are submitted to hot extraction with a Soxhlet apparatus or to cold extraction at room temperature; several solvents have been used: for example Me_2CO (Misso and Goad 1984; Sica et al. 1984); MeOH or EtOH (Palermo et al. 1984; Kac et al. 1984); Hexane or petrol (Artaud et al. 1984; Stankovic et al. 1984; Francisco et al. 1977; Combaut et al. 1976); $CHCl_3$ or CH_2Cl_2 (Fattorusso 1975, 1976; Kabore et al. 1983; Combaut et al. 1981 a, b). Popov et al. (1976) discussed the efficiency of the different solvents and precognized the $CHCl_3$/MeOH procedure (Kalinowska and Wojciechowski 1984; Sjöstrand et al. 1981; Combaut and Saenger 1984; Combaut et al. 1984). Extraction of fresh homogenized tissues is preferable in order to avoid any autoxidation. This is true for some marine sterols particularly desmosterol 2 o the amount of which is lower when the extraction is performed on air dried material. This is well illustrated by the results observed for the Rhodomelaceae *Halopytis pinastroides*: the ratio of desmosterol 2 o cholesterol 2 i is 0.5 (33/66) and 2 (66/33) respectively when air-dried and fresh algae are used for sterol extraction (Combaut, personal communication).

3.2 Free Sterols and (or) Sterols from Steryl-Esters

After extraction of lipids, the usual work-up includes basic hydrolysis (saponification). Some methods employ initial saponification of the fresh tissues followed by extraction of the nonsaponifiable lipids, but this procedure should be avoided (Popov et al. 1976). In some studies free sterols and sterols from steryl-esters are analyzed separately (Kokke et al. 1981; Rohmer et al. 1980). The composition of both fractions is often similar, differing only by the relative amounts of each sterol. But the results can be fundamentally different as it is the case with *Rissoella verruculosa*. Indeed this alga contains desmosterol 2 o in its esterified state (Artaud et al. 1980), whereas this sterol is quite absent in the free state (Kabore et al. 1983).

3.3 Purification of Sterolic Fractions

The digitonin fractional precipitation, often used for the isolation of 3β-hydroxysterols (Francisco et al. 1977, 1979; Combaut et al. 1976; Kabore et al. 1983; Sjöstrand et al. 1981) has been discussed by Popov et al. (1976). Usually TLC on silicagel with $CHCl_3$–Et_2O and $CHCl_3$–EtOH mixtures as developing solvents is used in sterol isolations (Artaud et al. 1984) and separations of 4,4-dimethyl sterols, 4-monomethyl sterols and 4-demethyl sterols (Misso and Goad 1984). TLC of the sterol acetates on silicagel with silver nitrate is one of the most often used techniques in the resolution of sterol mixtures (Popov et al. 1976; Misso and Goad 1984; Rohmer et al. 1980), together with reversed phase HPLC (Popov et al. 1976; Dow et al. 1983; Misso and Goad 1984; Stankovic et al. 1984; Rohmer et al. 1980; Artaud et al. 1980; Francisco et al. 1979). The different fractions thus isolated can be submitted to GC-MS analysis as free sterols or steryl-derivatives.

3.4 Derivatization

The pioneering works in combined gas chromatography-mass spectrometry used trimethylsilyl-derivatives (TMS). This type of derivatization was well refined by the Houston group of Horning (Vandenheuvel 1973). Acetates, trifluoroacetates, propionates, and methyl ethers have also been used (Knights 1973). However, when derivatization is required, acetates and trimethylsilyl-derivatives are usually preferred. Acetylation is used when the separation of sterols includes a preliminary $AgNO_3$-silicagel TLC (Misso and Goad 1984; Rohmer et al. 1980) and when HPLC is performed on steryl-acetates (Misso and Goad 1984).

Acetylating agents are generally anhydrides; the polarity of the hydroxyl group is reduced, and the volatility increased. Pyridine is added as solvent and also as a basic catalyst and basic acceptor of the acid derived from the derivatization reaction. Generally the free sterols are treated overnight at room temperature with acetic anhydride-pyridine (1:1) and the excess reagent is removed with a stream of N_2 (Rohmer et al. 1980).

Silylation consists of the substitution of an active hydrogen by a trimethylsilyl group $(CH_3)_3$ Si-yielding a TMS-derivative. The most general method to silylate all the hydroxyl groups has been pointed out by Horning (see literature cited by Prost et al. 1974). Sterols react overnight at 65 ° with bis (trimethylsilyl) acetamide (BSA) and trimethylchlorosilane (TMCS) (80:20, v/v). More rapid methods for usual sterols consist of silylation by bis (trimethylsilyl) trifluoroacetamide (BSTFA) and TMCS (80:20, v/v) at 65 ° during 1 h, or faster yet by hexamethyldisilazane (HMDS) and TMCS in pyridine (Artaud et al. 1984).

4 Characterization of Sterols

4.1 Characterization of Sterols by GC Data

Relative Retention Times (RR$_t$). Sterols can be initially identified by their mobilities in comparison with literature data of already known sterols. When isothermal chromatographic conditions are used, RR_t are often related to cholesterol or cholesterol-derivatives (Sica et al. 1984; Palermo et al. 1984; Artaud et al. 1984).

Kovats Indices (I). In programmed temperature GC separations, the Kovats retention indices (Knights 1973) apply, n alkanes being used as standards. For example one finds for cholesterol 2i-acetate $I^{SE-30}3185$ and for fucosterol 21-acetate $I^{SE-30}3370$ (Knights 1970).

Methylene Units (MU). The Horning's system differs by the position of the decimal point; for example on OV-1 2i-TMS, MU 31.81; 21-TMS, MU 33.70 (Francisco et al. 1979).

Itoh et al. (1982) listed the RR_t and MU values of 168 acetates of sterols and triterpene alcohols, most of which originated from higher plants. This work was performed on OV-1 (nonpolar) and OV-17 (slightly polar) glass capillary columns, and is a highly valuable reference for further plant sterol analysis. The RR_t values of steryl-acetates and those of free sterols being quite similar, it has been possible to apply these data to the particular case of the sterols of red algae belonging to the *Amansiae* tribe. Because of the small amounts of material of each alga available, separation of the sterol fraction was not feasible. Consequently, GC analysis of the crude lipid extracts was undertaken, and 22-dehydrocholesterol 2p (RR_t 0.91), cholesterol 2i (1.00), desmosterol 2o (1.08), brassicasterol 2m (1.11), 24-methylenecholesterol 2d (1.25), campesterol 2b (1.28) and sitosterol 2a (1.60) were identified. Eight different *Amansiae* species were analyzed and the amounts of the different sterols compared. A first group of these algae contains a majority of C-27 sterols, a second a majority of C-28, and a third equal amounts of C-27 and C-28 sterols (Combaut and Saenger 1984). The results support our presumption that the *Amansiae* contained species capable of alkylation at C-24 of the sterol side chain, and emphasizes the significance of this group of algae for sterol studies. At the present time those species which are an exception to the theory that red algae are incapable of alkylation at C-24, all belong to the *Amansiae* group.

4.2 Characterization of Sterols by MS Data

Mass spectra of the crude sterolic fraction can be of interest. For example, the mass spectrum of the sterols of the brown alga *Padina vickersiae* present two molecular ions at m/z (relative intensity) 412 (50) and 386 (100) confirming the surprising result that cholesterol 2i is the major sterol of this brown alga, fucosterol 21 being the minor one (Combaut et al. 1985). The mass spectrum of the sterol fraction extracted from the marine dinoflagellate *Amphidium carterae* (Withers et al. 1979a) shows molecular ions at m/z 426 (20), 412 (100), 400 (55) indicating the presence of a C-29 diunsaturated major sterol, a C-28 monounsaturated sterol and a C-30 diunsaturated sterol. The major sterol has been isolated by TLC on $AgNO_3$-Silicagel and identified by 360 MHz ^{1}H NMR as 4α-methyl-24-methylenecholest-8,14-en-3β-ol 4d (MW 412 RR_t 1.53). This sterol was later found to be widely distributed in the genus *Amphidium* (Withers 1983).

4.3 Characterization of Sterols by GC and MS Data

4.3.1 A Typical Analysis of 4-Demethyl and 4,4-Dimethyl Sterols from *Zea mays*

Misso and Goad (1984) analyzed Δ^{23}, $\Delta^{24(28)}$ and Δ^{25} sterols of *Zea mays* shoots.

The 4-demethyl sterols were isolated from 9-day-old maize shoots, acetylated and separated into eight bands by silver nitrate-silicagel TLC. GC-MS of these fractions identified the twenty two following sterols including four 5α-stanols, by their RR_t and mass spectra as steryl-acetates:

cholest-5-en-3β-ol 2i RR_t 1.00; MS m/z (rel int): 368 [M-AcOH]$^+$ (100), 353 (12), 260 (11), 255 (27), 253 (8), 247 (10), 213 (7)

5α-cholestan-3β-ol 1i RR_t 1.02; MS 430 [M]$^+$ (10), 370 (12), 355 (2), 335 (4), 318 (1), 275 (25), 215 (100)

24-methylcholest-5-en-3β-ol 2b and 2c RR_t 1.26; MS 382 [M-AcOH]$^+$ (100), 367 (19), 340 (2), 274 (14), 261 (14), 255 (16), 213 (15)

24-methyl-5α-cholestan-3β-ol 1b RR_t 1.27; MS 444 [M]$^+$ (28), 429 (5), 384 (27), 369 (21), 330 (2), 276 (33), 275 (28), 261 (7), 215 (100)

24-methylcholesta-5,22-dien-3β-ol 2m RR_t 1.14; MS 380 [M-AcOH]$^+$ (86), 365 (6), 337 (6), 282 (7), 267 (2), 255 (54), 253 (8), 228 (9), 213 (13)

24-methylcholesta-5,24(28)-dien-3β-ol 2d RR_t 1.35; MS 380 [M-AcOH]$^+$ (100), 365 (14), 296 (39), 281 (14), 259 (8), 255 (9), 253 (17), 228 (9), 213 (20)

24-methyl-5α-cholesta-7,24(28)-dien-3β-ol 3d RR_t 1.61; MS 440 [M]$^+$ (9), 425 (12), 380 (5), 365 (7), 356 (27), 342 (4), 313 (100), 296 (6), 273 (7), 255 (26), 227 (16), 213 (32)

24-methylcholesta-5,E-23-dien-3β-ol 2h RR_t 1.35; MS 380 [M-AcOH]$^+$ (61), 365 (5), 296 (5), 283 (67), 255 (7), 253 (27), 213 (8), 81 (87), 55 (100)

24-methyl-5α-cholesta-7,E-23-dien-3β-ol 3h RR_t 1.62; MS 440 [M]$^+$ (4), 425 (3), 394 (98), 380 (4), 379 (12), 343 (10), 313 (53), 283 (41), 273 (6), 255 (7), 253 (21), 227 (7), 215 (6), 213 (6)

24-methylcholesta-5,25-dien-3β-ol 2g RR_t 1.34; MS 380 [M-AcOH]$^+$ (100), 365 (12), 296 (11), 283 (7), 259 (9), 255 (12), 228 (7), 213 (16)

24-methylcholest-7-en-3β-ol 3b data not given

24-ethylcholest-5-en-3β-ol 2a RR_t 1.55; MS 396 [M-AcOH]$^+$ (100), 381 (17), 354 (2), 288 (12), 275 (12), 255 (15), 213 (4)

24-ethyl-5α-cholestan-3β-ol 1a RR_t 1.56; MS 458 [M]$^+$ (34), 398 (22), 383 (24), 344 (2), 276 (25), 275 (16), 257 (5), 215 (100)

24-ethyl-5α-cholest-22-en-3β-ol 1n RR_t 1.40; MS 456 [M]$^+$ (19), 442 (2), 396 (3), 353 (21), 344 (16), 315 (16), 275 (4), 257 (46), 255 (9), 229 (5), 215 (20)

24-ethyl-5α-cholest-7-en-3β-ol 3a RR_t 1.84; MS 456 [M]$^+$ (100), 441 (19), 396 (10), 381 (10), 315 (7), 273 (14), 255 (98), 229 (29), 213 (44)

24-ethylcholesta-5,22-dien-3β-ol 2n RR_t 1.38; MS 394 [M-AcOH]$^+$ (96), 379 (5), 351 (15), 282 (7), 255 (47), 253 (9), 228 (11), 213 (12)

24-ethylcholesta-5,Z-24(28)-dien-3β-ol 2e RR_t 1.80; MS 394 [M-AcOH]$^+$ (14), 379 (3), 296 (100), 281 (18), 253 (8), 229 (10), 213 (13)

24-ethylcholesta-5,25-dien-3β-ol 2j RR_t 1.63; MS 394 [M-AcOH]$^+$ (98), 379 (7), 315 (3), 296 (5), 281 (6), 255 (13), 253 (16), 228 (9), 213 (16), 55 (100)

24-ethylcholesta-5,24-dien-3β-ol 2f RR_t 1.87; MS 394 [M-AcOH]$^+$ (27), 379 (5), 296 (75), 281 (15), 255 (8), 253 (19), 288 (10), 213 (16)

24-ethyl-5α-cholesta-7,Z-24(28)-dien-3β-ol 3e RR_t 2.08; MS 454 [M]$^+$ (2), 439 (3), 379 (2), 356 (43), 341 (5), 313 (100), 296 (6), 288 (8), 281 (5), 273 (5), 255 (11), 253 (8), 227 (9), 213 (17)

24-methylcholesta-5,Z-23-dien-3β-ol 2k RR_t 1.36; MS 380 [M-AcOH]$^+$ (73), 365 (7), 296 (5), 283 (75), 253 (27), 227 (5), 213 (8)

24-ethylcholesta-5,E-24(28)-dien-3β-ol 2l RR_t 1.69; MS 394 [M-AcOH]$^+$ (44), 379 (7), 296 (100), 281 (19), 253 (14), 228 (10), 213 (11).

The 4,4-dimethyl sterols were acetylated and separated by silver nitrate-silica gel TLC to yield five bands, the constituents of which where identified by GC-MS.

Four 4,4-dimethyl sterols were identified, RR_t and MS of steryl-acetates are given:

Cycloartenol 5o RR_t 1.80; MS 468 $[M]^+$ (20), 453 (18), 408 (100), 393 (86), 365 (30), 357 (6), 339 (40), 286 (62), 297 (25)

Cyclolaudenol 5g RR_t 1.96; MS 482 $[M]^+$ (13), 467 (13), 422 (100), 407 (80), 379 (33), 357 (6), 353 (17), 300 (45), 297 (39)

24-methylenecycloartenol 5d RR_t 1.98; MS 482 $[M]^+$ (7), 467 (12), 422 (100), 407 (96), 379 (31), 357 (4), 353 (15), 300 (45), 297 (42)

Cyclosadol 5h RR_t 2.00; MS 482 $[M]^+$ (1), 467 (7), 422 (100), 407 (77), 379 (10), 353 (4), 325 (61), 300 (27), 297 (11).

As a general feature, it can be said that GC-MS is not able to distinguish the absolute configuration of C-24 alkyl sterols. However, it had been pointed out by NMR spectra that the major sterol of *Zea mays* was 24α-ethylcholest-5-en-3β-ol (2a; sitosterol) while 24-methylcholesterol (campesterol) was recognized to be a mixture of 24α-methylcholest-5-en-3β-ol 2b and 24β-methylcholest-5-en-3β-ol 2c (Scheid et al. 1982; Zakelj and Goad 1983).

4.3.2 Co-Occurrence of Δ^5 and Δ^7 Sterols in Tracheophytes

Artaud et al. (1984) observed the co-occurrence of Δ^5 and Δ^7 sterols in two *Gleditsia* species, *G. triacanthos* and *G. macracantha*, two species belonging to the plants rich in Δ^7 sterols. The Δ^5 sterol content ranged from 13 to 32%, showing therefore that the co-occurrence of Δ^5 and Δ^7 sterols in some Tracheophytes exists. Because no Δ^5 sterols or very small quantities of Δ^5 sterols are reported in plants rich in Δ^7 sterols, the authors reinvestigated the sterol composition of *Thea sinensis, Spinacia olearacea*, and *Medicago sativa*. The identity of the sterols was determined by comparison of RR_t with previously published retention data of silyl-, free-, and acetylated-standards (Artaud et al. 1980; Itoh et al. 1982).

The sterols were analysed by GC-MS as TMS derivatives. Two Δ^7 sterols were found in highest amounts, 24-ethyl-5α-cholest-7,trans 22-dien-3β-ol 3n (28–50%) and 24-ethyl-5α-cholest-7-en-3β-ol 3a (23–49%), two in low amounts, 24-methyl-5α-cholest-7-en-3β-ol 3b–3c (1.8–6.9%) and 24-ethylcholesta-7,Z-24(28)-dien-3β-ol 3e (2.2–6.8%). RR_t and MS of steryl-TMS are given:

24-ethyl-5α-cholest-7,trans 22-dien-3β-ol 3n RR_t 1.66; MS 484 $[M]^+$ (23.5), 469 (12.0), 441 (2.7), 394 (2.8), 372 (9.7), 357 (2.5), 343 (51.8), 329 (3.8), 318 (6.8), 255 (34.7), 230 (3.6), 229 (23.3), 213 (13.7), 201 (7.6)

24-ethyl-5α-cholest-7-en-3β-ol 3a RR_t 1.88; MS 486 $[M]^+$ (100), 471 (15.1), 396 (3.7), 381 (9.8), 345 (9.5), 303 (3.0), 255 (44.7), 230 (3.3), 229 (12.8), 213 (17.6), 201 (5.5)

24-methyl-5α-cholest-7-en-3β-ol 3b–3c RR_t 1.51; MS 472 $[M]^+$ (59.7), 457 (6.3), 367 (9.5), 343 (7.9), 269 (3.1), 255 (36.8), 237 (3.1), 229 (16.8), 213 (20.7)

24-ethylcholesta-7,Z-24(28)-dien-3β-ol 3e RR_t 2.07; MS data not given.

Concerning the Δ^5 sterols, 24-ethylcholest-5-en-3β-ol 2a was found in all species analysed and in higher content in *G. triacanthos* and *T. sinensis*. The other Δ^5 sterols identified were the sterols 2i, 2b–2c, 2n, 2l, 2e. RR_t and MS of steryl-TMS are given:

cholest-5-en-3β-ol 2i RR$_t$ 1.00; MS 458 [M]$^+$ (21.5), 443 (53), 370 (3.0), 368 (36.3), 353 (15.3), 329 (21.1), 275 (3.1), 255 (10.0), 247 (10.7), 233 (2.0), 217 (3.3), 213 (4.3), 129 (100)
24-methylcholest-5-en-3βol 2b–2c RR$_t$ 1.31; MS 472 [M]$^+$ (47.8), 457 (8.6), 384 (8.6), 382 (43.0), 367 (25.0), 343 (45.8), 269 (7.9), 255 (14.2), 217 (6.9), 213 (3.4), 129 (100)
24-ethylcholest-5,trans-22-dien-3β-ol 2n RR$_t$ 1.41; MS 484 [M]$^+$ (48.7), 469 (5.8), 394 (23.7), 379 (7.5), 356 (5.6), 355 (10.0), 330 (2.5), 282 (3.2), 271 (4.0), 255 (30.4), 239 (3.2), 227 (2.5), 213 (7.0), 129 (58.3)
24-ethylcholest-5-en-3β-ol 2a RR$_t$ 1.61; MS 486 [M]$^+$ (33.8), 471 (6.2), 396 (40.3), 381 (10.5), 357 (36.6), 355 (2.3), 275 (1.8), 255 (8.4), 217 (4.1), 213 (4.2), 129 (100)
24-ethylcholesta-5,E-24(28)-dien-3β-ol 21 (fucosterol) RR$_t$ 1.70; MS 484 [M]$^+$ (8.0), 469 (5.7), 386 (47.1), 371 (9.5), 343 (4.9), 296 (20.8), 281 (12.0), 258 (3.7), 257 (12.8), 255 (6.1), 243 (3.3), 227 (3.3), 211 (6.2), 129 (87.5)
24-ethylcholesta-5,Z-24(28)-dien-3β-ol 2e (Δ^5 avenasterol) RR$_t$ 1.78; MS 484 [M]$^+$ (5.6), 469 (3.4), 386 (60), 371 (9.8), 343 (3.0), 296 (30.6), 281 (20.7), 258 (4.2), 257 (17.1), 255 (7.4), 243 (3.3), 227 (4.7), 211 (9.6), 129 (76.9).

4.3.3 Side Chain Hydroxylated Sterols from Red Algae

The first dihydroxy steroid from a marine source was saringosterol 2q. Knights (1970) has shown that the diol 2q was an artifact arising during air-drying of the brown alga *Ascophyllum nodosum*, the fresh material containing only fucosterol 21. Since then Fattorusso et al. (1975) and Morisaki et al. (1976) described liagosterol 2r and cholesta-5,25-dien-3β,24-diol 2s. Francisco et al. (1979) found that the diols 2r and 2s, whose origin was considered as doubtful (Morisaki et al. 1976), result by autoxydation of desmosterol 2o. So, to avoid any autoxydation, the red alga *Asparagopsis armata* was harvested and immediately fixed on the seaside in methanol and after the usual work-up, sterols were submitted to GC-MS as TMS and acetate derivatives. Cholesterol 2i was found to be the major sterol with few amounts of brassicasterol 2m, desmosterol 2o and fucosterol 21. The interesting fact was the presence of three other sterols representing 12% of the total sterol fraction. The first one (MU 32.38 as TMS) displayed a molecular ion at mass 546. The presence in the side chain of a tertiary hydroxyl group was deduced from the fact that the acetylated sterol shows, by GC-MS, peaks at m/z 384 (M-AcOH) and 366 (M-AcOH-H$_2$O). This suggested a tertiary hydroxyl group in the side chain, not acetylated under mild conditions (Fattorusso et al. 1975). For the TMS derivative, the fragments at m/z 343 (80%, M-TMSOH-SC-2H) and m/z 145 (60%, cleavage of 23–24 bond) further confirmed that this sterol was 25-hydroxycholesterol 2t. By a similar reasoning, the fragmentation pattern of the second TMS derivative (MU 32.74, m/z 560 [M]$^+$, 343 M-SC-2H) and for the acetate m/z 380 (M-AcOH-H$_2$O), identified this sterol as 25-hydroxy-24-methylcholesterol 2u. The important fragments observed at m/z 386 and 296 are explained by elimination of TMSOH followed by McLafferty rearrangement in the side chain. The molecular ion (m/z 544) of the last TMS derivative (MU 34.30) pointed to a side chain of mass 199, thus containing a double bond and a TMS group. The base peak at m/z 143 (cleavage at allylically actived 23–24 bond) identified this sterol as cholesta-5,25-dien-3β,24-diol 2s. The identification was confirmed by the mass spectrum of the acetylated sterol: m/z 424 (M-AcOH), 364 (M-2AcOH). Another set of peaks, suggested that, as in the case of *Rhodymenia palmata* (Morisaki et al. 1976), the sterol 2s is accompanied by the isomeric liagosterol 2 r. The important feature is the presence, in the fresh alga *Asparagopsis armata*, of the hydroxy sterols 2 r and 2 s, which were also found in the tetrasporophyte *Falkenbergia rufolanosa* (Combaut et al. 1979). When one uses a same procedure including basic hydrolysis, neither the sterol 2 r nor 2 s were identified in the red alga *Rissoella verruculosa* (Kabore et al. 1983). Desmosterol 2 o is the major sterol (60%) with cholesterol 2 i (40%). However, when the free sterols are analyzed, after isolation by TLC of the crude lipid extract, desmosterol 2 o is absent and cholesterol 2 i (57%) is accompanied by three side chain oxygenated sterols (23%, 2%, 16% respectively). Two of them are the sterols 2 r and 2 s, the third (MU 34.00 as TMS derivative) is identified as 24,25-epoxycholesterol 2 v.

4.3.4 4-Methyl Sterols of Dinoflagellates

Dinoflagellates are important components of phytoplankton, being the foundation of many food chains, and are major producers of dietary sterols. Withers (1983) listed the numerous sterols isolated and identified from dinoflagellate species since the discovery of dinosterol (Shimizu et al. 1976). Relative retention times of 28 4-methyl sterols are given along with some mass spectra data. As a typical work using GC-MS, Nichols et al. (1983) recently detected eleven different sterols in the marine unicellular alga FCRG 51. Six 4-desmethyl sterols representing 59% of the total sterols were identified as sterols 2 i, 2 m, 2 d (major sterol), 1 d, 2 w, 1 w. Five 4-methyl sterols (41% of the total sterols) were present, of which only two were identified: the widely distributed dinoflagellate sterol 4α, 23,24-trimethyl-5α-cholest-22-en-3β-ol 6 w dinosterol (Withers 1983) and the other common dinoflagellate sterol (Withers 1983) 4α,24-dimethyl-5α-cholest-22 E-en-3β-ol 6 m. The latter is also present in the recently studied dinoflagellate *Prorocentrum cordatum* (Nichols et al. 1984). RR_t and MS as steryl-TMS are given:

24-methyl-5α-cholestan-24(28)-en-3β-ol 1 d RR_t 1.32; MW of TMS derivative 472. MS 457 (5), 388 (66), 345 (30), 255 (27), 215 (30), 75 (100)
23,24-dimethylcholesta-5,22 E-dien-3β-ol 2 w RR_t 1.39; MS 484 $[M]^+$ (4), 394 (3), 372 (5), 351 (4), 343 (10), 323 (4), 255 (16), 139 (16), 129 (13), 97 (16), 83 (23), 69 (100)
23,24-dimethyl-5α-cholest-22 E-en-3β-ol 1 w RR_t 1.42; MW of TMS derivative 486. MS 374 (6), 345 (16), 257 (18), 139 (11), 97 (24), 83 (25), 69 (100)
4α,24-dimethyl-5α-cholest-22 E-en-3β-ol 6 m RR_t 1.47; MS 486 $[M]^+$ (16), 471 (4), 388 (13), 359 (18), 271 (44), 229 (6), 69 (100)
4α,23,24-trimethyl-5α-cholest-22-en-3β-ol 6 w RR_t 1.79; MS 500 $[M]^+$ (3), 388 (7), 359 (19), 271 (21), 139 (10), 97 (24), 83 (25), 69 (100).

5 Conclusion

Combined GC-MS has been presented as a remarkable tool for the analysis of complex plant sterol mixtures. With selected examples from the literature it was seen how the use of retention times, methylene units, and mass spectral data afforded the identification of described sterols. Some research groups such as that of Djerassi have performed computer-based GC-MS analytical systems for structure analysis. In addition the use of reversed phase HPLC as a supplement to

AgnO$_3$-silicagel chromatography and high field NMR, has made possible the separation and structure elucidation of new sterols. Hence the literature is daily enriched with additional GC and MS data, especially from sterols isolated and identified in the vast area of marine organisms.

References

Artaud J, Iatrides MC, Tisse C, Zahra JP, Estienne J (1980) Études chromatographiques et spectroscopiques de sterols: application aux sterols d'algues. Analusis 8:277–286

Artaud J, Iatrides MC, Gaydou EM (1984) Co-occurrence of Δ^5 and Δ^7 sterols in two *Gleditsia* species. A reassessment of the sterol composition in oils rich in Δ^7 sterols. Phytochemistry 23:2303–2306

Brooks CJW, Middleditch BS (1973) Some aspects of mass spectrometry in steroid analysis. In: Heftmann E (ed) Methods of steroid analysis. Academic, London, pp 139–197

Brooks CJW, Middleditch BS (1979) Gas chromatography-mass spectrometry. In: The chemical society (ed) Specialist periodical report. Burlington House, London, pp 142–185

Brooks CJW, Horning EC, Young JS (1968) Characterization of sterols by gas chromatography-mass spectrometry of the trimethylsilyl ethers. Lipids 3:391–402

Combaut G, Saenger P (1984) Sterols of the *Amansiae* (*Rhodomelaceae: Rhodophyta*). Phytochemistry 23:781–782

Combaut G, Bruneau Y, Jeanty G, Francisco C, Teste J, Codomier L (1976) Contribution chimique à l'étude de certains aspects biologiques d'une Phéophycée de profondeur *Cystoseira zosteroides* (Turn) C Ag. Phycologia 15:275–282

Combaut G, Bruneau Y, Codomier L, Teste J (1979) Comparative sterols composition of the red alga *Asparagopsis armata* and its tetrasporophyte *Falkenbergia rufolanosa* Lloydia. J Nat Prod 42:150–151

Combaut G, Codomier L, Teste J (1981a) Seasonal chemical evolution of the alga *Cystoseira elegans*. Phytochemistry 20:2036–2037

Combaut G, Codomier L, Teste J, Pedersen M (1981b) The occurrence of C$_{28}$ sterols in red algae. Phytochemistry 20:1748–1749

Combaut G, Piovetti L, Kornprobst JM (1984) Étude de quelques algues rouges des côtes sénégalaises. C R Acad Sci Paris 299(II):433–435

Combaut G, Yacoubou A, Piovetti L, Kornprobst JM (1985) Sterols of the Senegalese Brown alga *Padina vickersiae*. Phytochemistry 24:618–619

Comita JJ, Klosterman HJ (1976) The sterols of normal and male-sterile maize tassels during development. Phytochemistry 15:917–920

Djerassi C (1984) Marine sterols. In: Krogsgaad (ed) Natural products and drug development. Munksgaard, Copenhagen, pp 164

Dow WC, Gebreyesus T, Popov S, Carlson RMK, Djerassi C (1983) Marine 4-methyl sterols: synthesis of C-24 epimers of 4α,24-dimethyl-5α-cholestan-3β-ol and 360 MHz ^{1}H NMR comparisons to the natural product from *Plexaura homomalla*. Steroids 42:217–230

Eneroth P, Hellström K, Ryhage R (1964) Identification of neutral fecal steroids by gas-liquid chromatography and mass spectrometry: studies of human excretion during two dietary regimens. J Lipid Res 5:245–262

Eneroth P, Hellström K, Ryhage R (1965) Identification of the neutral metabolites of stigmasterol found in human feces. Steroids 6:707–719

Fattorusso E, Magno S, Santacroce C, Sica D, Impellizzeri G, Mangiafico S, Oriente G, Piatelli M, Sciuto S (1975) Sterols of some red algae. Phytochemistry 14:1579–1582

Fattorusso E, Magno S, Santacroce C, Sica D, Impellizzeri G, Mangiafico S, Piatelli M, Sciuto S (1976) Sterols of mediterranean Florideophyceae. Biochem Syst Ecol 4:135–138

Francisco C, Combaut G, Teste J, Maume BF (1977) Étude des stérols d'algues brunes du genre *Cystoseira*. Identification par chromatographie Gaz-liquide couplée à la spectrométrie de masse. Biochim Biophys Acta 487:115–121

Francisco C, Combaut G, Teste J, Tarchini C, Djerassi C (1979) Side chain-hydroxylated sterols of the red alga *Asparagopsis armata*: significant products or artifacts due to autoxidation? Steroids 34:163–169

Goad LJ (1978) The sterols of marine invertebrates: Composition, biosynthesis, and metabolites. In: Scheuer PJ (ed) Marine natural products II. Academic, New York, pp 75–172

Hartmann MA, Normand G, Benveniste P (1975) Sterol composition of plasma membrane enriched fractions from maize coleoptiles. Plant Sci Lett 5:287–292

Ikekawa N, Kenkyusho R, Morisaki N, Tsuda K, Yoshida T (1968) Sterol compositions of some green algae and brown algae. Steroids 12:41–48

Itoh T, Shimizu N, Tamura T, Matsumoto T (1981) 24-methyl-E-23-dehydrolophenol, a new sterol and two other 24-methyl-E-Δ^{23} sterols in *Zea mays* germ oil. Phytochemistry 20:1353–1356

Itoh T, Tani H, Fukushima K, Tamura T, Matsumoto T (1982) Structure-retention relationship of sterols and triterpene alcohols in gas chromatography on a glass capillary column. J Chromatogr 234:65–76

Kabore SA, Combaut G, Vidal JP, Codomier L, Passet J, Girard JP, Rossi JC (1983) Sterols of the red alga *Rissoella verruculosa*. Phytochemistry 22:1239–1240

Kac D, Barbieri G, Falco MR, Seldes AM, Gros EG (1984) The major sterols from three species of Polyporaceae. Phytochemistry 23:2686–2687

Kalinowska M, Wojciechowski ZA (1984) Sterol conjugate interconversions during germination of white mustard (*Sinapis alba*). Phytochemistry 23:2485–2488

Kemp RJ, Mercer EI (1968) The sterol esters of maize seedlings. Biochem J 110:111–118

Kemp RJ, Goad LJ, Mercer EI (1967) Changes in the levels and composition of the esterified and unesterified sterols of maize seedlings during germination. Phytochemistry 6:1609–1615

Kemp RJ, Hamman ASA, Goad LJ, Goodwin TW (1968) Studies on phytosterol biosynthesis: observations on the esterified sterols of higher plants. Phytochemistry 7:447–450

Knights BA (1967) Identification of plant sterols using combined GLC/mass spectrometry. J Gas Chromatogr 5:273–282

Knights BA (1970) Sterols in *Ascophyllum nodosum*. Phytochemistry 9:903–905

Knights BA (1973) Qualitative and quantitative analysis of plant sterols by gas-liquid chromatography. In: Heftmann E (ed) Method of steroid analysis. Academic, London, pp 103–138

Knights BA, Brooks CJM (1969) Isomers of 24-ethylidenecholesterol: Gas chromatographic and mass spectrometric characterization. Phytochemistry 8:463–467

Knights BA, Smith AR (1976) Sterols of male and female compound inflorescences of *Zea mays* L. Planta 133:89–93

Kokke WCM, Withers NW, Massey IJ, Fenical W, Djerassi C (1979) Isolation and synthesis of 23-methyl-22-dehydrocholesterol a marine sterol of biosynthetic significance. Tetrahedron Lett 38:3601–3604

Kokke WCM, Fenical W, Djerassi C (1981) Sterols with unusual nuclear unsaturation from three cultured marine dinoflagellates. Phytochemistry 20:127–134

Matsumoto G, Torii T, Hanya T (1982) High abundance of algal 24-ethylcholesterol in Antarctic lake sediment. Nature 299:52–54

Misso NLA, Goad LJ (1984) Investigation on the Δ^{23}, $\Delta^{24(28)}$ and Δ^{25} sterols of *Zea mays*. Phytochemistry 23:73–82

Morisaki M, Kidooka S, Ikekawa N (1976) Studies on steroids. XXXIX. Sterol profiles of red algae. Chem Pharm Bull 24:3214–3216

Nes WR, Krevitz K, Behzadan S (1976) Configuration at C-24 of 24-methyl and 24-ethylcholesterol in tracheophytes. Lipids 11:118–126

Nes WR, Krevitz K, Joseph J, Nes WD, Harris B, Gibbons GF (1977) The phylogenic distribution of sterols in tracheophytes. Lipids 12:511–527

Nichols PD, Volkman JK, Johns RB (1983) Sterols and fatty acids of the marine unicellular alga, FCRG 51. Phytochemistry 22:1447–1452

Nichols PD, Jones GJ, de Leeuw JW, Johns RB (1984) The fatty acid and sterol composition of two marine dinoflagellates. Phytochemistry 23:1043–1047

Palermo JA, Seldes AM, Gros EG (1984) Free sterols of the red alga *Gigartina skottsbergii*. Phytochemistry 23:2688–2689

Popov S, Carlson RMK, Wegmann A, Djerassi C (1976) Minor and trace sterols in marine invertebrates. Steroids 28:699–732

Prost M, Maume BF, Padieu P (1974) Étude des stérols de la glande surrénalienne de Rat. Identification de phytosterols, d'oxo- et d'hydroxycholestenols par chromatographie Gaz-Liquide couplée à la spectrométrie de masse. Biochim Biophys Acta 360:230–240

Rohmer M, Kokke WCMC, Fenical W, Djerassi C (1980) Isolation of two new C_{30} sterols, (24 E)-24-n-propylidenecholesterol and 24-n-propylcholesterol, from a cultured marine chrysophyte. Steroids 35:219–231

Scheid F, Benveniste P (1979) Ergosta-5,23-dien-3β-ol and ergosta-7,23-dien-3β-ol, two new sterols from *Zea mays* etiolated coleoptiles. Phytochemistry 18:1207–1209

Scheid F, Rohmer M, Benveniste P (1982) Biosynthesis of $\Delta^{5,23}$ sterols in etiolated coleoptiles from *Zea mays*. Phytochemistry 21:1959–1967

Shimizu Y, Alam M, Kobayashi A (1976) Dinosterol, the major sterol with a unique side chain in the toxic Dinoflagellate, *Gonyaulax tamarensis*. J Am Chem Soc 98:1059–1060

Sica D, Piccialli V, Masullo A (1984) Configuration at C-24 of sterols from the marine phanerogames *Posidonia oceanica* and *Cymodocea nodosa*. Phytochemistry 23:2609–2611

Sjöstrand U, Kornprobst JM, Djerassi C (1981) (22E)-ergosta-5,22,25-trien-3β-ol and (22E,24R)-24,26-dimethylcholesta-5,22,25(27)-trien-3β-ol. Two new marine sterols from the sponge *Pseudaxinella lunacharta*. Steroids 38:355–364

Stankovic S, Bastic MB, Jovanovic JA (1984) Composition of the sterol fraction in horse chestnut. Phytochemistry 23:2677–2679

Suemitsu R, Yoshikawa T, Tsuji T, Sakata K (1977) Isolation of three sterols from Dentcorn, *Zea mays* L var *indentata*, Silage. Agric Biol Chem 41:211–212

Swenson W, Tagle B, Clardy J, Withers NW, Kokke WCMC, Fenical W, Djerassi C (1980) Peridinosterol. A new Δ^{17} unsaturated sterol from two cultured marine algae. Tetrahedron Lett 21:4663–4666

Vandenheuvel WJA, Smith JL, Albers-Schönberg G, Plazonnet B, Belanger P (1973) Derivatization and gas chromatography in the mass spectrometry of steroids. In: Heftmann E (ed) Methods of steroid analysis. Academic, London, pp 199–219

Varkony TH, Smith DH, Djerassi C (1978) Computer-assisted structure manipulation. Studies in the biosynthesis of natural products. Tetrahedron 34:841–852

Withers NW (1983) Sterols of dinoflagellates. In: Scheuer PJ (ed) Marine natural products V. Academic, New York, pp 87–130

Withers NW, Goad LJ, Goodwin TW (1979a) A new sterol 4α-methyl-5β-ergosta-8(14),24(28)-dien-3β-ol, from the marine dinoflagellate *Amphidinium carterae*. Phytochemistry 18:899–901

Withers NW, Kokke WCMC, Rohmer M, Fenical WH, Djerassi C (1979b) Isolation of sterols with cyclopropyl-containing side chains from the cultured marine algae *Peridinium foliaceum*. Tetrahedron Lett 38:3605–3608

Zakelj M, Goad LJ (1983) Observations on the biosynthesis of 24-methylcholesterol and 24-ethylcholesterol by *Zea mays*. Phytochemistry 22:1931–1936

GC-MS Methods for Terpenoids

L. WITTE

1 Introduction

Terpenoids are widely distributed in living systems and they occur predominantly as plant products with an immense number of members. Despite their structural diversity, many of them are chemically and physically closely related to each other. Sophisticated techniques are therefore required to separate mixtures of them and to identify the particular components. Terpenoids are, with only a few exceptions, volatile compounds. Thus gas chromatography is the analytical method of choice and has the great advantage that it can easily be combined with mass spectrometry.

Complete chromatographic separation is almost impossible, due to the high complexity of essential oils and their composition of structurally similar, or even isomeric compounds, with rather small variations in physical parameters of the particular members of mono-, sesqui-, di-, and triterpene groups. Even using long and efficient capillary columns many compounds are still either not or at least only incompletely resolved. The use of columns with different liquid phases of different polarity usually solves most of the separation problems, but uncertainty in identification often remains. Unequivocal structure determination of the many different terpenoids based on only a single analytical method, such as gas chromatographic retention data, is questionable.

Likewise definitive identification on the basis of mass spectral data is difficult, especially with stereoisomers or other structurally related compounds, by no means an infrequent problem with terpenoids. The solution of these difficulties lies in the combination of two different physical methods, capillary gas chromatography and mass spectrometry, resulting in one of the most powerful analytical techniques for volatiles currently available. Mass spectral fragmentation patterns combined with retention data render the possibility of rapid and reliable identification of individual components of complex mixtures.

Compound determination by comparing the gas chromatographic and mass spectrometric behavior of individual substances with authentic samples yield the most credible results. If authentic material is not available, and there is no way of synthesizing it from related identified compounds, or of comparing it with results of essential oils of known composition, literature data must be used. This will be necessary in many cases. Complete structure assignment of hitherto unknown compounds is more difficult to carry out with GC-MS information alone. This may be achieved only in the most favorable cases. Normally other analytical techniques, such as nuclear magnetic resonance, must also be used (Witte et al. 1983).

2 Isolation Methods

Commonly used techniques for isolating essential oils from natural sources are extraction, steam distillation, vapor collection by absorption or cryogenic concentration. Extraction seems to be a good method that avoids artifact formation. Comparison of extracted terpenes with the steam-distilled oil from the same material shows significant differences (Weston 1984). A remarkably lower content of monoterpene hydrocarbons can be found in the solvent extract compared with the steam distillate. Since steam distillation promotes dehydration of labile alcohols, the monoterpene hydrocarbons are probably artifacts. In contrast to other methods extraction allows the determination of less- or nonvolatile compounds such as di- and triterpenes as well. This technique is usually performed using a Soxleth-apparatus and sometimes long extraction times of several days are reported (Epstein and Gaudioso 1984). Solvents with low boiling points such as pentane or trichlorofluoromethane (MacLeod and Pieris 1982) are preferably used, since enrichment of the extract by evaporation is usually necessary and losses of volatile components can be avoided. For the same reason, concentration is best performed using a gentle stream of nitrogen to evaporate the solvent or low temperature vacuum distillation. Highly volatile compounds are then collected in a trap cooled with liquid nitrogen (MacLeod and Cave 1975).

Conventional steam distillation procedures likewise result in a relatively large amount of essential oil containing solvent, and again concentration steps, with possible losses, are necessary. To avoid this problem, micro methods which require no further enrichment by evaporation have been developed. Godefroot et al. (1981) described a modified Likens and Nickerson apparatus in which the volatiles are collected in only 1 ml solvent during the continuous steam distillation and extraction. One to 15 g of plant material, the amount depending on the content of essential oil, are required and a complete analysis, including gas chromatographic separation, can be performed in less than 4 h.

Another micro-scale apparatus for steam distillation has been developed by Bicchi et al. (1983). It requires very small amounts of plant material with less than 1 g being normally sufficient. The material is suspended in 50 ml water and distilled for half an hour. The essential oil is collected with cooling in 100 µl of the low boiling alkanes pentane or hexane. The complete analysis takes only about 2 h.

Trapping of volatiles on adsorbents such as Tenax GC is a further isolation and concentration method (Yokouchi et al. 1981), particularly if high volatile compounds from gaseous samples have to be analyzed. A specific volume of the volatile containing gas is passed through a tube filled with the adsorbent. Heat desorption with the carrier gas transfers the sample onto the column for separation and detection. This procedure is prone to artifact production with sensitive components, especially when working in the ppm range.

The headspace technique is another very useful auxiliary sampling method in essential oil evaluation. It offers several advantages for volatile component analysis. Time-consuming isolation procedures such as extraction, distillation, adsorption, or other methods can be avoided. Beside the simplicity and rapidity of the method, its nondestructive character is particularly important. Sample composition and the structures of substances are not altered during investigation, due to the mild conditions. A further advantage is that transfer of solvents or nonvolatiles into the chromatographic system, which causes destruction of the liquid phase and contamination of the injection port, is avoided. Automatic injection enables exact dosing and reproducible sample volumes, an important fact during quantitative evaluation. Peak areas in headspace analysis are dependent on equilibrium time and temperature, volume of the vial used, the injected sample amount and instrumental parameters such as split ratio and detector sensitivity.

A typical analysis consists of the following steps: Five to 10 g of the material are ground, Gabri and Chialva (1981) used a modified coffee grinder for this procedure, then the sample containing vial is heated to 40 °–60 °C under thermostat control for 30 to 60 min for equilibration. About 1 ml of the vapor phase is then injected, with the syringe being heated to the temperature of the sample and a low split ratio of about 1:5 is normally used. Following this procedure, good results are only achieved for the most volatile components of an essential oil, namely the monoterpenes. Recovery of the highly volatile monoterpene hydrocarbons are even better than with other sample preparations (Chialva et al. 1982; Bicchi et al. 1984). Monoterpene esters and sesquiterpenes need higher equilibrium temperatures of more than 100 °C (Hiltunen et al. 1984).

All the above sampling methods cannot give results for all the terpenoids actually present in the sample. The headspace technique cannot detect compounds with low volatility, but its use is preferable for highly volatile constitutents, which are usually diminished or not recovered with steam distillation or extraction procedures in which concentration steps are involved. Extraction has the advantage of yielding results also for the less volatile diterpenes and higher terpenoids, and the compounds are not thermally stressed. During steam distillation, and particularly with adsorption methods, sensitive components may undergo modifications. Thus, the best sample preparation technique can only be chosen after consideration of all aspects of the particular analytical problem.

3 Prefractionation and Ancillary Reactions

Despite the very high separation efficiency of capillary gas chromatography, complete separation of all components of an essential oil cannot always be achieved, as terpenes belong to a group of natural substances containing an immense number of compounds. Many of them have nearly identical structures and therefore closely related retention behavior. Various attempts have been made to prefractionate the complete sample in order to achieve a better gas chromatographic separation and therefore an easier identification afterwards.

Stepwise chromatography using different principles, such as adsorption and distribution, often facilitates the separation of compounds, which are difficult to be separated with only one of these methods alone. The simplest, but very efficient, procedure for prefractionating terpenoids is their adsorbtion chromatography on silica gel. Using a pure alkane such as pentane as solvent, only terpene hydrocarbons are eluted while more polar solvents, such as ethers or chlorinated hydrocarbons, elute oxygenated terpenoids. Collection of a number of small fractions (Scheffer and Baerheim Svendsen 1975) leads to a separation or an enrichment of several compounds in different fractions. Gradient elution using increasing levels of diethyl ether in pentane (Scheffer et al. 1977) leads to a separation of particular groups of oxygen-containing compounds. Esters were first eluted, followed by aldehydes and ketones, and finally alcohols.

Care must be taken during this type of separation as isomerization of certain terpenes is possible (Scheffer et al. 1976). These modifications may be caused by traces of metal ions present in the adsorbent. Hence silica gel should be stepwise pretreated with hydrochloric acid and ammonia. After washing with distilled water and drying at 105 °C, 5% water should be added to deactivate the adsorbent.

One disadvantage of such prefractionation procedures is the recovery of diluted solutions which are normally too dilute for direct injection into the gas chromatographic system. A micro method is often sufficient or even better (Jennings and Shibamoto 1980). Here fractionation of terpenoids is achieved using a micro column consisting of a thin-walled Teflon tube, 1 mm wide and 10 cm long, packed with silica gel. One or two microlitres of the terpene mixture are added to one end of the column, which is then placed loaded end down in a test tube containing 1 ml dichloromethane. The test tube is stoppered and, when the solvent has reached the top, the column is removed and cut into sections. Monoterpene hydrocarbons will be found at the top of the column, sesquiterpene hydrocarbons are in the central section and the bottom contains the oxygenated terpenes. Each section is washed with a minimum quantity of diethyl ether or dichloromethane, which is then injected directly for their analysis.

Prefractionation with Girard-T reagent, trimethylaminoacetohydrazide chloride, has been used in terpenoid analysis to separate carbonyl compounds (Debrauwere and Verzele 1975). These are converted into water-soluble derivatives, which can be separated from noncarbonyl compounds by partitioning between water and organic solvents. The Girard-T derivatives are then hydrolyzed and the carbonyl compounds recovered.

A typical procedure is as follows: The terpenoid mixture and approximately the same quantity of Girard-T reagent are refluxed in acetic acid ethanol 1:10 as a 10% solution for 2 h. After cooling, the mixture is poured into water and the noncarbonyl compounds are then extracted with diethyl ether. Sodium chloride can be added to avoid emulsion formation. The aqueous solution is then treated with hydrochloric acid and the recovered carbonyl compounds are extracted with diethyl ether and washed with sodium bicarbonate solution.

A method for the separation of saturated and unsaturated terpenoids of an essential oil is described by Stahl (1983). The separation is based on the fact that unsaturated compounds form adducts with mercury (II) salts which are practically insoluble in apolar organic solvents in which the saturated terpenoids are then extracted. Regeneration of the unsaturated compounds is achieved using ammonium rhodanate and hydrochloric acid. The strong acidic conditions can lead to artifact formation.

A variety of other functional group reactions have been employed to terpenoid mixtures to gain additional component information. Often the chemical nature of components can be deduced by the disappearance of peaks and appearance of new ones in the gas chromatogram. Helpful reactions are hydrolysis of esters, dehydration or acetylation of hydroxyl compounds and reduction of carbonyl compounds (Vostrowsky et al. 1981). Catalytic hydrogenation often leads to a simpler gas chromatogram and to simpler mass spectra. Complete hydrogenation to the fully saturated system often yields information about the carbon skeleton of the treated compounds. Results using this technique are especially useful for revealing the type of sesqui- and diterpenes investigated.

Various groups of terpenoids, such as gibberellines, cannabinoids (Harvey 1977) and abscisic acid (Netting et al. 1982), require derivatization to give sufficient volatility and to prevent thermal decomposition, such as decarboxylation of acids. Diazomethane reacts easily with acid groups to form the corresponding methyl esters, and silylating agents are often used to form derivatives of alcoholic and carboxylic groups of these terpenoids. Combination of these two derivatizations have been successfully used with gibberellines (Ingram et al. 1984; Heupel et al. 1985).

With all these methods compound modification is possible and often occurs. Thus care must be taken with their interpretation and the results must be critically compared with the original extracts and additional information should be considered.

4 Gas Chromatography

Capillary columns should be used for gas chromatographic analysis of complex mixtures. They are especially suited to this type of analysis as they have a high resolving power, and a higher sensitivity and shorter analysis time than packed columns. The choice of a suitable liquid phase is important. Various phases ranging from nonpolar ones such as SE-30 and OV-101 to polar ones such as Carbowax 20 M have been used successfully.

Compounds are eluted in the order of their boiling points from nonpolar columns and a separation is achieved according to the compound type. Thus monoterpenes are eluted first followed by sesqui-, di-, and triterpenes. In each class oxygenated compounds are retained longer than pure hydrocarbons. The high thermal stability of these phases has several advantages in GC-MS-work. Column bleeding is low and the background in the mass spectra is negligible even at elevated temperatures. Thus compounds with high boiling points, such as di- and triterpenes can easily be analyzed using this type of liquid phase (Hooper et al. 1982; Witte et al. 1983). Insensitivity to moisture and oxygen results in a reasonably long column life time. Problems may arise for the separation of structurally related terpenes, as differences in their boiling points are often very small leading to similar retention times.

The polar liquid phase Carbowax 20 M has different separation characteristics and is one of the most widely used phases in mono- and sesquiterpene analysis. Difficult separations on nonpolar phases are often resolved with polar ones. The use of different liquid phases increases the number of compounds which can be separated and hence allows more structure assignements. The disadvantages of Carbowax 20 M is its oxygen and moisture sensitivity, and its relatively low thermal stability. Thus high boiling terpenoids, such as the majority of diterpenes and all the triterpenes, cannot be analyzed using this polar liquid phase.

Good results can often be obtained on semi-polar phases, which are a compromise between the ones discussed above. Thus the selection of the suitable liquid phase depends on the particular chromatographic problem, which has to be solved.

5 Retention Data

As indicated above, terpenes cannot be unambiguously identified by their mass spectra alone. Mass intensity differences due to instrument parameters, such as ion source temperature, are often greater than the intensity differences which distinguish structurally related compounds, especially stereoisomers. However, precise retention data can often distinguish between isomers in such situations.

The Kováts retention index system (Kováts 1958) is the preferred method of reporting gas chromatographic retention data. In this system retention behavior is reported relative to that of the n-alkanes. The resulting values are only dependent on the stationary phase used and, to a lesser extent, on the column temperature. Column length and diameter, inlet pressure and flow rate are unimportant factors, and it does not matter if packed or capillary columns are used. Other methods of reporting retention data, such as retention times or relative retention times (Lemberkovics 1984), are far more influenced by the gas chromatographic conditions and they do not have the accuracy needed for terpenoid analysis. They may be helpful when the system is fixed for a long period, but their inflexibility limits their usefulness.

Retention indices are obtained with the greatest precision if they are determined isothermally. The relationship between the retention times of n-alkanes is then logarithmic, so that the indices can be determined precisely. For analysis of mixtures over a wide boiling range, several isothermic runs are required or programmed temperature procedures are necessary. In programmed runs, interpolation between consecutive members of the n-alkanes is required to assign the retention index of the compound in question (Majlát et al. 1974). This is preferably done graphically and results in a slightly S-shaped curve for the complete chromatogram. Program rates should be kept low, as rapid temperature changes leads to decreased accuracy. Precision of retention indices, determined during programmed temperature runs, are between ± 10 index units (Benecke et al. 1982). Such an approximate value is normally sufficient to confirm structure assignments derived from mass spectral data, especially if the possible structures suggested by mass spectrometry differ widely in their retention indices.

Kováts retention indices for a given compound vary with changes of the column temperature, and a reversal of the elution sequence is possible (Roberts 1962;

Karlsen and Siwon 1975). Temperature differences of 20 °C may result in a change of more than 40 index units (Perry and Weavers 1984) with polar liquid phases and polar substances show greater effects than nonpolar ones. Temperature accuracy and stability are therefore very critical for exact reproducibility of the retention indices.

The polarity of the column, the most important cause of retention behavior, may change during its life span. This effect is more noticeable with polar liquid phases, which are oxygen-sensitive. Degradation of the liquid phase increases the column polarity and hence changes the retention characteristics.

Many retention indices of terpenoids have been published in the literature during the last two decades. However, the multitude of stationary phases used causes difficulties in using the data, and the problems are enhanced by the habit of manufacturers of using their own brand names for the same phase. Thus methylsilicone phases comparable to the well-known phases SE-30, OV-1, OV-101 are sometimes called SP-2100, SF-96, CP Sil 5, RSL-150 etc. while DB-1, SPB-1, and CP Sil 5 CB are the chemically bonded phases. In the following only publications are listed, in which retention indices were obtained on the two most commonly used phases Carbowax 20 M and phases comparable to SE-30.

An extensive compilation of indices determined on these two phases is presented by Jennings and Shibamoto (1980). Saeed et al. (1979) reported retention indices of 20 monoterpene hydrocarbons on the same two phases, and Andersen and Falcone (1969) compiled indices of 55 sesquiterpenes on different phases. Perry and Weavers (1984) presented Kováts indices for more than 25 diterpene hydrocarbons. Several other authors (Michaelis et al. 1982; Weyerstahl et al. 1983; Witte et al. 1983; De Pooter et al. 1985) have reported various retention indices of different terpenoids.

6 Mass Spectrometry

In the following the demands on the mass spectrometric side of a capillary GC-MS-system for terpenoid analysis are discussed. Many terpenoids are thermally labile compounds, and hence a high degree of inertness of the GC-MS-system is essential in order to avoid degradation or rearrangement. This is especially important for the transfer lines from the gas chromatograph to the ion source of the mass spectrometer as the constant high temperatures of this interface promote degradation reactions. Fused silica capillary columns solve many of the deactivation problems. They are sufficiently flexible and can be linked directly through the interface with the ion source.

Another important point for labile components is the temperature of the ion sources itself. At elevated temperatures, molecular ions of various terpene alcohols are not detectable. This dehydration process and other thermal degradations can be diminished if the ion source temperature is kept low. High temperatures are not necessary, as only volatiles enter the ion source.

The time which is required to scan the mass range greatly influences the quality of the mass spectra. Intensities of masses acquired during the rising part of a

GC-peak are not present in their correct proportions. If the mass spectrometer scans from high to lower masses, high fragment ions are underrepresented and low mass ions overaccentuated, caused by the increasing concentration of the compound in the ion source.

The opposite happens during the falling part of the peak with the decreasing concentration resulting in a de-emphasis of later acquired masses. Fast scan rates lessen this effect and make it possible to obtain data at the top of the peak with only small changes in the concentration, which results in a fairly representative spectrum.

For this reason instruments with fast scanning capabilities are preferred for capillary GC-MS. Quadrupole instruments fulfill the above requirement and are therefore those mostly used in essential oil analysis. Another advantage of this instrument type are the relatively low costs compared with magnetic sector field instruments as the high resolution option and high mass capabilities of the latter are usually unnecessary in terpenoid analysis.

Complex mixtures can only be analyzed if the GC-MS-system is equipped with a data system. The computer system can store thousands of acquired mass spectra on disc and software capabilities greatly facilitate identification of compounds. Beside the total ion current chromatogram, a form of reconstructed GC-trace, mass chromatograms of interesting ions can be build up. Thus particular compounds can be located in mixtures that are normally difficult to analyze and computer libraries of mass spectra can be used to identify unknowns.

Mass spectra of terpenes show various characteristic fragment ions. Several fragments and losses from the molecular ion are very common. Prominent fragments, which are present in nearly all terpenoid mass spectra, result from loss of a methyl group yielding an $(M-15)^+$-ion and the loss of C_3H_7, normally the isopropyl group, producing an $(M-43)^+$-ion. The latter is a characteristic peak in the mass spectra of cyclic terpene hydrocarbons. Unsaturated acyclic terpenoids form the fragment m/z 69 as an intense ion by allylic bond cleavage. Limonen shows a characteristic base peak at m/z 68 resulting from the retro-Diels-Alder decomposition of the cyclohexene ring. Hydroxylated terpenes decompose very easily with loss of water. The molecular ions of these compounds are generally very small or sometimes completely absent. They can then be recognized by the occurrence of an $(M-15)^+$ combined with the $(M-18)^+$-ion. Terpene esters also produce only very weak molecular ions, as they readily eliminate, by analogy with the hydroxylated compounds, their carboxylic acid moiety as a neutral fragment. Consequently their low mass fragmentation is very similar to those of the respective alcohols. A detailed discussion of fragmentation pathways is presented by Enzell et al. (1972), Enzell and Wahlberg (1980) and in the literature cited therein.

Compilations of mass spectra, including mono- and sesquiterpenes, are published in the *Registry of Mass Spectral Data* (Stenhagen et al. 1974) and by Jennings and Shibamoto (1980). Further mass spectra of monoterpene hydrocarbons are presented by Ryhage and von Sydow (1963), Thomas and Willhalm (1964), and von Bünau et al. (1969). Various spectra of monoterpene alcohols, aldehydes ketones, and esters are published by von Sydow (1963–1965). Sesquiterpene mass spectra are reported by Maurer and Grieder (1977), Williams et al. (1981), and

Shieh and Matsubara (1981). Mass spectra of diterpenes are compiled in several publications of Enzell, Ryhage, and Wahlberg (Enzell and Ryhage 1965, 1967a; Enzell 1966; Enzell and Ryhage 1967b; Enzell and Wahlberg 1969). Publications of Hill et al. (1968), Papageorgiou (1980), Mazza (1983), and Witte et al. (1983) contain several mass spectra of different terpenoids. Another large compilation of tobacco terpenoid mass spectra is found in a review of Enzell et al. (1984). The above enumeration is not complete, as further mass spectra are widely dispersed in the literature and cannot be listed here.

As discussed earlier, identification of terpene alcohols and esters is sometimes difficult, as their molecular ions are often not present in the acquired mass spectra. Chemical ionization, the best "soft ionization" technique for GC-MS work, permits molecular weight estimation of many labile compounds. Various reactant gases can be used to form quasi molecular ions with particular classes of compounds, although several other classes are not susceptible to this treatment. Thus the widely used reactant gases methane and isobutane fail to form good quasi-molecular ions with terpene hydrocarbons, alcohols, and esters. Molecular weight determination is only possible with aldehydes and ketones.

Negative chemical ionization (NCI), using nitrous oxide and methane in a 1:1 mixture as reactant gas, is a useful method for obtaining quasi-molecular ions of alcohols and esters (Hendriks and Bruins 1980). The generated $(OH)^-$-ions abstracts a proton from these compounds and the resulting $(M-H)^-$ is, in most cases, the base peak of the mass spectra. A further advantage is that esters produce a second intense ion, the carboxylate anion, which facilitates their assignment. Mass chromatograms of these indicate the kind of compound present in the mixture (Bruins 1979). Unsaturated terpene hydrocarbons yield a quasi-molecular ion of relatively low intensity, as protons of these compounds are more difficult to abstract. Their NCI-spectra are further complicated by reaction of their $(M-H)^-$-ion with neutral N_2O from the reactant gas. Saturated terpenoids are not ionized by this technique (Hendriks and Bruins 1983). Thus it is advantageous to divide the complete terpenoid mixture into a hydrocarbon and an oxygenated compound fraction by column chromatography. The hydrocarbon fraction can then be subjected to electron impact-GC-MS and negative chemical ionization is applied to the oxygenated terpenoid mixture.

References

Andersen NH, Falcone MS (1969) The identification of sesquiterpene hydrocarbons from gas-liquid chromatography retention data. J Chromatogr 44:52–59
Benecke R, Thieme H, Nyiredy SZ (1982) Retentionsverhalten der Hauptkomponenten der Lavendel- und Lavandinöle an verschiedenen stationären Phasen in Abhängigkeit von der Säulentemperatur. J Chromatogr 238:75–87
Bicchi C, D'Amato A, Nano GM, Frattini C (1983) Improved method for the analysis of small amounts of essential oils by microdistillation followed by capillary gas chromatography. J Chromatogr 279:409–416
Bicchi C, D'Amato A, Nano GM, Frattini C (1984) Capillary GLC controls of some alpine artemisiae and of the related liqueurs. Chromatographia 18:560–566

Bruins AP (1979) Negative ion chemical ionization mass spectrometry in the determination of components in essential oils. Anal Chem 51:967–972

Bünau v G, Schade G, Golnick K (1969) Massenspektrometrische Untersuchungen von Terpenen. III. Kohlenwasserstoffe der Caran- und der Menthanreihe. Z Anal Chem 244:7–17

Chialva F, Gabri G, Liddle PAP, Ulian F (1982) Qualitative evaluation of aromatic herbs dy direct headspace GC analysis. Applications of the method and comparison with the traditional analysis of essential oils. HRC & CC 5:182–188

Debrauwere J, Verzele M (1975) New constituents of the oxygenated fraction of pepper essential oil. J Sci Food Agric 26:1887–1894

De Pooter HL, Nor Omar M, Coolsaet BA, Schamp NM (1985) The essential oil of greater galanga (alpinia galanga) from malaysia. Phytochemistry 24:93–96

Enzell CR (1966) Mass spectrometric studies of diterpenes. 4 Aromatic diterpenes. Tetrahedron Lett 1966:2135–2143

Enzell CR, Ryhage R (1965) Mass spectrometric studies of diterpenes. 1 Carbodicyclic diterpenes. Ark Kemi 23:367–399

Enzell CR, Ryhage R (1967a) Mass spectrometric studies of diterpenes. 3 Aromatic diterpenes. Ark Kemi 26:425–434

Enzell CR, Ryhage R (1967b) Mass spectrometric studies of diterpenes. 5 Aromatic diterpenes. Ark Kemi 27:213–229

Enzell CR, Wahlberg I (1969) Mass spectrometric studies of diterpenes. 6 Aromatic diterpenes. Acta Chem Scand 23:871–891

Enzell CR, Wahlberg I (1980) Terpenes and terpenoids. In: Waller GR, Dermer OC (ed) Biochemical applications of mass spectrometry (1st suppl vol). Wiley-Interscience, New York, pp 311–406

Enzell CR, Appleton RA, Wahlberg I (1972) Terpenes and terpenoids. In: Waller GR (ed) Biochemical application of mass spectrometry. Wiley-Interscience, New York, pp 351–385

Enzell CR, Wahlberg I, Ryhage R (1984) Mass spectra of tobacco isoprenoids. Mass Spectrom Rev 3:395–438

Epstein WW, Gaudioso LA (1984) Volatile oil constituents of sagebrush. Phytochemistry 23:2257–2262

Gabri G, Chialva F (1981) Qualitative evaluation of aromatic herbs by direct head space (GC) analysis. Methodology and some preliminary applications. HRC & CC 4:215–217

Godefroot M, Sandra P, Verzele M (1981) New method for quantitative essential oil analysis. J Chromatogr 203:325–335

Harvey DJ (1977) Cyclic alkylboronates as derivatives for the characterization of cannabinolic acids by combined gas chromatography and mass spectrometry. BMS 4:88–93

Hendriks H, Bruins AP (1980) Study of three types of essential oil of valeriana officinalis L. s. 1 by combined gas chromatography-negative ion chemical ionization mass spectrometry. J Chromatogr 190:321–330

Hendriks H, Bruins AP (1983) A tentative identification of components in the essential oil of cannabis sativa L. by a combination of gas chromatography negative ion chemical ionization mass spectrometry and retention indices. BMS 10:377–381

Heupel RC, Phinney BO, Spray CR, Gaskin P, MacMillan J, Hedden P, Graebe JE (1985) Native gibberellins and the metabolism of [^{14}C]gibberellin A_{53} and of [17-^{13}C, 17-$^3H_2$] gibberellin A_{20} in tassels of zea mays. Phytochemistry 24:47–53

Hill HC, Reed RI, Robert-Lopes MT (1968) Mass spectra and molecular structure. Part I. Correlation studies and meta-molecular structure. Part I. Correlation studies and metastable transitions. J Chem Soc C 1968:93–101

Hiltunen R, Vuorela H, Laakso I, v. Schantz M (1984) Application of headspace gas chromatography in essential oil analysis (IV). Quantitative analysis of α-bisabolol in volatiles of matricaria recutita L. Farm Tijdschr Belg 61:354–355

Hooper SN, Chandler RF, Lewis E, Jamieson WD (1982) Simultaneous determination of sonchus arvensis L. triterpenes by gas chromatography-mass spectrometry. Lipids 17:60–63

Ingram TJ, Reid JB, Murfet IC, Gaskin P, Willis CL, MacMillan J (1984) Internode length in pisum. The le gene controls the 3β-hydroxylation of gibberellin A_{20} to gibberellin A_1. Planta 160:455–463

Jennings W, Shibamoto T (1980) Qualitative analysis of flavor and fragrance volatiles by glass capillary gas chromatography. Academic, New York

Karlsen J, Siwon H (1975) Elution sequence as a function of temperature in the gas-liquid chromatography of monoterpene hydrocarbons. J Chromatogr 110:187–189

Kováts E (1958) Gas-chromatographische Charakterisierung organischer Verbindungen Teil 1: Retentionsindices aliphatischer Halogenide, Alkohole, Aldehyde und Ketone. Helv Chim Acta 41:1915–1932

Lemberkovics É (1984) Gas chromatographic characterization of frequently occurring monoterpenes in essential oils. J Chromatogr 286:293–300

MacLeod AJ, Cave SJ (1975) Volatile flavor components of eggs. J Sci Food Agric 26:351–360

MacLeod AJ, Pieris NM (1982) Analysis of the volatile essential oils of murrays koenigii and pandanus latifolius. Phytochemistry 21:1653–1657

Majlát P, Erdös Z, Takács J (1974) Calculation and application of the retention indices in programmed-temperature gas chromatography. J Chromatogr 91:89–103

Maurer B, Grieder A (1977) Sesquiterpenoids from costus root oil (saussurea lappa Clarke). Helv Chim Acta 60:2177–2191

Mazza G (1983) Gas chromatographic-mass spectrometric investigation of the volatile components by myrtle berries (myrtus communis L.). J Chromatogr 264:304–311

Michaelis K, Vostrowsky O, Paulini H, Zintl R, Knobloch K (1982) On the essential oil components from blossoms of artemisia vulgaris L. Z Naturforsch 37c:152–158

Netting AG, Milborrow BV, Duffield AM (1982) Determination of abscisic acid in eucalyptus haemastoma leaves using gas chromatography-mass spectrometry and deuterated internal standards. Phytochemistry 21:385–389

Papageorgiou VP (1980) GLC-MS computer analysis of the essential oil of Thymus capitatus. Planta Med Suppl 1980:29–33

Perry NB, Weavers RT (1984) Kovats indices of diterpene hydrocarbons on fused-silica capillary columns. J Chromatogr 284:478–481

Roberts JB (1962) Elution sequence as a function of temperature in gas chromatography. Nature 193:1071–1072

Ryhage R, v. Sydow E (1963) Mass spectrometry of terpenes I. Monoterpene hydrocarbons. Acta Chem Scand 17:2025–2035

Saeed T, Redant G, Sandra P (1979) Kovats indices of monoterpene hydrocarbons on glass capillary columns. HRC & CC 2:75–76

Scheffer JJC, Baerheim Svendsen A (1975) Improved gas chromatographic analysis of naturally occurring monoterpene hydrocarbons following pre-fractionation by liquid-solid chromatography. J Chromatogr 115:607–611

Scheffer JJC, Koedam A, Baerheim Svendsen A (1976) Occurrence and prevention of isomerization of some monoterpene hydrocarbons from essential oils during liquid-solid chromatography on silica gel. Chromatographia 9:425–432

Scheffer JJC, Koedam A, Schüsler MThIW, Baerheim Svendsen A (1977) Improved gas chromatographic analysis of naturally occurring oxygen-containing monoterpenes following prefractionation by liquid-solid chromatography. Chromatographia 10:669–677

Shieh B, Matsubara Y (1981) Gas chromatography-mass spectrometry to the study of the sesquiterpenoids on the series of longifolanes, caryophyllanes, cadinanes, cedranes, longipinanes and thujopsanes. Shitsuryo Bunseki 29:97–111

Stahl E (1983) Pre-chromatographic procedure using mercury(II) acetate to separate saturated from unsaturated components in essential oil analysis. GIT Suppl 3:69–72

Stenhagen E, Abrahamsson S, McLafferty FW (1974) Registry of mass spectral data. Whiley-Interscience, New York

Thomas AF, Willhalm B (1964) Les spectres de masse dans l'analyse 3. Les spectres de masse des hydrocarbures monoterpeniques. Helv Chim Acta 47:475–488

v Sydow E (1963) Mass spectrometry of terpenes. II. Monoterpene alcohols. Acta Chem Scand 17:2504–2512
v Sydow E (1964) Mass spectrometry of terpenes. III. Monoterpene aldehydes and ketones. Acta Chem Scand 18:1099–1104
v Sydow E (1965) Mass spectrometry of terpenes. IV. Esters of monoterpene alcohols. Acta Chem Scand 19:2083–2088
Vostrowsky O, Brosche T, Ihm H, Zintl R, Knobloch K (1981) On the essential oil components from artemisia absinthium L. Z Naturforsch 36c:369–377
Weston RJ (1984) Composition of essential oil from leaves of eucalyptus delegatensis. Phytochemistry 23:1943–1945
Weyerstahl P, Kaul VK, Meier N, Weirauch M, Marschall H (1983) Volatile constituents of plectranthus rugosus leaf oil. Planta Med 48:99–102
Williams HJ, Strand MR, Bradleigh Vinson S (1981) Synthesis and purification of the allofarnesenes. Tetrahedron 37:2763–2767
Witte L, Berlin J, Wray V, Schubert W, Kohl W, Höfle G, Hammer J (1983) Mono- and diterpenes from cell cultures of thuja occidentalis. Planta Med 49:216–221
Yokouchi Y, Fujii T, Ambe Y, Fuwa K (1981) Determination of monoterpene hydrocarbons in the atmosphere. J Chromatogr 209:293–298

GC-MS of Auxins

L. Rivier

1 Introduction

The frequent lack of correlation between endogenous levels of auxins and selected physiological processes which has often been reported in the literature suggests the conclusion that such correlations do not often exist. However, from a critical evaluation of the methods used for such studies, it is apparent that most of the qualitative and quantitative determinations made for auxins in general did not comply with good analytical practice. For example, some mass spectra (MS) of both reference and putative IAA derivatives that have been recently published in the literature contain gross errors (i.e., Arteca et al. 1980). In addition, many of the quantitative calculations have been based on nonspecific ions or detector signals.

Clearly, accurate and precise measurements of auxin levels have first to be obtained to establish meaningful physiological correlations. It is, for example, mandatory to correct for all losses which occur during handling and analyzing of the sample. This can most easily be done by the use of a well-chosen Internal Standard (IS). Mass spectrometry, when coupled with gas chromatography (GC-MS), offers the potential not only for unequivoqual identification of the auxins, but also for very specific and sensitive detection of trace amounts of selected endogenous compounds with auxinic activity.

With the recently introduced new generation of small GC-MS equipment controlled with micro-processors and micro-computers, the initial cost of the necessary equipment is relatively low. The high quality of the results obtained with GC-MS justifies the investment. When using these small machines, the running costs are similar to those encountered with other GC or HPLC analyses using selective detectors. The operator must be trained primarily to recognize very rapidly the significance of the data obtained from the computer. These new microcomputer-based machines can be operated, therefore, by most laboratory personel.

At the present time, it is difficult to give an over-all definition of the term auxin. Historically, the auxins are represented by the activity of the various compounds on the *Avena* coleoptile curvature or elongation test (see Schneider and Wightman 1978).

Indole-3-acetic acid (IAA, Fig. 1) has been considered as the most important auxin found in plants. Extensive coverage of the occurrence, biochemistry, and some aspects of the physiology of IAA and related auxins are to be found in recently published treaties (Letham et al. 1978; MacMillan 1980; Bandurski 1984; Cohen and Bialek 1984; Scott 1984).

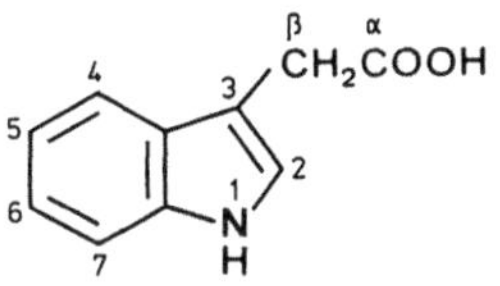

Fig. 1. Structure of indole-3-acetic acid and numbering of the ring's atoms

In 1978, the general guidelines of a GC-MS technique for endogenous IAA analyses were described in a text book (McDougall and Hillman 1978). Since then, many important improvements of the technique, as well as new methods, have been published. The present chapter aims to offer an in-depth description and critical assessment of the most recent GC-MS techniques used in practical endogeneous IAA analysis cases and based on the author's own experiences in this field.

2 The Compounds Involved

The current status of the auxins which have been identified by reliable physicochemical means has been reviewed recently by Bearder (1980). Since then, very few new auxins have been identified: the following list presents the enumeration of published works from 1980 on the characterization of these auxins by reliable physicochemical methods.
- IAA has been found in a large number of plant species and tissues;
- Indole-3-propionic acid (IPA) has been found by GC-MS to be present in *Cucurbita pepo* seedlings (Segal and Wightman 1982).
- Indole-3-carboxylic acid (ICA) has been identified and quantified in *Pinus sylvestris* seedlings by using GC-MS (Sandberg et al. 1984).
- Indole-3-ethanol (IEtOH) or tryptophol, which was found also in cucumber seedlings (Rayles and Purves 1967) has now been proven to be an endogenous compound in *Pinus sylvestris* seedlings and not to arise from the artifactual degradation of endogenous IAA during work-up of the sample (Sandberg 1984).
- For the first time, indole-3-methanol (IMeOH) has been found to be an endogenous compound in etiolated seedlings of Scots pine by GC-MS (Sundberg et al. 1985).
- The presence of phenyl-acetic acid (PAA) in tobacco shoots is now established (Wightman and Lichty 1982).
- The chlorinated indoles, 4-chloro-indole-3-acetic acid (4-Cl-IAA) and its methyl ester have been detected in seven Viciae species (Fabiaceae) analyzed so far (Engvild et al. 1981; Pless et al. 1984; Schneider et al. 1985).
- Precursors of IAA like tryptophane (Tryp) have been identified in plants (Allen and Baker 1980), but it is most probable that these compounds show auxin activity because they are transformed in vivo into IAA itself.

- Catabolites of IAA are many (Nonhebel et al. 1983), but only 2-ox-indole-3-acetic acid (OxIAA) and 7-hydroxy-2-ox-indole-3-acetic acid-7'-O-β-D-glucopyranoside in *Zea mays* caryopses and shoots have been fully characterized so far (Reinecke and Bandurski 1983; Nonhebel and Bandurski 1984). None of these appears to have any auxinic activity.
- The occurrence of indole bases, namely tryptamine and derivatives, has been reviewed (Smith 1977). Some of them do have physiological activities in plants (Rivier and Pilet 1971). These bases can easily undergo reactions under "physiological conditions" with aldehydes for ring closure of their side-chain giving tricyclic-3,4-b-indoles or β-carbolines: their occurrence in plants in a few instances has been questioned in a review (Allen and Holmstedt 1980).
- A number of more complex indolic molecules showing auxin activity are the IAA-conjugates. These complex molecules are formed by covalent linkage of IAA to small molecules, such as glucose, myo-inositol, aspartic acid, glucoproteins, or polypeptides. Esters formed with several IAA molecules per myo-inositol have also been characterized (see Cohen and Bandurski 1982). They release IAA upon acid hydrolysis. Indole glucosinolates – found only in some Brassicaceae species (see Schneider and Wightman 1978) release indole-3-acetonitrile (IAN) under similar conditions. These complex substances do not appear to have auxinic activities per se, but, because of their metabolic instability, they give IAA during the biotest, and thus have some auxin activity as a result of hydrolysis to IAA (Bialek et al. 1983).

In conclusion, auxin is a generic term and there must still be considerable effort made to demonstrate the presence of these compounds as well as new, as yet unidentified auxins in various plant tissues. GC-MS techniques will be instrumental in the continued development of such knowledge.

3 Reference Compounds

For the majority of GC-MS determinations it is necessary to have access to reference compounds, mainly to calibrate the instrument responses. For auxins, many indoles and related compounds and the stable isotope-labeled molecules which are the best suited Internal Standards in MS, readily available from commercial sources, are shown in Table 1 (the list is not exhausive as new indolic compounds frequently appear on the market).

To our knowledge, few attempts to synthesize ^{15}N or ^{13}C labeled indoles have been made, and it is not possible to find commercial sources of IAA with such label. J. Cohen and coworkers have obtained ^{13}C$_6$-[benzene ring]-indole-3-acetic acid starting from the recently available and highly enriched ^{13}C$_6$-analine for use as a new internal standard for quantitative mass spectral analysis of IAA in plants (Cohen et al. 1986). Such labeled compounds might be very useful for feeding experiments, too, although nothing is known yet about the isotopic effect or discrimination of the new molecule. Further synthesis of these valuable chemicals is to be expected in the near future for unrestricted availability.

Table 1. Chemicals related to auxins, available from commercial sources for GC-MS analysis

Chemical	Purity	Source[a]
Indole-3-acetic acid		Sigma
Methyl indole-3-acetate		K. & K./Fluka
Ethyl indole-3-acetate		Calbiochem/Research Organics
n-Hexyl indole-3-acetate		Vega
Indole-3-methanol		Sigma
Indole-3-ethanol (Tryptophol)		Regis
Indole-3-pyruvic acid		Regis
Indole-3-acrylic acid		CSIRO, Nutritional Biochemical Corp.
Indole-3-lactic acid		CSIRO, Nutritional Biochemical Corp.
Indole-3-acetyl-L-Alanine		Research Organics Inc.
Indole-3-acetyl-DL-Aspartic acid		Research Organics Inc.
Indole-3-acetyl-Glycine		Research Organics Inc.
Indole-3-acetyl-L-Phenylalanine		Research Organics Inc.
Indole-3-aldehyde		Research Organics Inc.
Indole-3-butyric acid		Research Organics Inc.
Indole-3-glyoxylic acid		Research Organics Inc.
Indole-3-propionic acid		Research Organics Inc.
Indoxyl-1,3-diacetate		Research Organics Inc.
Many indole synthetic derivatives such as:		Laboratoires Plan S.A.
Tryptamines, substitued tryptophanes and		Laboratoires Plan S.A.
derivatives, gramines, auxin derivatives,		Laboratoires Plan S.A.
carboxy indoles, indolic esters, substituted		Laboratoires Plan S.A.
indoles, indole-3-carboxaldehydes, indoxyls		Laboratoires Plan S.A.
derivates, indole-3-amides, indole-3-		Laboratoires Plan S.A.
nitrovinyl,		Laboratoires Plan S.A.
Indole-3-acetonitrile		Laboratoires Plan S.A.
Indole-3-carbinol		Laboratoires Plan S.A.
Indoyl-β-D-glucoside (Plant indican)		Laboratoires Plan S.A.
4-Chloro-indole		Aldrich
6-Chloro-indole		Aldrich
7-Chloro-indole		Aldrich
5-Bromo-indole		Aldrich
Phenyl-acetic acid		Aldrich
Cinnamic acid		Eastman
Phenyl-acetaldehyde		K. & K. Lab.
Phenyl-propionic acid		Pfalz and Bauer
p-Hydroxy-phenyl-propionic acid		Pfalz and Bauer
$^2H_2$-Tryptophan-3,3		Merck, Sharp, and Dohme, Canada, Ldt.
$^2H_3$-Tryptophan-2',3',3'		Merck, Sharp, and Dohme, Canada, Ldt.

[a] The addresses of some of the above-mentioned manufacturers are:
- Isocommerz, Lindenberger Weg 70, DDR-1115 Berlin, German Democratic Republic,
- K.O.R. Isotopes, 56 Rogers Street, Cambridge, MA 02142, USA
- Merck, Sharp, and Dohme, Canada Ldt, distributed in Europe by IC Chemikalien GmbH, Sohnckestrasse 17, D-8000 München 71, Fed. Rep. Germany
- Research Organics Inc., 4353 East 49th Street, Cleveland, OH 44125, USA
- Stohler Isotope Chemicals, 49 Jones Road, Waltham, MA 02154, USA, and
- Laboratoire Plan S.A., Chemin des Sellières, CH-1211 Aïre-Genève, Switzerland

Table 1 (continued)

Chemical		Purity	Source[a]
$^2H_5$-L-Tryptophan-2,4,5,6,7		97%	Merck, Sharp, and Dohme, Canada, Ldt.
$^2H_8$-Tryptophan		98%	Merck, Sharp, and Dohme, Canada, Ldt.
$^2H_{12}$-DL-Tryptophan	Nucleus: chain:	50–70% 20%	Isocommerz
$^2H_2$-Tryptamine-β,β		98%	Merck, Sharp, and Dohme
$^2H_4$Tryptamine		98%	Merck, Sharp, and Dohme
$^2H_2$-IAA-2'2'		97%	Merck, Sharp, and Dohme
$^2H_5$-IAA-2,4,5,6,7		97%	Merck, Sharp, and Dohme
$^2H_7$-IAA-2,4,5,6,7-β,β		97%	Merck, Sharp, and Dohme
$^2H_2$-Indole-3-acetonitrile		98%	Merck, Sharp, and Dohme
$^2H_7$-Indole			Merck, Sharp, and Dohme
$^2H_4$-5-Methoxy-N,N-dimethyl-tryptamine			Merck, Sharp, and Dohme
$^2H_6$-5-Methoxy-N,N-dimethyl-tryptamine			Merck, Sharp, and Dohme
$^2H_2$-5-Methoxy-IAA			Merck, Sharp, and Dohme
$^2H_4$-5-Methoxy-Tryptamine			Merck, Sharp, and Dohme
$^2H_2$-5-Hydroxy-IAA-2',2'		98%	Stohlers Isotope Chemicals
$^2H_4$-5-Hydroxy-Tryptamine		98%	Stohlers Isotope Chemicals
^{15}N-Indole		99%	Stohlers Isotope Chemicals
^{15}N-L-Tryptophan (side chain)		99%	Stohlers Isotope Chemicals
^{15}N-L-Tryptophan-UL		95–98%	Stohlers Isotope Chemicals
^{15}N-DL-Tryptophan (side chain)		95%	Isocommerz
^{15}N-DL-Tryptophan-(ring)		95%	Isocommerz
^{15}N-L-Tryptophan-(ring)		95%	Isocommerz
^{15}N-L-Tryptophan-(ring)		95–99%	K.O.R. Isotopes
$^{13}C_1$-Indole		90%	K.O.R. Isotopes
$^{13}C_6$-Analine			K.O.R. Isotopes
$^{13}C_1$-DL-Tryptophan-1'		90%	Merck, Sharp, and Dohme
$^{13}C_1$-DL-Tryptophan-2'		90%	K.O.R. Isotopes
$^{13}C_1$-L-Tryptophan-(ring 3)		90%	K.O.R. Isotopes

The isotope purity of some of these labeled compounds listed above is often low, and caution should be used depending on the GC-MS application. For deuterium-labeled compounds in particular, the labeled proton should remain in the position where placed during synthesis, and if such conditions are fulfilled, these stable labeled molecules are extremely useful tools for quantitative GC-MS determinations.

4 Extraction

Extraction of auxins from plant tissues for GC-MS determinations must be carried out immediately after harvesting under reduced temperature (around 0 °C) as for other techniques in order to reduce possible enzymatic degradative activities. If this cannot be done, the tissue must be immediately frozen and stored at as low a temperature as possible to prevent any further changes to the original composition of the sample. For long-term storage, the sample can best be freeze-dried and stored at $-30\ °C$. Losses of IAA under such conditions do not exceed 1% per month on average (Rivier, unpublished data). IAA has been reported to sublime from ether solutions under vacuum at room temperature (Mann and Javorski 1970). When IAA is still within the tissue, however, freeze-drying over 20 h did not change the amount of endogenous IAA found in maize roots (Rivier and Saugy, unpublished).

The extraction should be performed in a room specially reserved for such purposes: the risk of contamination of the working bench by spitting micro-droplets of concentrated standard solutions should not be underestimated. Further, all the glassware used for extraction and purification of auxins should be rigorously cleaned and kept exclusively for such use. This is specially recommended for quantitative trace analyses. The procedure for cleaning the glassware is similar to that used for radio-active contamination: the use of special detergents is necessary. Then, for removing all traces of soap, citric acid can be added at $1\ g\,l^{-1}$ concentration to the rinsing demineralized water. Finally, it is good practise to rinse all the glassware with MeOH and dry it at $120\ °C$ after silylation (when necessary).

The deactivation of the surface of the glass is recommended when polar auxins are present in very low quantities and the analyst wishes to avoid their irreversible adsorption onto the polar sites of the glass surfaces. This is done by silylation of the glassware: all the surfaces are immersed for 15 min in a 5% dimethyldichlorosilane solution in dry toluene. The material is then rinsed with MeOH and dried. Such deactivation is not usually necessary for IAA, but we observed that it can significantly improve the recovery of IAA and other auxins for an inexperienced analyst. As an alternative to deactivated glass vials, the use of Teflon labware is also recommended, especially for hydrolysis of auxin conjugates (J. Cohen, pers. comm.).

The solvents most often used for the extraction of auxins are MeOH, EtOH, acetone, and diethyl ether. Alcohols and acetone provide the most thorough extraction of IAA (the so-called free-IAA). This might not be the case for other auxins, and the efficiency of the extraction should be checked systematically. This is done by re-extracting the solid residue from the first extraction: if a significant amount of the auxin is still present in the second extract, the first extraction was either incomplete or some bound IAA was slowly released during that time. The use of 80% acidic MeOH at $0\ °-4\ °C$ is believed to inhibit degradative enzymes.

Careful selection of the extraction solvent(s) is important, to minimize contributions of unwanted endogenous compounds or impurities from the solvent itself. Phthalates, present in plastics (especially in soft materials such as Tygon or Sylastic tubing, and Parafilm; use of these materials should be stricly avoided!), are omnipresent even in water. The importance of these impurities for the analysis depends on the system of detection: when scanning the whole mass range of the GC-MS, it is clear that all compounds will be detected. On the contrary, when using the MS as selected ion detector, impurities might not be "seen" at all even if present in large amounts. However, they can significantly degrade the chromatographic performances, even to the point of altering retention times during HPLC and GC.

Generally, it is safe to use freshly distilled solvents or chemicals which have been "stabilized" (e.g., di-ter-butyl-2,6-methyl-4 phenol or butyl-hydroxy-toluene-BHT is often added to diethyl ether and chloroform contains a few % of EtOH). Such additives might change to solubilization properties of the pure solvent. If special mixed solvent systems have to be used, it is necessary to establish their efficiency. The addition of anti-oxidants is considered to improve the yield of extraction of IAA (Iino et al. 1980; Iino 1982) even when very rapid and simple procedures are used (see below), and to reduce the easy conversion of indole-3 pyruvic acid to IAA (Atsumi et al. 1979). BHT, ascorbic acid, Santoquin or CO_2 have been used, but only the first two are still employed, BHT for lipophillic solvents and ascorbic acid for aqueous solvents (Yokota et al. 1980).

Acetone has been used extensively for the recovery of sugar esters of IAA (Espstein et al. 1980). For free IAA, MeOH, EtOH, acetone, diethyl ether, chloroform, and ethyl acetate have been reported to be equivalent (McDougall and Hillman 1978). IAA, however, becomes esterified in contact with MeOH or EtOH after extraction in acidic medium and during the final phase of the evaporation of the solvent (Allen et al. 1982 a). The importance of such artifactual esterification has been shown to be higher than 10% with MeOH, and not more than 3% with EtOH. For these reasons, iso-propanol has been proposed as alternative alcohol (Elliott and Stowe 1971), but may not be as effective as an extraction solvent.

Acetone can react with tryptamines during extraction to give Schiff bases called β-carbolines (Allen and Holmstedt 1980). Finally, chloroform is sometimes considered more as a reagent than a solvent as upon storage it may rapidly form significant amounts of $COCl_2$, which can react quantitatively with IAA. 1-butanol was also used to extract chlorinated IAA (Engvild et al. 1981).

5 Purification

The purification of the auxin(s) is a crucial step in GC-MS anylses. The goal is to obtain an extract pure enough for the signals of the MS detector for these target compounds to be free from signals originating from other substances. Such purification can be obtained in numerous ways, the choice of which depends on the nature of the tissue and the level of the endogenous auxins. It is practically impossible to draw specific guidelines for insuring successful purification, but the successive approximation approach described by Reeve and Crozier (1980) can be used for quantitative determination for validation of the method. A trial and test procedure is recommended when working on new material, as published procedures might not be suitable for a particular plant or tissue, as they might contain different "impurities".

After extraction of the plant tissue with a water – miscible solvent and removal of the solids by either filtration and/or centrifugation, a water phase is obtained after evaporating the organic solvent under reduced pressure. Some precipitation of undefined compounds usually occurs at this stage. The remaining water solution cannot, however, be injected into the GC-MS. It is therefore necessary to transfer the target compounds into an organic phase by partition at a suitable pH. As a general rule for the selection of the partition system, it is advisable to determine the behavior of the compounds expected to be isolated beforehand whenever possible: low extraction yields may partly explain why analysts have sometimes reported the absence of an auxin in an amount of tissue sufficient to yield detectable quantities of that auxin. Another method may involve the use of C_{18} preparative columns, when the IAA is retained on the column by using a very dilute alcohol solution (e.g., converting the 80% methanol to <10% methanol by adding H_2O), the IAA then being eluted with a pure alcohol. However, more polar auxins or IAA conjugates may not be retained quantitatively even in dilute solutions and the preparative C_{18} columns must be checked with appropriate standard compounds (R. P. Pharis, pers. comm.).

A classical purification sheme consists of partitioning buffer at pH 7–8 with peroxide-free diethyl ether or dichloromethane. Then the aqueous phase is acidified to pH 3 with 1 N HCl, and the indole acids like IAA are extracted with ethyl acetate. It should always be kept in mind that the efficiency of such partitioning is questionable, as the partitioning coefficient of the various auxins is very different. For these reasons, alternative methods have been used: gel permeation columns, ion exchange procedures, adsorption and partition chromatographies, to mention just a few. A recent review giving many practical details of these procedures can be obtained elsewhere (Sandberg et al. 1986).

6 Columns for GC

In the past, most of the GC separations of auxin extracts have been done on conventional, packed glass columns with inner diameters of 2–6 mm and with the stationary phase coated onto relatively inert supports. However, in recent years, there has been increasing interest in capillary gas-liquid chromatography, as auxin analysts have become more and more aware of the dramatic increase in separation power of the capillary columns over conventional, packed columns. As an illustration, Table 2 shows a comparison of the characteristics of the capillary and packed columns. Capillary columns most often used are wall-coated open tubular columns (WCOT) with mid-polarity bonded phase. Besides superior separation power, the capillary GC column – with the stationary phase chemically bonded to the internal surface of the tubing – is also claimed to offer better sensitivity, lower noise level and increased inertness, resulting in minimal sample adsorption and decomposition on the columns (see below). These improvements are highly beneficial for auxin analysis. In addition, instrumentation for capillary GC is commercially available, and recent developments and innovations in column technology – like the very stable chemically bounded or, also called cross-linked stationary phase, fused silica WCOT columns – make this technique quite suit-

Table 2. Characteristics of capillary and packed GC columns

	Capillary	Packed
Length (m)	5–100	1–5
Internal diameter (mm)	0.2–0.7	2–4
Flow (ml min)[1]	0.5–15	10–60
Pressure drop (psi)	3–40	10–40
Total effective plates	150,000 (50 m)	5,000 (2 m)
Effective plates/meter	3,000 (i.d. 0.25 mm)	2,500 (i.d. 2 mm)
Capacity	50 µg/peak	10 µg/peack
Liquid film thickness (µm)	0.05–3	1–10

able for routine work. The following paragraphs relate to practical conditions of capillary GC usually encountered in auxin analysis.

In the early days of capillary GC, columns were usually drawn from glass. Besides being fragile, it was soon discovered that these columns were not sufficiently inert toward the various substances to be chromatographied. This appeared to be due to the presence of alkali metal oxides and other metal ions (Al and Fe) in most glass preparations and to surface silanol groups. Several approaches have been tried to overcome these problems and the use of fused silica as column material has evolved as the best alternative. Fused silica columns are made from amorphous silica, which is virtually free of metal oxides (less than 1 ppm) and gives rise to columns that are extremely flexible (due to the polyimide external coating) and durable. Sometimes these columns are also called fused quartz, but it should be realized that the latter is produced from naturally occurring silica deposits that contain relatively high amounts of metal ions.

Fused silica columns still have active silanol groups at their surfaces, so that deactivation remains necessary. Various procedures are used, depending on the manufacturer, to influence the adhesion of the stationary phase to the column material. The upper temperature limit of the column is determined by the agent that has the lower limit, either the liquid phase or the deactivation agent (e.g., polyethylene glycol or polysiloxane gums). Moreover, the presence of this deactivation agent may influence the retention characteristics of the column. Various methods have been described for the coating of capillary columns. The reader is referred to Freeman (1979) and Jennings (1980) for further details and references.

The preparation of capillary columns requires experience and special equipment. Capillary columns are therefore usually obtained from commercial manufacturers. This cost (at least $ 200) may be a handicap to those with limited budgets. However, the working life of one of these columns is quite long. Finally, it should be noted that columns of the same type, but from different manufacturers, may show different properties and performances. It is generally assumed that this is due to the use of different deactivation procedures, details of which are usually not available. When the chromatographic performances of the bonded phase capillary column gradate, it is possible to restore them by washing out the unvolatiles residues remaining in the tubing. This is obtained in discon-

Table 3. Gas chromatographic retention time (expressed in min) of authentic indoles and indole derivatives on a 15 m × 0.25 mm (i.d.) cross-linked DB-1 fused silica column (J&W Scientific) with a film thickness of 0.25 μm. Injector temp.: 250 °C, Oven temp. program: 180 °C to 220 °C in 20 min. Detector temp.: 300 °C. (E. Jensen, pers. comm.) A: underivatized standards; B: methylated standards; C: silylated standards; D: methylated and silylated standards

Compound	A	B	C	D
Indole-3-methanol	2.9		4.7	
Indole-3-acetonitrile	4.2		5.1	
Indole-3-aldehyde	4.4		5.2	
Indole-2-carboxylic acid		2.9	5.7	4.2
Indole-3-ethanol	3.9		5.9	
Indole-3-acetic acid		4.5	6.8	5.6
Indole-3-carboxylic acid	5.9	4.8	7.9	6.1
Indole-3-propionic acid		5.6	9.1	7.0
Indole-3-acetamide	9.1		9.8	
Indole-3-butyric acid		7.3	11.6	8.8
5-hydroxyindole-3-acetic acid		9.4	12.8	11.1
L- and D-tryptophan		8.1	13.0	10.8
N-acetyl-L-tryptophan			18.6	
Indole-3-pyruvic acid		12.2		16.5

necting the column from the oven and sucking 2–5 ml of methanol through the cold column at room temperature by a suitable vacuum pump.

To illustrate the usefulness of capillary columns in auxin separation, a few data on the Rt of several indoles are given in Table 3.

7 Injection Techniques

Capillary columns have a low flow rate, a small gas hold-up volume and a low sample capacity. The low sample capacity applies not only to the amount of solutes that a column with such thin stationary phase can accommodate (no peak distortion), but also to the volume of the sample injected.

In the direct injection mode, the sample is first vaporized in the inlet and then passed onto the column. Even 1 μl of liquid results in a relatively large amount of vapor. If the direct mode is used in capillary GC, only very small volumes can be injected at high flow rates. Evidently, alternative modes have to be used: initially, stream splitters (also called split-injection) were introduced. Then splitless inlet systems were developed and, more recently, on-column injection systems have been described. Purcell (1982) has surveyed the potential and limitations of these techniques for their application to quantitative analyses. For qualitative analyses, such as auxin identification in a biological sample, split injection does not lend itself too well to trace analysis, because solutes present at relatively low concentrations may easily go undetected at a split ratio of 1:50 to 1:500 which is used for capillary GC.

With on-column injection, the sample is directly injected on the capillary column at a column temperature below the boiling point of the solvent. This results in a "cold trapping" of the sample at the top of the column. Then the temperature of the column is raised rapidly to the level suitable for chromatography. Although there are some advantages, such as minimal sample decomposition, no septum, and quantitative recovery of all solutes, there are limitations as well. On-column injection cannot be automated, only a limited volume can be injected, and non-volatile material may bring deterioration of the top of the column. The latter is particularly true with plant extracts.

Requirements of Splitless Injection. In splitless injection – sometimes referred to as the Grob technique – the sample is injected through the septum in the inlet liner with a relatively large volume (1–3 µl). Due to the high temperature of the injector, the sample evaporates, resulting in a large volume of gas. Then the solute must be reconcentrated the top of the column. This can be achieved by the so-called solvent effect, or by cold trapping.

In cold trapping, the column is held at about 150 °C below the boiling point(s) of the solute(s), so that they condense in the first part of the column.

For an optimum solvent effect, the column temperature during the injection has to be held at about 20 °–40 °C below the boiling point of the solvent: the solvent liquifies in the column and forms a film at the head of the column. When the temperature of the oven is then rapidly increased, evaporation of this liquid film starts at the rear (injector end) and the wet zone will be "rolled up" from the rear of the column to the front (Grob effect). Solutes dissolved in the liquid film tend to stay therein and are also carried toward the front, so that concentration into a small zone takes place.

After the sample has entered the column, the liner of the injector is back-flushed to avoid tailing of the solvent. For both cold trapping and the solvent effect, the column temperature has to be low at injection. Thus a temperature program has to be carried out to elute the solutes.

Different factors influence the reconcentration effect and thus the amount and distribution of the solutes on the column:

1. The solvent has to fulfill the following conditions: it must sufficiently dissolve the solutes; it must interact with the liquid phase without damaging it; it must be of high purity; and its boiling point must not be too low. Alcohols will dissolve many auxins, but they are too polar and carry the risk of column damage. Most ketones are lacking in purity. We have found ethyl acetate to be sufficiently pure and have sufficient polarity to dissolve most underivatized auxins. For derivatives of indoles, nonpolar and dry solvents like hexane, cyclohexane, or toluene are recommended.

2. A minimum volume of injection is required, usually between 0.5 and 2 µl.

3. The carrier gas flow rate and the splitless period have to be experimentally determined. Since in capillary GC the carrier flow rate is low (0.5–15 ml min^{-1}) and the volume of the liner large (0.1–1 ml), it will take some time before all sample has been transferred onto the column. Due to diffusion and turbulence, a 100% transfer cannot be achieved. In experiments with 0.32 mm column, a car-

rier flow of about 2 ml min^{-1} and a liner volume of 0.2 ml, a splitless period of 30 s was found suitable, as it corresponds to a transfer greater than 95%.

4. The injector temperature may be lower than with split injection, since the sample can remain in the inlet for a relatively longer period. Solute stability is also to be considered. Therefore, we prefer an injection port at as low of a temperature as possible. For screening purposes and identification, we set the injection port at 250 °–270 °C. When using ethyl acetate as solvent, the initial column temperature can be set at 60 °C, and the following temperature program may be used: 2 min at 60 °C, 25 ° min^{-1} to 190 °C, 10 ° min^{-1} to 260 °C, and hold for 5 min at 260 °C. The interface between the GC and the MS can be kept at 260 °C. An excellent all-purpose column can be, for example, a 5% phenylmethyl silicone (like SE-54) thick-coated 0.31 mm i.d. fused silica 25 m long WCOT capillary column from Hewlett-Packard. For simple methyl esters a mid-polarity column such as CP-Sil 19 CB from Chrompack will decrease problems with tailing and column memory seen with less polar phases.

8 Derivatisation

GC is a simple separation technique which is based on partition principles between the stationary liquid phase and the mobile gas phase. Solutes should then pass alternatively to the liquid and gas phase. As the acidic and phenolic auxins, as well as auxin conjugates are compounds usually of high polarity (due to hydrogen bonds), their volatility is limited and high temperatures can cause decomposition. Derivatization of the polar moieties is therefore needed to increase their vapor pressure. Basic and neutral indoles can usually be injected into a GC column without modification. Separation of both the native auxins and/or their derivatives can be modulated by temperature programming during the GC run.

Derivatization can serve a number of different purposes. Most commonly, it is used to improve the chromatographic characteristic of the analyte. For example, even though IAA-Me can be chromatographed, the free amine moiety causes peak tailing and adsorptive losses on most nonpolar chromatographic columns. These effects become most evident when subnanogram quantities of the methyl esters of auxins are analyzed. Derivatization is also used to enhance the detector response. The following paragraphs will show that the choice of derivative and method of ionization can have a great impact on the intensity of the ion current response, and hence the sensitivity of the assay. Indeed, the selection of the derivative must go hand in hand with the selection of the method of ionization. For example, a derivative that has suitable EI characteristics may be poorly ionized by negative chemical ionization (NCI), or give an NCI mass spectrum that is useless for quantitative measurements.

A large number of derivatives of IAA and related indoles have been tested on various GC systems. The choice of the most suitable derivative is determined by several factors:

1. the ease of preparation,
2. the yield of the reaction when using biological sample (and not only pure standards),
3. the stability of the derivative upon storage and further purification,
4. the selectivity of the detector response of the derivative, and
5. the extent of the fragmentation of the derivative in the MS source.

Methyl esters of IAA and other indole acids have often been used for qualitative (Magnus et al. 1980; Segal and Wightman 1982) and quantitative analyses (Pengelly et al. 1982; Rivier and Pilet 1983). Methylation is usually accomplished using diazomethane. The reaction (Fig. 2) is fast (a few minutes) and the reagent is easily prepared in a diethyl ether solution in small quantities without distillation (Cohen 1984).

Ethyl esters have been obtained from diazoethane. The reaction is slower but as simple as with diazomethane. It is thus possible to differentiate between the endogenous IAA-Me and IAA, and also any IAA-Me which could originate from transesterification from conjugates or indiscriminate use of MeOH as solvent during extraction and/or purification (see above; Allen et al. 1982a). Use of ^{13}C-diazomethane for such a purpose, however, is preferred, and ^{13}C-diazomethane precursors are now easily available. When using a diazoalkane reagent, the phenolic moieties can also react slowly (as for 5-hydroxy-IAA for example).

The esterification of IAA greatly increases the stability of the molecule, and it also considerably reduces its binding to the glass surface. However, IAA-Me is subject to some "memory effects" when injected (Epstein et al. 1980). This effect is characterized by the retention in the GC system – usually in the injector, the septum, or even the syringe – of a very small part of the sample. It is demonstrated by injecting immediately after the sample containing IAA-Me a large quantity of solvent and by observing the detector response at the retention time of IAA-Me: if a signal occurs, it is an indication that the system had kept part of the IAA-Me previously injected. The accuracy of the quantitative determinations will, of course, be affected and detection of auxin in sample suspect. Injections of large amount of solvents between samples has been proposed as a remedy (Epstein et al. 1980). Not only are these procedures times-consuming, but also full control of the accuracy of the determinations cannot be guaranteed. Also, if IAA is not subject to sublimation, IAA-Me is under reduced pressure (care is recommended when reducing these methylated solutions to dryness). It would be better to select other columns or derivatives for the auxins, which show no memory effect and are less volatile. In my experience, a memory effect with IAA-Me can be avoided by the further derivatization of the amine moieties or the use of mid-polarity-bonded phase capillary columns. An additional and general precaution consists in avoiding the injection of unnecessarily large quantities of standards or samples.

Fig. 2. Scheme of reaction for IAA with diazomethane

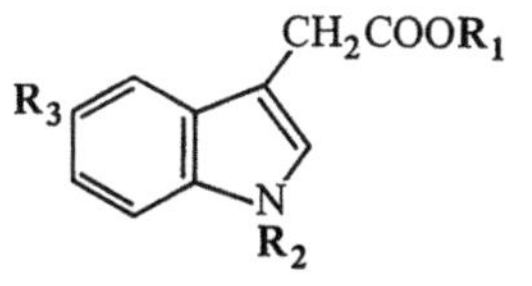

$$CH_2COOR_1$$ on indole with substituents R_2 (on N) and R_3

Name	R_1	R_2	R_3
Indol–3yl acetic acid (IAA)	H	H	H
Indol–3yl acetic acid methylester (IAA–Me)	CH_3	H	H
Indol–3yl acetic acid bis–Trimethyl silyl derivative (IAA–TMS)	$(CH_3)_3Si$	$(CH_3)_3Si$	H
5–Fluoro–indol–3yl acetic acid methylester (5–F–IAA–Me)	CH_3	H	F
Indol–3yl acetic acid methylester trifluoroacetyl–1 (IAA–Me–TFA)	CH_3	F_3CCO	H
Indol–3yl acetic acid pentafluoro-benzylester (IAA–PFB)	$F_5C_6CH_2$	H	H
Indol–3yl acetic acid methylester heptafluorobutanoyl–1 (IAA–Me–HFB)	CH_3	F_7C_3CO	H
Indol–3yl acetic acid pentafluoro-propionylester pentafluoropro-pionyl–1 (IAA–PFP)	$F_5C_2CH_2$	F_5C_2CO	H

Fig. 3. Developed formula and structure of IAA and its derivatives used for GC-MS determinations

The molecular weight derivative of IAA can be greatly augmented by the use of silylated or fluorinated reagents. A selection of possible derivatives are illustrated in Fig. 3 and are discussed below.

Trimethylsilyl (TMS) derivatives have been useful for GC-MS and structure determinations of IAA and IAA-conjugates as silylation reagents are able to react both with acidic, basic (primary and secondary amines) and phenolic moieties (Bandurski 1979). TMS derivatization under mild conditions [e.g., with hexamethyldisilizane (HMDS) and trimethylchlorosilane (TMCS) in pyridine (1:1:1) at 60 °C for 45 min produces mainly the TMS ester. Depending on temperature and time of reaction, the bis-TMS derivative can also be obtained. When using a more powerful reagent [e.g., bis-trimethylsilylacetamide (BSA)] in pyridine (2:1) at 60 °C for 30 min (McDougall and Hillman 1980), or better with bis-trimethylsilyl-trifluoroacetamide (BSTFA) containing 1% of TMCS in pyridine (1:5) at 75 °C for 45 min, the bis-TMS-IAA derivative is obtained quantitatively. The determination of yields of these derivatives has often be made on a large amount of standard solutions (mg levels) and is thus a poor guide for trace quantities often encountered in biological samples. TMS derivatives are sensitive to water and must be stored in an excess of reagent until injected into the GC-MS. Moreover, the bis-TMS-IAA derivative has shown memory effects on packed columns (McDougall and Hillman 1980). This can be circumvented by injecting large amounts of the TMS reagent between samples.

TMS derivative of IAA-Me has been also proposed in the general screening for plant growth substances (Martin et al. 1982; Savidge and Wareing 1983): its stability is greatly improved over bis-TMS-IAA derivative, and its chromatographic behavior is excellent.

Perfluorinated IAA derivatives are usually very stable. They are extremely volatile even though their molecular weights are much larger than the parent compounds. Further purification can be performed on them, either by partition (Allen et al. 1982a) and/or by HPLC (Epstein and Cohen 1981). By selecting the suitable reagent, it is possible to add from 3 to 10 fluorine atoms to IAA (Rivier and Saugy 1986):

1. trifluoroacetylation is quantitatively obtained by heating trifluoroacetylanhydride (TFAA) and IAA-Me at 60 ° for 3 h in a micro-vial (Seeley and Powell 1974),
2. pentafluorobenzyl esterification was proposed by Epstein and Cohen (1981),
3. acylation with heptafluorobutyrylimidazole (HFBl) has been critically evaluated (Allen et al. 1982a), and,
4. pentafluoropropionylation of the carboxyl and the amine moieties was made directly by using a mixture of pentafluoropropionyl anhydride (PFPA) with the corresponding alcohol (PFPOH) in a 4 to 1 ratio (Allen and Baker 1980).

When selecting the most suitable derivative of auxins of GC and GC-MS analyses, it should be remembered that the best selectivity is obtained only when the reactions are linked specifically to the indole nucleus. For example, silylation is not selective, as the vast majority of polar components of the sample will react also. In contrast, after methyl esterification (not selective), a further reaction of alkylation on the nitrogen of the indole ring is highly selective. It has been possible to detect and quantify IAA in crude extracts by obtaining the methyl ester 1-heptabutyryl derivative of IAA (Fig. 4; Rivier and Pilet 1974; Hofinger 1980).

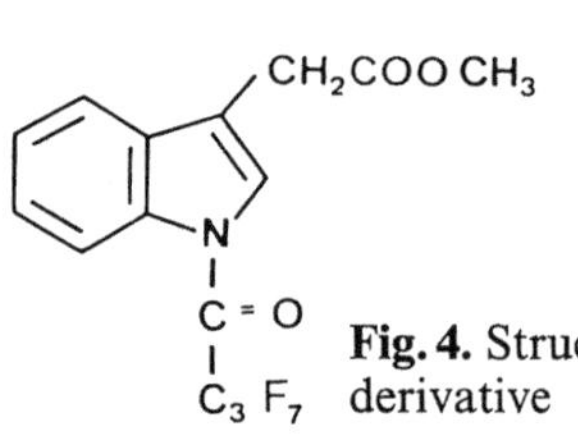

Fig. 4. Structure of indole-3-acetic acid methyl ester 1-heptafluorobutyryl derivative

9 Interface Between GC and MS

The pressure difference between the outlet of the GC (1 atm) and the mass spectrometer source (10^{-6} Torr range) requires a pressure reduction system, which transfers the effluents from the GC to the ionization chamber. Three basic systems are used in GC-MS: direct connection, effluent split and molecular separation (McFadden 1973).

1. Modern MS instruments have efficient differential pumping systems which can cope with the flow of at least 2 cm^3 min^{-1} and thus allow direct connection of capillary columns without losses. The end of the column is progressively exposed to reduced pressure so that the chromatographic properties of a coupled capillary column are somewhat different from those of an independent one.
2. Reduction of the gas flow can be achieved by a split of the GC effluent. An open split system has the advantage that the complete separation system is at atmospheric pressure. This maintains the integrity of the chromatography, but results in the loss of over 90% of the sample. Such a loss makes the device not suitable for trace auxin analyses.
3. Several types of device built for the interfacing of a packed column to a MS have been devised. The most commonly available from MS manufacturers is the Ryhage jet separator. In this device, helium is pumped away from heavier molecules as they pass at high speed between openings of two critically aligned jets. The much greater momentum of the sample molecules carries them across the gap while helium atoms are pumped away. The efficiency of the jet separator is satisfactory for auxin analyses with a sample yield of up to approximately 80%.

Since the establishment of GC-MS as a powerful analytical tool, much effort has been applied to the direct linking of liquid chromatographs to MS (McFadden 1973). The combination appeals to the plant chemist seeking a separation and inlet system for underivatized, polar, and thermally labile compounds. However, the problem is considerably greater than for GC-MS, as liquid flow of 1 ml min^{-1}, commonly used for HPLC separation, would yield up to 1000 cm^3 min^{-1} of vapor to be disposed of. To date, three devices have been developed: the moving belt, the direct liquid and the thermospray interfaces. None of them has yet been evaluated for auxin analyses.

10 Mass Spectrometer

Instruments may be grouped according to the method of mass analysis that they use. Three designs, operating on a wide range of principles, have been used for plant hormone analysis. They are both magnetic-sector and quadrupole low-resolution instruments, and high-resolution double-focusing instruments.

Low-resolution spectrometers yield spectra calibrated to the nearest whole mass. The resolution of an instrument is defined somewhat differently for different types of instrument, but basically measures the ability to resolve ions differing by unit mass: an instrument operating at resolution 1000 would resolve ions of masses up to 1000 a.m.u. to a predefined standard.

Magnetic Sector Instruments. A beam of ions is accelerated from an ion source through several kV and thence through a magnetic field to a detector. A beam of ions of mass/charge ratio (m/z) is deflected on passage through the magnetic field according to the equation:

$$m/z = B^2 r^2 / 2V,$$

where B is the strength of the magnetic field, r the radius of the circular path into which the ions are deflected and V the voltage used to accelerate the ions out of the source. Ions are usually detected by an electron-multiplier tube, placed behind a narrow slit through which the ions must pass. A spectrum is produced by scanning the magnetic field, so that ions of successive m/z value achieve the correct path through the collector slit, the accelerating voltage being kept constant. Alternatively, the magnetic field can be held constant and the accelerating voltage scanned, but this method is not commonly used, as sensitivity decreases with decreasing accelerating voltages, and the important high-mass end of the spectrum would be recorded at considerably reduced sensitivity. The resulting spectra are of the form of intensity plotted against m/z value (and not mass).

Quadrupole Mass Filters. This is a mass analyzer (Dawson 1976) with filter in the form of four parallel rods whose cross-section is hyperbolic – opposite rods being electrically connected and a combined d.c. and r.f. voltage applied between the two pairs. Ions passing through the filter follow a complex path along the axis of the rods. For a given combined voltage, only ions in a small mass window will have a stable path between the rods. A mass spectrum is thus obtained by sweeping the voltages at a constant d.c.–r.f. ratio.

Recent quadrupoles have equivalent performances in terms of resolution and sensitivity to magnetic intruments. They are more tolerant of high source pressure, because high accelerating voltages are not needed. Interfacing a computer is simpler because mass scale is linear. Scanning can also be faster, as hysteresis effects in the magnet preclude fast cycling times for sector instruments. No metastable ions, important in the IAA conjugates identifications, however, can be detected by these intruments.

Perhaps the greatest advantage of quadrupoles lies in the ability to switch quickly and easily from monitoring one particular chosen ion to another in selected-ion-monitoring (SIM) – the technique used to provide selective and sensitive detection of a particular compound (see below). This can be achieved over the whole mass range, as only voltages have to be changed. Similar detections are usually obtained on magnetic intruments by keeping the magnetic field constant and switching between precalculated accelerating voltages to bring the different ions to detection; however, it is not recommended to scan more than 10 to 20%

away from the highest mass, as otherwise serious loss of sensitivity at high mass will be experienced.

Double Focusing Instruments. These use two analyzing fields to measure m/z ratios to high precision. An energy-focusing electrostatic analyzer is used to provide a more nearly monoenergetic beam of ions for analysis in a magnetic sector: resolution of over 100,000 can be achieved and unambiguous assignments of elementary formulas are possible. However, the higher resolution gained over a single-focusing magnetic sector instrument is made at the expense of sensitivity, because only a limited number of ions in a narrower energy band are analyzed. These instruments have less value for routine capillary GC-MS determinations, where fast scan rates are employed, although much improvement in this direction has been recently obtained.

Other types of instrument involve time-of-flight, ion trap or Fourier-transform mass spectrometers. All of these machines could be proven to be of utility for auxin analyses, although they have not yet been tested for such a purpose.

11 Data Systems

A computerized data system (Chapman 1978) is now considered an essential part of GC-MS instrumentation. A basic data system provides for the collection, storage, and calibration of spectra. Some computers are programmed in a way to control fully any parameters of the GC-MS system allowing fully automated data acquisition and reduction of several samples. Tabulation and graphical output of individual spectra, substraction of background spectra, plotting of total-ion currents and/or individual on intensities through a GC run are minimum requirements. Provision for the comparison of unknown spectra with a library of reference spectra is usually held in the computer memory. Newer systems utilize powerful microcomputers and offer a considerable price-saving over the minicomputers once used on many machines.

12 Ionization

Until 1980, the only widely used method of ionizing auxins as well as other plant hormones in a mass spectrometer was by electron impact (EI). Now other ionization methods have been evaluated (Rivier and Saugy 1986) and will be developed further. However, EI is still the mode most commonly used in obtaining and interpreting the spectrum of an unknown compound.

In an EI source, a beam of electron is drawn from a heated filament by a voltage, usually set at 70 V. Such an electron beams has an energy of 70 eV, which is in excess of the ionization potentials of organic molecules, these lying typically between 10 and 20 eV. Collision between sample molecules and the well-defined electron beam result in ionization of the sample in a well-defined region of the source. A small voltage is produced on the "repeller plates" to expel the produced ions from the source. Radical anions, the molecular ion of the compound, are formed, but the excess of beam energy over ionization potential will appear in the molecule as internal energy, which may initiate fragmentation of the molecular ion to produce positively charged and neutral fragments. Further fragmentation of the anions also occurs spontaneously in the source, prior to separation in the analyzer.

Chemical ionization (CI) (Munson 1977; Harrison 1983) uses a reagent gas, e.g., methane, iso-butane, ammonia. The gas is introduced into the source at a pressure of about 1 torr (about 133 Pa). The reagent gas is ionized by the electron beam to produce positive ions as in the EI source. But when a sample is introduced into the source, its molecules are ionized by reaction with the reagent gas ions present in very large quantities. This ionization occurs by transfer of either a positive charge or a charged fragment from the reagent gas ions to the unknown molecule. Because less energy is transferred to the sample molecule during ionization, less fragmentation occurs, and the resulting CI spectrum is much simpler than the EI spectrum.

Field desorption (FD) is used to volatilize substances with very low vapor pressure (molecules with very high molecular weight and high polarity). Fast atom bombardment (FAB) is also a soft ionization technique using Argon atoms as reagent gas. Both are especially suitable for obtaining useful information from high molecular compounds with very low volatility. Most of the time it gives a strong molecular ion with little fragmentation. They cannot be used with chromatographic separation, as the total sample has to be placed on a special holder directly into the ion source. The sample to be analyzed should be essentially in a pure form for obtaining good quality mass spectra. Unfortunately, this rarely applies to plant extracts.

Until recently, analytical mass spectrometry has concentrated on the study of positive-ion spectra: under EI conditions, the yield of positive ions is several orders of magnitude greater than the yield of negative ions. Efficient negative ionization requires the production of thermal electrons, which are available under certain CI condition. Sample molecules with positive-electron affinities will under such circumstances produce negative ions (it is rarely clear whether ionization involves electron or anions species produced from the reagent gas – often both are involved). Spectrometers capable of CI can be cheaply and easily modified for anion studies. Magnetic sector intruments require the reversing of magnetic field and accelerating voltage, while quadrupole instruments require the reversal of voltages only. The technique may offer several advantages over cation mass-spectrometry: sensitivity, selectivity, and fragmentation, the details of which will be briefly discussed below.

13 GC-MS Strategy for Auxin Analysis

Mass spectrometry, linked directly to a gas chromatograph, has been used since 1970 as the ultimate tool for the identification of most the endogenous auxins isolated so far. Used in the scanning mode, GC-MS served to establish the proof of identity of a previously detected chromatographic peak, as for example obtained by HPLC (see Reeve and Crozier 1980). The sensitivity of such a mode varies from instrument to instrument and from compound to compound: 40 ng of IAA-Me, the base peak of which at m/z 130 in the EI mass spectrum corresponds to 12% of the total ion current (TIC), gives a readable spectrum just above the background. On the other hand, using the same instrument and the same scan settings, 5 ng only of IAA-Me-HFB derivative are sufficient for recording a spectrum of similar quality. This is due to the 44% of the TIC from the base peak at m/z 326.

A minimum of basic chemical knowledge is necessary to evaluate the quality of MS data.

First, the signal corresponding to the molecular ion of the detected compound should be present in the recorded spectrum.

Second, the selection of a particular derivative of the auxins implies to choice of specific ions. For example, all TMS derivatives give abundant nonspecific ions at m/z 73 and 75. They are of poor informative value. Better is the ion at m/z 130, which is characteristic of the stable quinolinium ion (Fig. 5) of the majority of the indoles. This ion is often the base peak of the spectrum (Jamieson and Hutzinger 1970).

Third, no major ion should be visible in the EI spectrum at higher values than the molecular ion and the naturally occurring isotopes. Of course this is not the case with CI spectra.

Fourth, logical losses should be observed from the molecular ion – or what is believed to be the molecular ion. For example, a loss of 9 a.m.u. is obviously im-

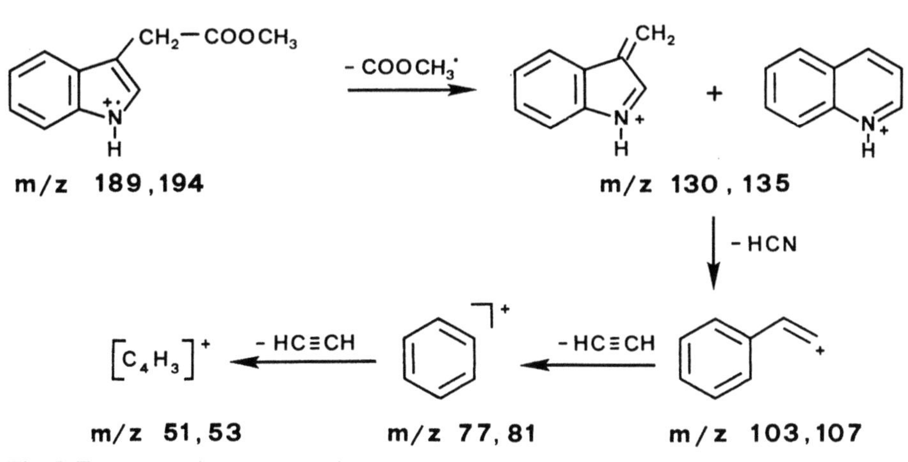

m/z 189,194   m/z 130, 135

m/z 51,53   m/z 77,81   m/z 103,107

Fig. 5. Fragmentation pattern of IAA methyl ester and its deuterated analogue under electron impact at 70 eV. The m/z values represent the mass to charge ratio of the ions of IAA-Me and the deuterated analogue, respectively

Table 4. Mass spectra of indoles and indole derivatives in tabulated form

Compound	MW	Base	2nd	3rd	4th	5th	6th	7th
		Peak						
N-Acetyl-5-methoxy-tryptamine	232	160	173	232	145	117	174	161
N-Acetyl-5-hydroxy-tryptamine	218	146	159	218	160	147	43	81
4-Chloro-indole-3-acetic acid methyl ester	223	164	166	223	128	225	165	101
4-Chloro-indole-3-acetic acid methyl ester HFB derivative	419	360	163	69	128	169	419	362
4-Chloro-indolyl-3-aspartate	352	164	191	128	129	352	193	355
N-(p-coumaryl)-tryptamine 2 TMS derivative	450	215	73	202	143	216	450	130
N-(p-coumaryl)-tryptamine 3 TMS derivative	522	73	215	202	219	320	522	450
N,N-Dimethyltryptamine	188	58	188	130	59	42	143	129
N,N-Diethyltryptamine	216	86	30	58	130	87	77	42
N,N-Dimethyltryptophan methyl ester	246	116	130	187	246	117	144	
Ethyl indole-3-acetate	203	130	203	131	103	102	77	204
Ethyl indole-3-carboxylate	187	143	187	115	89	144	116	188
N-Ferrulyltryptamine 2 TMS	480	215	73	143	202	130	480	249
N-Ferrulyltryptamine 3 TMS	552	73	215	202	249	350	480	522
5-Fluoro-indole-3-acetic acid	207	148	207	101	149	120	127	74
5-Hydroxy-N,N-diemthyl-tryptamine	204	58	204	146	160	161	205	
5-Hydroxyindole-3-acetic acid	191	146	191	147	130	57	145	117
Hydroxyphenylacetic acid (meta)	152	107	152	77	39	108	51	79
Hydroxyphenylacetic acid (para)	152	107	77	51	39	53	78	50
Hydroxyphenylacetic acid (ortho)	152	78	134	106	51	77	39	40
4-Hydroxyphenyl-3-propionic acid	166	107	166	77	39	108	45	65
Indole	117	117	90	89	118	116	63	59
Indole-3-acetaldehyde	159	130	159	77	64	131	103	48
Indole-3-acetamide	174	130	174	77	131	103	102	129
Indole-3-acetic acid	175	130	175	77	131	103	102	129
Indole-3-acetic acid	175	130	45	131	159	77	62	43
Indole-2-acetic acid	175	130	131	77	103	175		
Indole-1-acetic acid	175	130	175	103	77	131		
Indole-3-acetic acid-t-butyl-dimethylsilyl derivative	247	130	73	247	131	75	232	129
Indole-3-acetic acid bis-TMS	319	202	73	203	319	204	304	75
Indole-3-acetic acid bis-TMS	319	202	73	75	319	130	129	
Indole-3-acetic acid D2-TMS	321	204	73	205	321	206	322	306
Indole-1-acetic acid TMS	247	130	73	247	202	77	131	103
Indole-3-acetic acid methyl ester TMS	261	202	261	145	137	170	73	75
Indole-3-acetic acid Me-HFB derivative	385	326	385	129	69	102	169	76
Indole-3-acetic acid Me-TFA derivative	285	226	285	198	129	227		
Indole-3-acetic acid PFB derivative	355	130	181	77	103	355		
Indole-3-acetic acid PFP derivative	453	276	453	129	277	454		

Table 4 (continued)

Compound	MW	Base Peak	2nd	3rd	4th	5th	6th	7th
2-O-(Indole-3-acetyl)-D-glucose O-methyloxime 4 TMS	654	73	130	157	103	147	217	205
2-O-(Indole-3-acetyl)-D-glucose O-methyloxime 5 TMS	726	229	73	202	103	147	246	319
4-O-(Indole-3-acetyl)-D-glucose O-methyloxime 4 TMS	654	157	130	73	217	205	654	494
4-O-(Indole-3-acetyl)-D-glucose O-methyloxime 5 TMS	726	229	73	202	217	147	160	319
6-O-(Indole-3-acetyl)-D-glucose O-methyloxime 4 TMS	654	73	130	157	290	147	217	404
6-O-(Indole-3-acetyl)-D-glucose-O-methyloxime 5 TMS	726	229	73	202	362	217	160	476
1-DL-1-o-(Indole-3-acethyl)-myo-inositol 6 TMS	769	229	73	202	147	157	217	318
2-O-(Indole-3-acethyl)-myo-inositol 6 TMS	769	229	73	202	147	217	191	318
1-O-(Indole-3-acethyl)-β-D-glucopyranose 5 TMS	697	202	73	361	217	147	316	271
Indole-3-aldehyde	145	144	145	116	89	63	146	90
Indole-2-carboxylic acid	161	115	143	117	89	161	90	63
Indole-3-carboxylic acid	161	144	161	89	116	117	63	115
Indole-4-carboxylic acid	161	161	144	116	89	117	63	90
Indole-5-carboxylic acid	161	161	144	116	89	63	117	115
Indole-6-carboxylic acid	161	116	161	144	89	117	90	63
Indole-7-carboxylic acid	161	161	143	116	89	115	117	114
Indole-3-glyoxylic acid methyloxime methyl ester	246	142	246	143	115	173	247	116
Indole-3-glyoxylic acid quinoxalinol-TMS deriv.	333	216	73	144	217	318	333	246
	333	216	73	144	217	318	333	246
Indolyl-3-glyoxyl-dimethylamide	216	144	216	116	89	72	63	0
Indolyl-3-N,N-diethylglyoxamide	244	144	72	145	116	89	29	100
Indolyl-3-N,N-dipropyl-glyoxamide	272	144	43	100	86	116	145	128
Indole-3-ethanol	161	130	161	143	131	103	77	115
Indole-3-lactic acid	205	130	44	205	77	129	131	103
Indole-3-lactic acid	205	130	131	205	129	77	43	41
Indole-3-pyruvic acid	203	130	129	45	77	203	157	102
Indole-3-pyruvic acid methyl oxime methyl ester	246	130	155	215	246	156	77	103
Indole-3-pyruvic acid methyl oxime ethyl ester	260	130	155	229	260	156	77	44
Indole-3-pyruvic acid quinoxalinol-TMS deriv.	421	73	147	347	404	419	260	314
Indole-3-pyruvic acid quinoxalinol-TMS deriv.	421	73	347	202	130	230	332	257
5-Methoxy-tryptamine	190	160	161	190	145	117	30	146
5-Methoxy-N,N-dimethyl-tryptamine	218	58	218	160	59	44	219	77
2-Methyl-indole	131	130	131	77	103	132	66	51
Methyl indole-3-acetate	189	130	189	131	79	190	52	103

Table 4 (continued)

Compound	MW	Base Peak	2nd	3rd	4th	5th	6th	7th
Methyl indole-3-acetate	189	130	189	77	103	65		
Methyl indole-3-butyrate	217	130	217	143	131	186	218	144
Methyl indole-3-carboxylate	175	144	175	116	89	145	176	117
Methyl 5-methoxyindole-3-acetate	219	160	219	161	220	145	74	69
2-Methyl-5-methoxyindole-3-acetic acid methyl ester	233	174	233	175	159	131	130	234
5-Methylindole-3-acetic acid TMS derivative	333	216	73	217	333	334	318	218
Methyl phenylacetate	150	91						
N-Methyl-tryptophan methyl ester	232	130	232	173	115	103	77	102
Oxindole	133	104	105	78	52	77	134	53
OxIAA-PFB derivative	371	181	146	145	369	116		
OxIAA-Methyl ester	205	145	144	203	117	116	172	81
OxIAA-Methyl ester	205	145	117	146	128	77	51	205
Phenylacetic acid	136	91	136	92	65	39	63	45
Phenylacetone	134	43	91	134	92	65	39	63
Phenylacetonitrile	117	117	116	90	89	51	77	118
Psilocin	204	58	42	204	30	146	117	130
Psilocin-TMS	276	58	276	218	261	0	0	0
Psilocybin	284	58	204	59	146	159	205	160
Tryptamine	160	130	131	36	77	108	65	160

possible as it corresponds to no possible atom weight or combinations of atom weights.

It is not intended here to discuss fragmentation mechanisms for the various indoles studied so far. The interested reader is referred to the few publications on the topic (Budzikiewicz et al. 1964; Powers 1968). Formation of the ions can be established by the used of labeled molecules. In Fig. 5, the fragmentation of IAA-Me and the pentadeuterated analog is illustrated. Some indole derivatives have been studied by GC-MS and their spectra published. A few examples are reproduced here in the Table 4. By comparing directly the ion values and abundance of a reference mass spectrum with an unknown, it is possible to identify the unknown compound.

Selected Ion Current Monitoring (SICM) offers greater sensitivity than the full scanning mode, as more time is spent for measuring the signals from fewer ions. The sensitivity gain is about 500- to 1000-fold, depending on the compounds and conditions. Single Ion Monitoring (SIM) is a particular case of SICM where one ion only is measured during the chromatographic run. Ultimate sensitivity is thus achieved, as full instrument time is devoted for the detection of the one ion. This mode of detection is more specific than the full scan mode and could be better compared to electron capture detection by GC (ECD) and specific detection of nitrogen- or phosphorus-containing compounds (NPD). Another advantage of SIM over ECD or NPD is that such detection can be applied to any organic compound volatile enough to enter the MS source with a significant vapor pressure. The best specificity is obtained by using at least two ions per compound to be detected. This mode is then called Multiple Ion Monitoring (MIM) or Selected Ion Monitoring (SIM) or sometimes Mass Fragmentography (MF). The sensitivity of the detection is lowered by the square root of the number of ions. With these techniques there is always a compromise between sensitivity and specificity.

Such techniques are useful to establish that no other substances are emerging at the same Rt in the GC run. When two or more ions are recorded for the auxin derivatives, the ratio of the intensities of these ions at the retention time (Rt) corresponding to the standard should be within 2–4% deviation of that obtained with the standard run under the same conditions. There is an extremely low probability that a contaminant behaving during the GC separation in exactly the same way as the auxin will have the same MS ions at the same intensities.

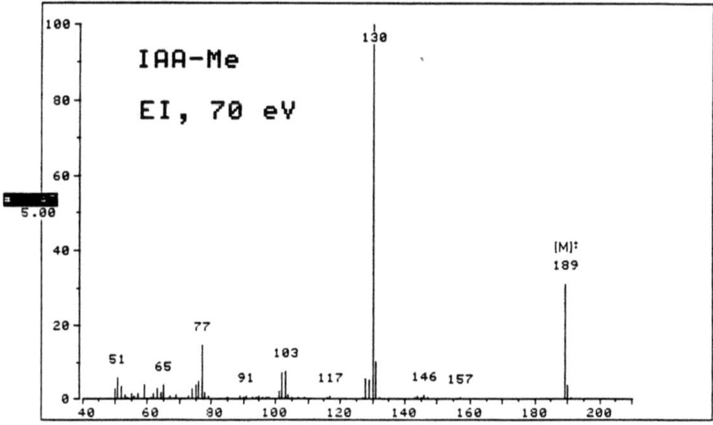

Fig. 6. Electron impact mass spectrum of indole-3-acetic acid methyl ester at 70 eV

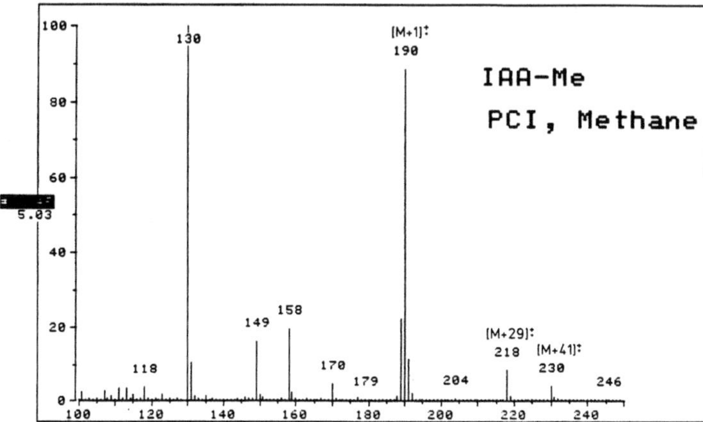

Fig. 7. Positive chemical ionization mass spectrum of indole-3-acetic acid methyl ester using methane as reagent gas at 0.5 Torr ion source pressure

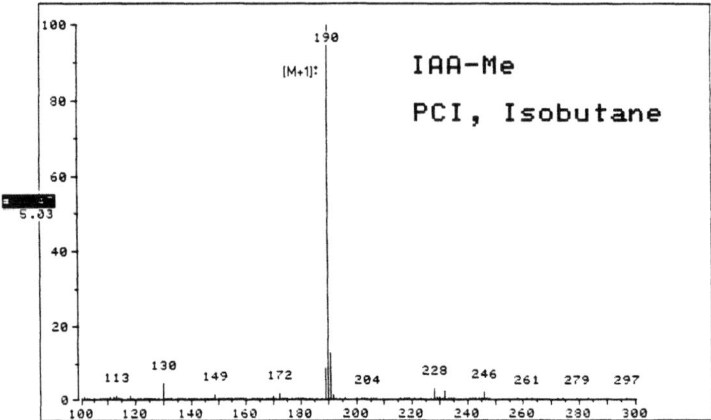

Fig. 8. Positive chemical ionization mass spectrum of indole-3-acetic acid methyl ester using isobutane as reagent gas at 0.5 Torr ion source pressure

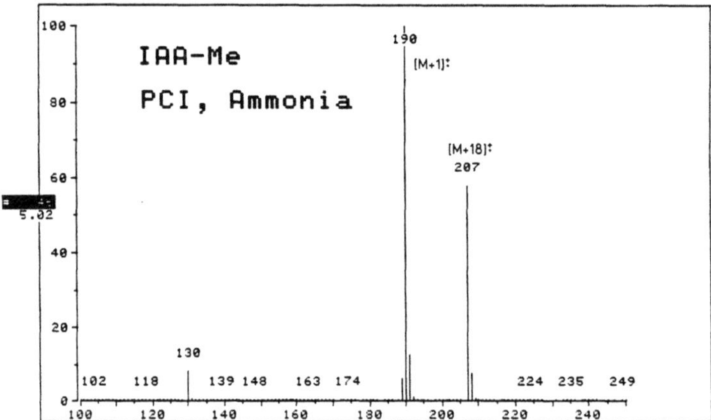

Fig. 9. Positive chemical ionization mass spectrum of indole-3-acetic acid methyl ester using ammonia as reagent gas at 0.5 Torr ion source pressure

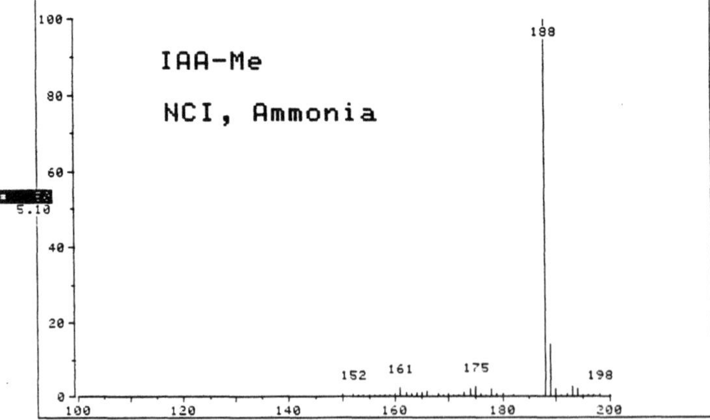

Fig. 10. Negative chemical ionization mass spectrum of indole-3-acetic acid methyl ester using ammonia as reagent gas at 0.5 Torr ion source pressure

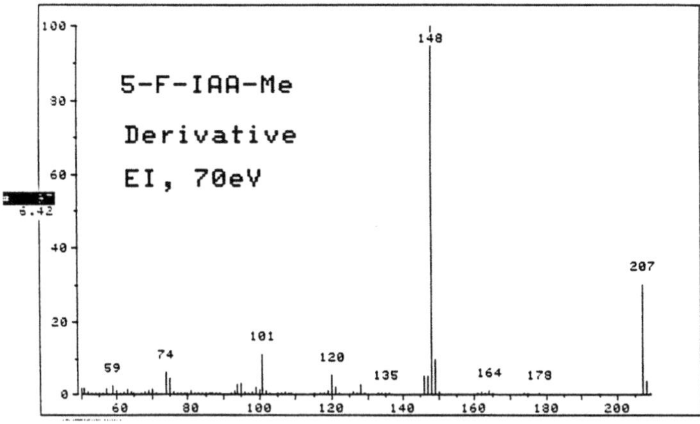

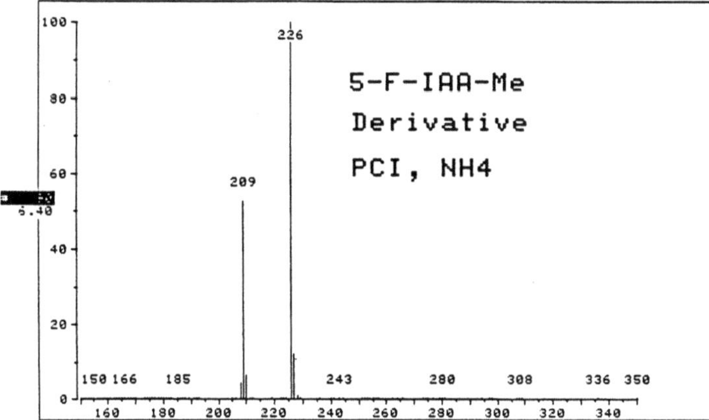

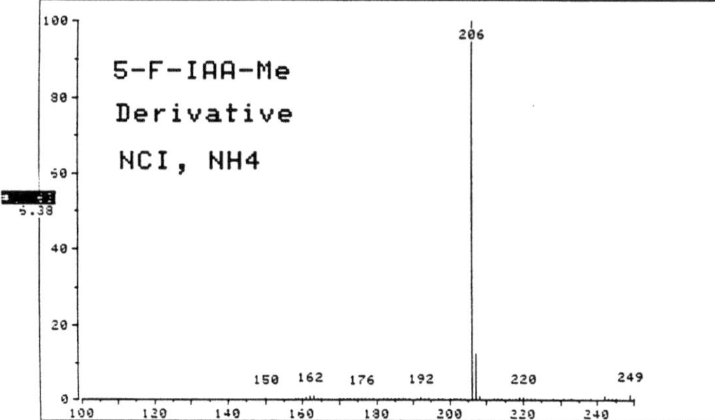

Fig. 11. Electron impact, Ammonia positive chemical ionization and negative ionization mass spectra of 5-fluoro-indole-3-acetic acid methyl ester

                                                        L. Rivier

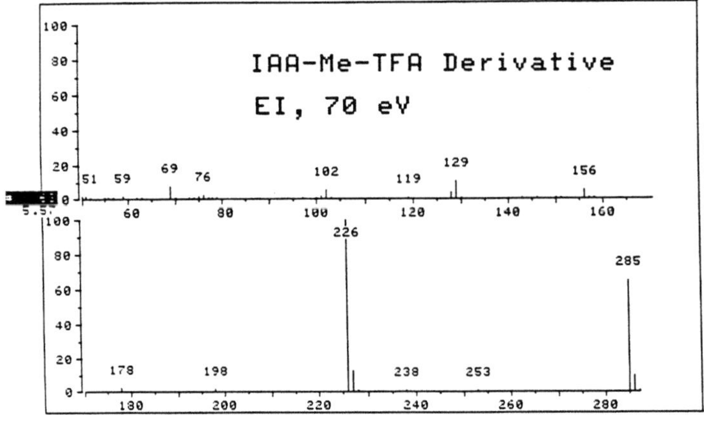

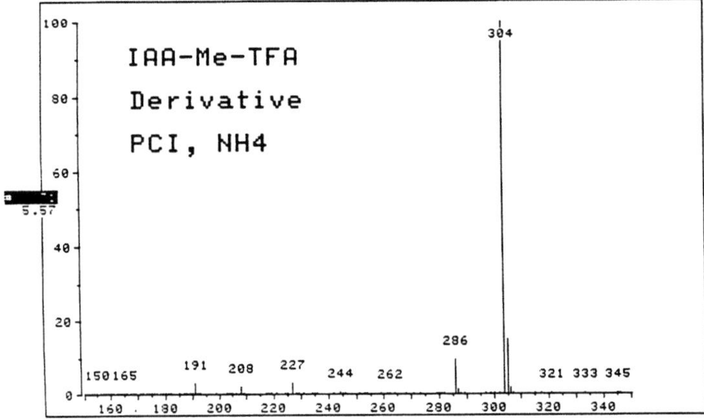

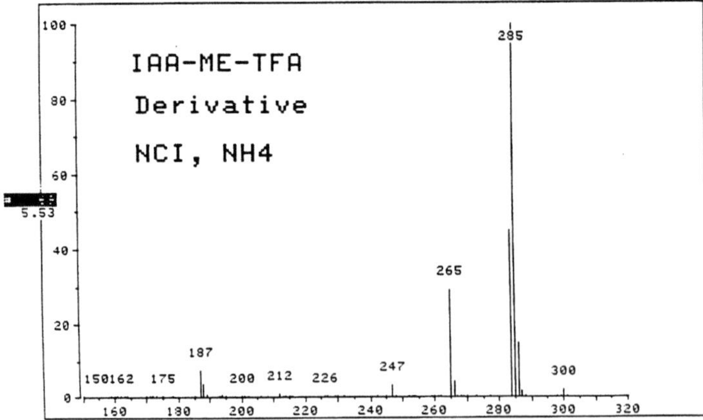

Fig. 12. Electron impact, Ammonia positive chemical ionization and negative ionization mass spectra of indole-3-acetic acid methyl ester trifluoroacetyl derivative

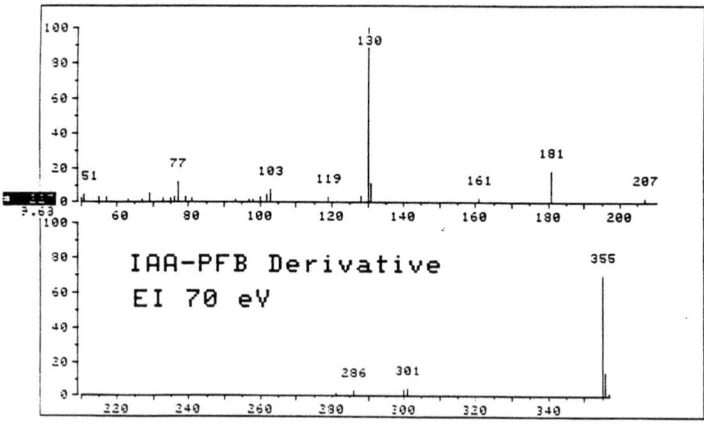

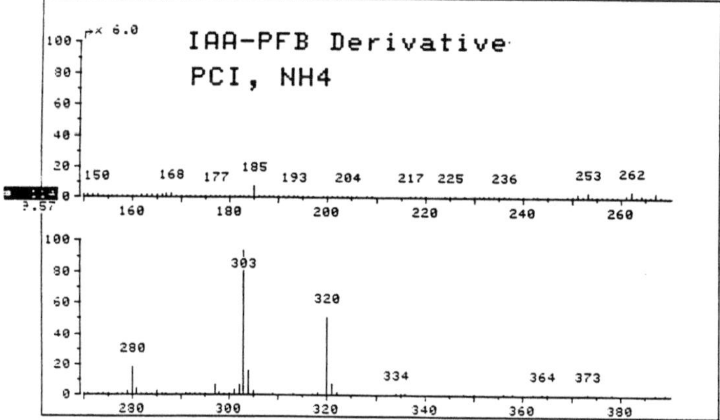

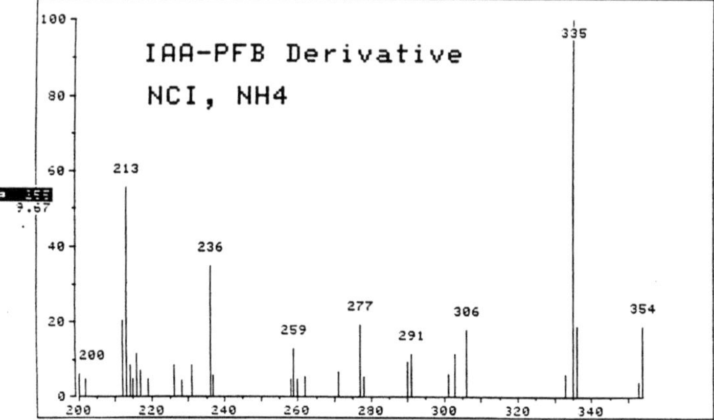

Fig. 13. Electron impact, Ammonia positive chemical ionization and negative ionization mass spectra of indole-3-acetic acid pentafluorobenzyl ester

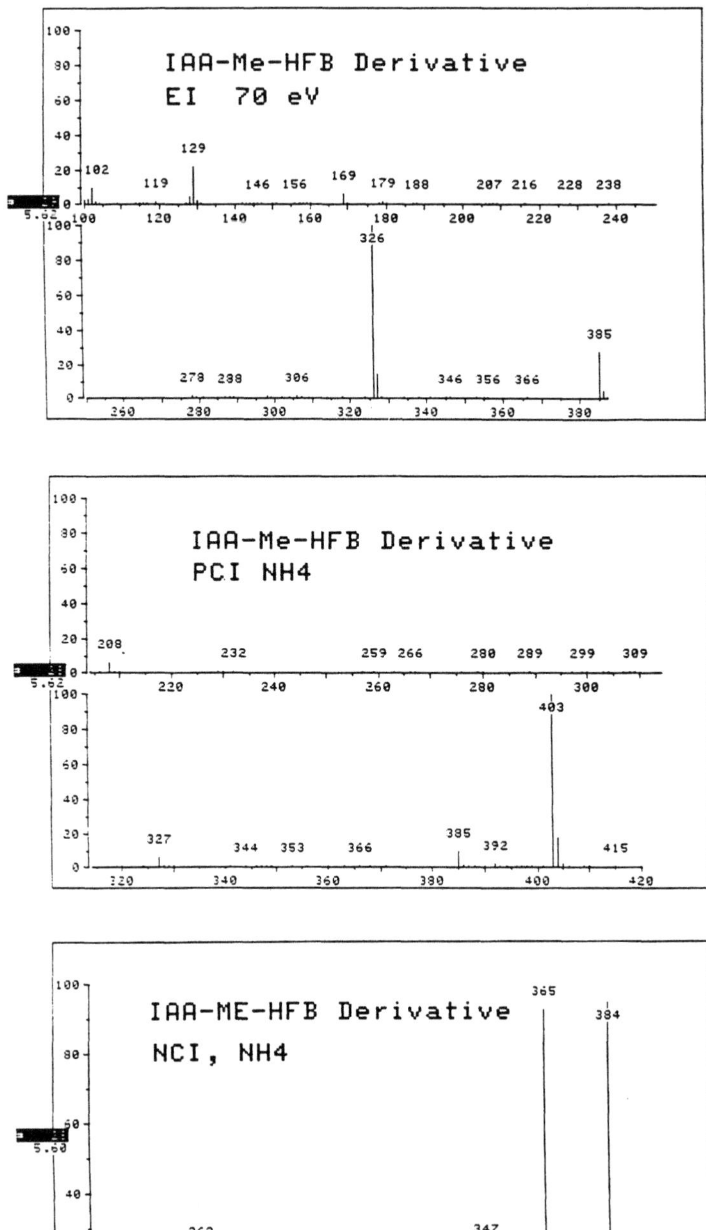

Fig. 14. Electron impact, Ammonia positive chemical ionization and negative ionization mass spectra of indole-3-acetic acid methyl ester heptafluorobutanoyl-1 derivative

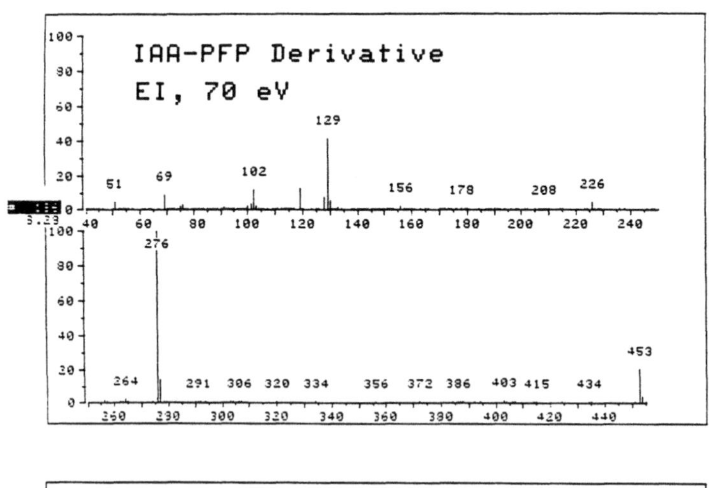

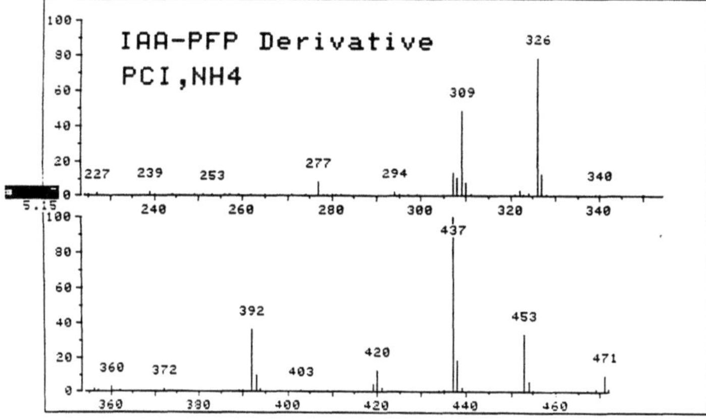

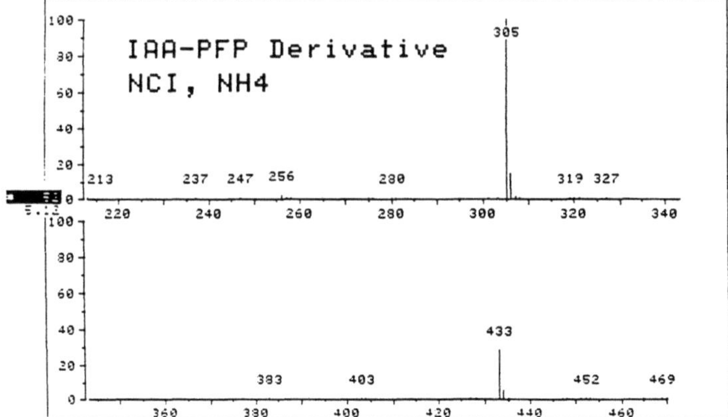

Fig. 15. Electron impact, Ammonia possive chemical ionization and negative ionization mass spectra of indole-3-acetic acid pentafluoropropionyl ester pentafluoropropionyl-1 derivative

The totality of the GC-MS analyses of endogenous auxins published so far have been performed by EI. Potential advantages of Cl in the positive and negative modes have recently been investigated (Rivier and Saugy 1986).

The EI spectrum of IAA-Me is not very informative at the molecular ion region and the m/z 130 is common to most indoles (Fig. 6). When using methane PCI (Fig. 7), isobutane PCI (Fig. 8) or ammonia PCI (Fig. 9), mass spectra are quite different. With the same compound, the ammonia NCI mass spectrum is still different (Fig. 10). It is thus possible to make use of these differences for collecting evidence of identification. Sensitivity is also modified, as ion responses are higher with Cl with lower background so that for IAA-Me a 20- to 50-fold increase in detection sensitivity is obtained with NCI compared to EI.

NCI is based partially on an electron-capturing process. It is reasonable to speculate that improved sensitivity could be obtained if electron-capturing atoms like halogens are added to the auxin molecules. The number of fluorine atoms which can be added to IAA ranges from 1 to 10 (see Fig. 3, above). The derivatives obtained so far give informative spectra (Fig. 11–15). As could be expected, the PFP derivative of IAA which contained 10 fluorines gave the most intense response (Rivier and Saugy 1986). Pulsed PCI and NCI on IAA-PFB derivative has also given higher sensitivity compared to EI (Epstein and Cohen 1981). Much further improvement is to be expected from such highly selective ionization modes. Not only is it possible to accumulate new evidence for the identification of an endogenous auxin by using the same instrument, but also to extend the detection to a lower limit.

14 Quantification

Even with very specific and sensitive detection, efficient extraction and effective purification of the plant material are necessary. Even then, loss of auxins inevitably takes place. Use of [^{3}H] IAA (16.7 Ci mmol^{-1} from Amersham, or 10.0 Ci mmol^{-1} from Israeli AEC) can be sought as an internal standard up to the stage of actual injection on the GC-MS. This necessitates the help of a radioactivity detector, on-line (HPLC radio-monitor) or off-line (scintillation counter) and two measurements have to be made, one from the detector for unlabeled auxin and another for the radiolabeled molecules. By contrast, a stable isotope standard is required for the GC-MS or SCIM steps as internal standard. The main advantage of using such a combination compared to others is that all determinations are made on the same apparatus almost simultaneously on the same aliquot of extract.

Quantitative determinations will be precise only if full correction for these losses is made. It has been advocated with good reason that GC dosages can be imprecise due to detector instabilities between and within runs. To alleviate such drawbacks, a double isotope dilution method for IAA has been proposed (Cohen and Schulz 1981). It allows the direct determination of the specific activity of the radiolabeled auxin by GC and liquid scintillation counting and it is recommended to those who have no access to a GC-MS. It requires the availability of the radio-labeled Internal Standard (IS) and also a chemically related radiolabeled compound to serve as a second IS. This second IS serves to calibrate the GC detector. Both a GC and a scintillation counter are needed and the measurements are done separately. It is a relatively complicated method.

As MS measures the mass of molecules, the use of auxins labeled with stable isotopes provides with the best compromise between accuracy and simplicity. It is called the stable isotope dilution method and at least one book has been devoted to the subject (Millard 1978). The stable isotope dilution technique consists in measuring by MIM the ratio of heavy to normal isotopes in the finally purified compound. Knowing the amount of heavy isotopes added at the beginning of the analysis, the absolute amount of auxin in the plant extract is then obtained.

Practically, using GC-MS, the measurements of both types of molecule population are obtained by selecting at least two ions per molecule (i.e., 4 channels minimum) and determining the peak areas at the Rt expected for the auxin. The stable isotope labeled molecule usually emergies a few tenths of a second before the natural one. Any drift in the instrument settings will affect all ions equally. It is rather paradoxal to realize practically that such an expensive and complex instrument can be used so simply to obtain highly reliable quantitative data.

It consists of measuring by mass spectrometry the ratio of normal to heavy isotope in the finally purified compound. Knowing the amount of heavy isotope added, the absolute amount of auxin in the crude extract is then obtained. The isotope dilution equation used by Magnus and co-workers (1980) for IAA determinations with stable isotopes is:

$$Y = (C_i/C_f - 1)X,$$

where Y = amount of compound in the extract,
$\quad\quad\quad C_i$ = the initial percentage of deuterated auxin in the IS
$\quad\quad\quad C_f$ = the percentage of deuterated auxin found in respect to the total auxin
$\quad\quad\quad\quad\quad$ (both labeled and nonlabeled), and
$\quad\quad\quad X$ = the amount of labeled auxin added.

A number of procedures have been described for calculations using the isotope dilution equation (Millard 1978). It is, however, complicated to apply them on a routine basis, and it is simpler to establish all these values experimentally and to run a series of standard solutions containing various amounts of the auxin with the same amount of IS. These solutions have to be extracted and derivatized in the same way as the plant extracts to which the same amount of IS is also added. After measuring by GC-MS the signal ratio between the two ions selected (each corresponding to one of the molecule species), a calibration curve is constructed. This curve compensates for all errors, losses, or isotope effects and instrument variations which might occur during the analyses. The first critical requirement lies on the reproducible pipeting of the solutions performed at the beginning of the procedure. It is my experience that instrument variations are not significant with modern apparatus. If they occur, they are more important during CI than EI modes, and the pressure of the reagent gas should be monitored more carefully for PCI than NCl.

Usually, at least duplicates of five concentrations should be used to buildup the curve. Mathematical manipulations – such as linear regression fitting – can be applied and standard errors calculated (Bush and Trager 1981). Any abnormal distribution between ions from the same molecule will indicate possible co-elution of foreign compounds.

It is clear that the establishment of such a curve requires additional laboratory work, but if rapid clean-up procedures can be used, the confidence in the measured data is such that it compensates largely for the effort. If one single sample or 30 to 50 samples together have to be determined, the calibration curve should be established each time. In practice, the calibration curve has to be determined for each day of GC-MS analysis. Within the correct range of concentrations, the curve is usually a straight line and the calculated ratio of ions in the sample gives directly the amount of auxin in the starting material. From the many IS available, it is important to select the best suited to the problem. This is discussed in the next paragraph.

15 The Internal Standard

There are three types of Internal Standard that can be used for SIM assays: Type 1 is a stable isotope analog of the auxin to be quantified; type 2 is a homologuous compound whose mass spectrum contains a reasonably intense ion at the same mass as the ion being monitored for the auxin; and type 3 is a compound that has similar physical properties to the auxin being measured, but whose mass spectrum has no significant ions in common with the mass spectrum of the auxin. There are very few reports in the literature in which the accuracy and precision of an assay are related to the different type of IS used, and they are in the area of pharmaceuticals. However, it can be considered that either type 1 or 3 IS appear to produce the lowest variance factors due to long-term instrumental and sample manipulation errors. With type 2 IS, mass switching can be avoided and the accuracy and precision should be improved. Thus, the limited studies to date support the intuitive decision made by a majority of mass spectroscopists to use a stable isotope labeled analog when available instead of a structural analog.

The selection of the suitable IS for GC-MS determinations is not only a matter of the final measurements, but also depends strongly on the extraction and purification procedures. For example, it is obvious that type 2 or 3 IS are separated from the auxin of interest during HPLC or TLC and should be added at the correct time of the analytical procedure. Even type 1 IS are separated from their unlabeled homologs. For example, $[^2H_5]$IAA separates clearly from IAA on isocratic C_{18} HPLC runs at Rt > 30 min, and on some GC capillary columns, wheatheras $[^{13}C]$IAA not (R. P. Pharis, pers. comm.).

Of concern to many analysts is the uncertainty of the extraction of an auxin from an actual plant matrix. The recovery of the auxin in an experimental sample is estimated from the recovery obtained by spiking various known amounts of the auxin and IS into the sample plant material. Such an estimate may be inaccurate because the composition of most plant matrix depends on the time when the specimen was obtained, the individual source of the specimen and storage and handling of the specimen. In this regard, an isotope dilution procedure (type 1 IS) provides the most reliable data.

The most desirable stable isotope analog of the auxin is one that is multilabeled with 99% enriched ^{13}C and/or ^{15}N atoms because these analoges more closely resemble the auxin to be quantified and are less likely to undergo isotopic exchange biologically, during the extraction and purification, and during the GC-MS determinations. To date, most IS available are labeled with deuterium because of the commercial availability of highly enriched synthetic precursors and chemical reagents (see Table 3 above). The EI mass spectra from several methylated IS used so far for the dosage of endogenous IAA are given in Fig. 16–19. These spectra have been obtained by quadrupole GC-MS at 70 eV ionization. Deuterated IAA are type 1 IS, 5-Me-IAA is of type 3. The type 2 has never been used for auxins.

When using molecules labeled with deuterium, some special precautions are necessary. Depending on their position on the molecule, some deuterons might exchange with protons of the solvent. For IAA analysis, a number of procedures have been described using side-chain $^2H_2$-IAA (Allen and Baker 1980). The labeled ion will be two a.m.u. higher than the naturally occurring ion. This latter is formed by a certain number of carbon atoms, the natural abundance of ^{13}C can be calculated. The probability of having a naturally occurring molecule containing two ^{13}C atoms can also be calculated, and for molecules with 10 C atoms, this number is around 1.4%. The mass value of this ion will be the same as the ion of the IS, that is at m/z 191 for IAA (see Figs. 6 and 17). As the low resolution MS cannot differentiate between the deuterium labeled and the heavy carbon labeled molecules, determinations can be inaccurate depending on the relative

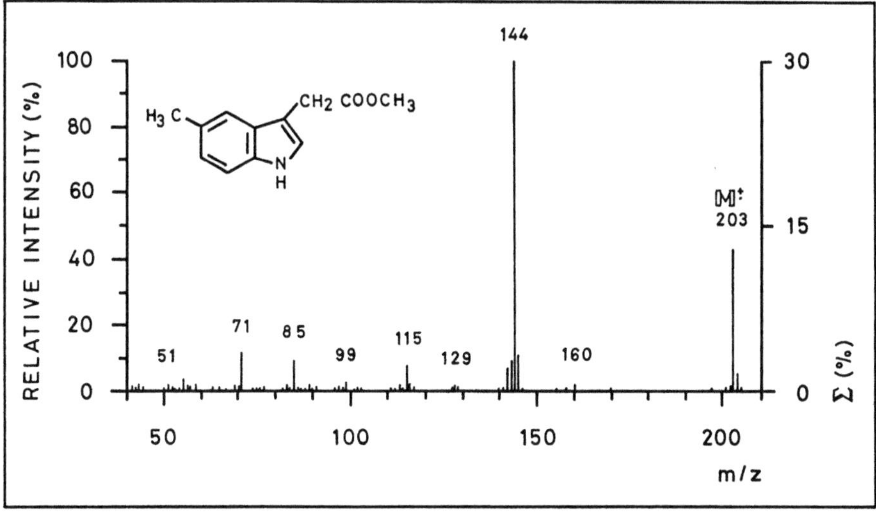

Fig. 16. Electron impact mass spectrum of 5-methyl indole-3-acetic acid methyl ester

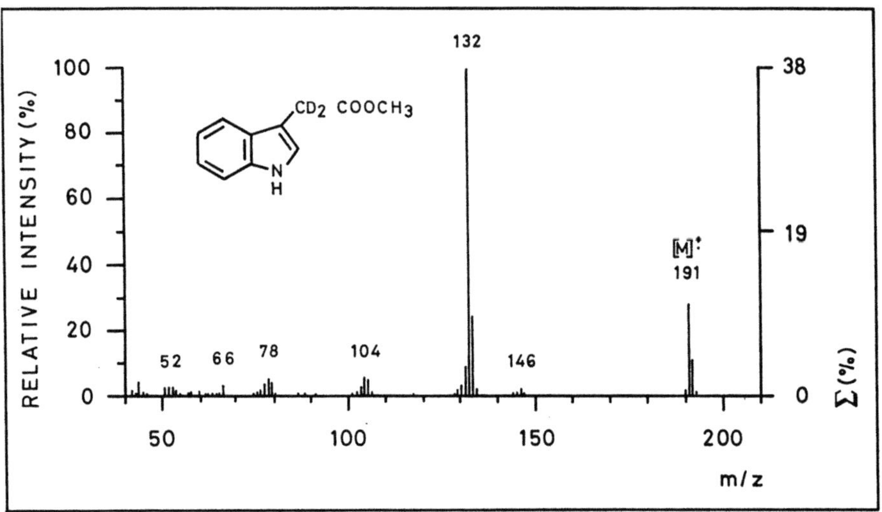

Fig. 17. Electron impact mass spectrum of $^2H_2$-$\beta\beta$-indole-3-acetic acid methyl ester

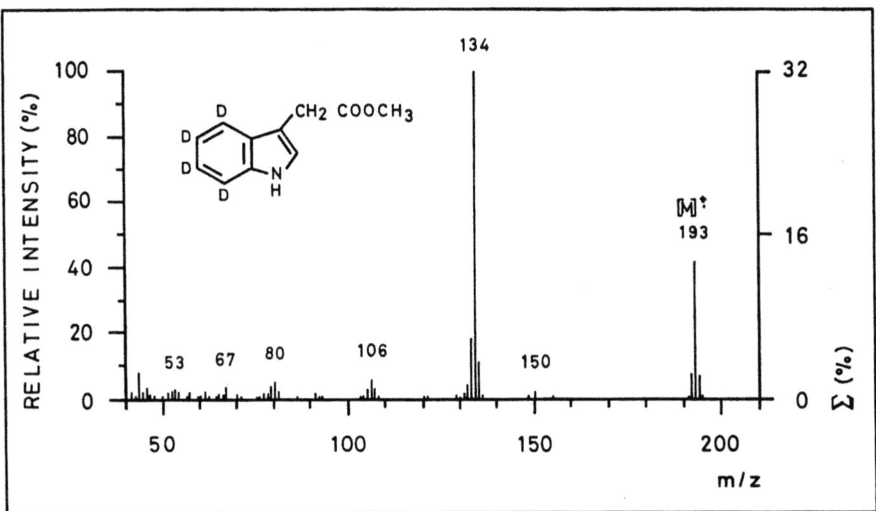

Fig. 18. Electron impact mass spectrum of 4,5,6,7-$^2H_4$-indole-3-acetic acid methyl ester

amount of IS added compared to the endogenous auxin level. As a rule, the amount of IS used should be similar to that of the endogenous auxin.

A number of assays have been described using side-chain ^2H-IAA as IS for plant material and careful selection of the procedure is mandatory (Allen et al. 1982b):

Base catalyzed $^1H/^2H$ exchange in basic medium can be carried out with relative ease. Under such alkaline conditions, IAA is sufficiently stable and soluble to allow a high yield of incorporation to be obtained. The resulting product is pre-

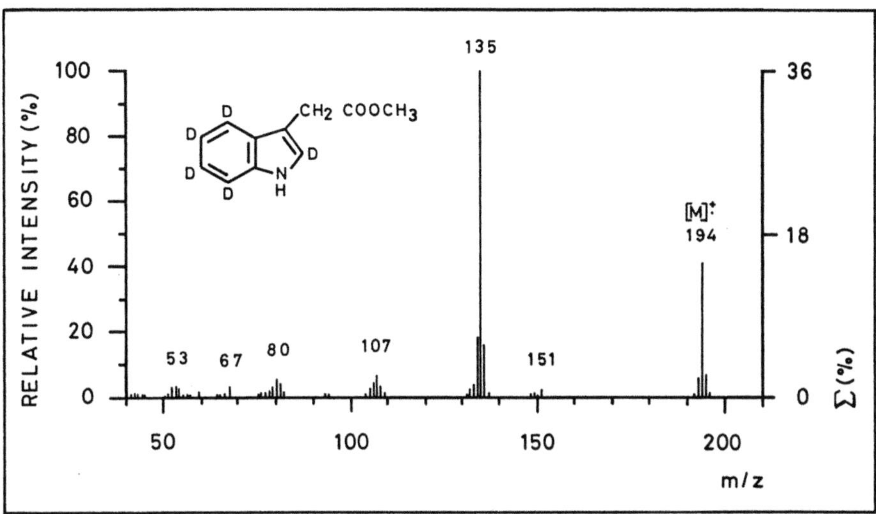

Fig. 19. Electron impact mass spectrum of 2,4,5,6,7-$^2H_5$-indole-3-acetic acid methyl ester

dominantly labeled on the methylene carbon of the side-chain, with small amounts of deuterium incorporated at the C-2 position of the indole ring. Using $^2H_2$-IAA obtained this way as IS for GC-MS, a small (0.1–0.4%) amount of interference of the $^2H_0$-IAA from $^2H_2$-IAA standard at m/z 130 or m/z 189 is therefore observable (Fig. 17; Allen and Baker 1980). Extremes of pH produce exchange of hydrogen during storage and sample manipulation, and consequently $^2H_2$-IAA is not recommended. Great care is necessary if this IS has to be used anyway. The principal advantage of using $^2H_2$-IAA as IS for GC-MS is the ease of preparation and low cost.

The use of acid catalyzed deuterium exchange for IAA is unsatisfactory because IAA is virtually insoluble in acidic solvents; it also decomposed and polymerizes.

Magnus et al. (1980) have described the preparation of ring-labeled IAA. The isotope purity of the $^2H_4$- and the $^2H_5$-IAA thus obtained is typically 70–82% of all deuterated IAA molecules treated (Figs. 18 and 19). The nonlabeled species is usually less then 0.1%, giving almost no detectable interference with the unlabeled endogenous IAA molecules. The principal advantage of using the ring-labeled IAA is stability of the label during alkaline hydrolysis when analyzing the conjugates of this auxin. However, the labeled at ring-2 position can move and exchange in either 1 N NaOH solution (Magnus et al. 1980) or during derivatization using acid anhydrides (Allen et al. 1982b). Thus caution should be taken when using $^2H_5$-IAA under extreme conditions. $^2H_4$-IAA is not commercially available, but has been shown to be the most suitable deuterated IS for IAA quantification (Magnus et al. 1980).

Recently, J. Cohen and coworkers have synthesized large amounts of $^{13}C_6$-[benzene ring]-indole-3-acetic acid using highyl enriched $^{13}C_6$-analine (Cohen et al. 1986). Such labeled compound is certainly now the most suitable stable labeled IS available for GC-MS: the labels are stable and the polarity of the molecule is almost identical to that of IAA. It is to be expected that the partitioning and the chromatographic behaviors on GC and HPLC are the same for both molecules

To illustrate the use of stable labeled IS, the following examples are given:

- Leaves, phloem and xylem saps of *Ricinus communis* were prepared for IAA analysis using $^2H_2$-IAA (Allen and Baker 1980),
- $^2H_3$-Methyl-IAA was the IS used for the quantification of endogenous IAA after methylation with diazomethane of the extracts from *Pinus contorta* (Savidge and Wareing 1983) and from *Phaseolus vulgaris* plants (Knox and Wareing 1984),
- *Zea mays* mesocotyls were analyzed for endogenous IAA (free and esters) using $^2H_4$-IAA as IS (Pengelly et al. 1982),
- Pea and maize seedlings were analyzed with $^2H_5$-IAA on various occasions in Lausanne (Allen et al. 1982; Pilet and Rebeaud 1983; Saugy and Pilet 1984; Pilet and Saugy 1985),
- Douglas Fir seedlings have been analyzed for their endogenous IAA content using $^2H_5$-IAA by Caruso and Zeisler (1983).
- Endogenous IAA has been measured using $^2H_5$-IAA in stem of tobacco by Noma and coworkers (1984).

Practical analytical details for each specific plant material can be found in these publications. In the next paragraph, the reader will find the description of the procedure used routinely for over 7 years by the author. Such a procedure is valid for the material tested here and must be adapted consequently for other plant parts or species.

16 Experimental Procedure

Freeze-dried maize or pea seedlings (0.05 to 0.3 g dry wt.) or equivalent of fresh weight tissue are placed in a 15-ml round-bottomed screw-cap glass tube (Sovirel No 20). 10 to 50 ng $^2H_5$-IAA (Merck, Sharp and Dohme Isotopes, Canada LDT) – but preferably $^{13}C_6$-[benzene ring]-IAA, which was obtained as a generous gift from J. Cohen – were added in minimum ethanolic solution with 500 ng indole or 5-Methyl-IAA (Aldrich). Both analogs act as carriers. Three ml of cold Na acetate buffer (0.1 M, pH = 4.0) and 5 ml of ethyl acetate (redistilled prior use) were added. The tissue does not need to be ground, as extraction is quantitative in the present case whatever the size of the tissue pieces (this point must be checked for other tissues). The glass tubes are capped with PTFE-lined screw tops and tumbled for 20 min by slow rotation at 8–10 rpm. After standing for 10 additional min and low speed centrifugation, the ethyl acetate phase is transferred with a Pasteur pipette into conical 10-ml glass tubes. The organic phase is reduced to ca. 0.5 ml volume under reduced pressure (50 mm Hg at 40 °C) in a vortex evaporator (Buchler, Fort Lee, USA). The residues are esterified by incubation in an ice-H_2O bath with 0.15 ml ethanol (redistilled before used) and 1.5 ml etheral diazoethane for 15 min. The solution is carefully reduced to dryness under vacuum (50 mm Hg, 22 °C) and taken up in 100 µl ethanol for preparative TLC. If the TLC step is omitted, the residue is dried more thoroughly by adding a few drops of toluene for subsequent derivatisation with HFBI.

For extraction of 1–3 g fr. wt. pigmented plant tissues, methanol was used after grinding the material. Centrifugation and evaporation of the methanol leave a residue which is purified by partition, first at pH = 3 with diethyl ether, then at pH = 8.6 with K_2HPO_4 and finally at pH = 3.0 again with diethylether. The crude acid-ether fraction is esterified with diazoethane and subjected to preparative TLC.

The ethanolic solutions of the alkylated phases are loaded as 15 mm streaks onto a 250 µm silica gel F-254-366 TLC plates (Woelm, Eschwege, or Merck, Darmstadt, Germany – these latter are more difficult to scrape off) after having added 10 µl of BHT ethanolic solution (100 mM) to each of them. On both sides of the plate, 500 ng of 5-Me-IAA-Et are loaded as TLC markers. The plate is immediately developed in either n-hexane : ethyl acetate (1 : 1) or $CHCl_3$: methanol (93 : 7). Up to six samples can be run simultaneously on the same TLC plate. After 20–30 min, the solvent front reaches 10–15 cm height. The markers, running at similar Rfs as IAA-Et (Rf = 0.6 ± 0.1) are visualized by UV_{254}. The corresponding zones of the extracts are scraped off and eluted immediately with 1.5 ml methanol in a Pasteur pipette containing a cotton wool plug previously cleaned with 2 ml of methanol. The eluates are collected in a 10-ml all-glass conical vial and reduced to dryness under vacuum (50 mm Hg, 40 °C).

To the dried residue in the glass-stoppered tube (PTFE material should be avoided as artifactual peaks are produced during the reaction), 50 µl dry pyridine and 50 µl heptafluorobutyrylimidazole (HFBI) (Pierce Eurochemie, Amsterdam, The Netherlands) are added sequentially. After incubation for 3 h at 85 °C, the tubes are chilled, 400 µl n-hexane and 1.0 ml 0.5 N H_2SO_4 are added in turn and the mixture was vortexed for 20 s. Following low speed centrifugation, the aqueous phase is removed by a Pasteur pipette and discarded. The hexane phase is washed with 1.0 ml H_2O and transferred to a 1.5-ml glass vial. Prior to analysis,

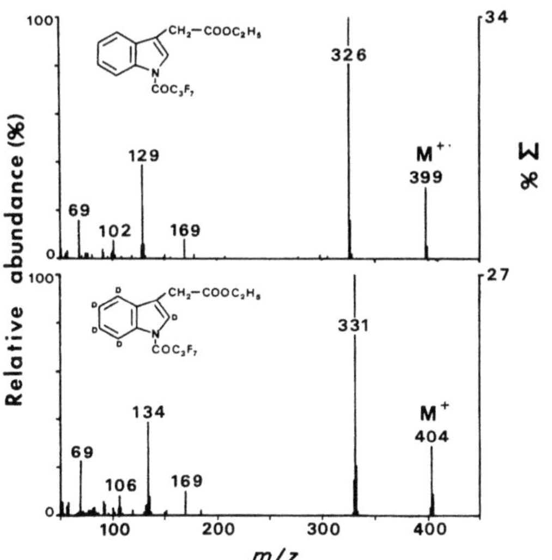

Fig. 20. Electron impact mass spectrum of the heptafluorobutyryl ethyl ester derivative of indole-3-acetic acid and 2,4,5,6,7-$^2H_5$-indole-3-acetic acid

the solution is reduced to around 40 µl under vacuum. The derivative of IAA remains stable for at least 10 days in the dark at -25 °C. Its MS was obtained from standards and the most representative ions are selected for SIM (Fig. 20).

GC-MS analysis is carried out with splitless injections of 30 s duration on a 25 m × 0.32 mm (i.d.) fused silica capillary column coated with 5% phenylmethyl silicone (like SE 54) cross-linked stationary phase. The column is connected directly to the MS ion source. The conditions of the run are such that the IAA derivative has a retention time between 6 and 7 min. The IS emerges from the column a few tenths of a second earlier, and the carrier 1.5 min later (Fig. 21).

Calibration is obtained on a day-to-day basis in the following way: quantification of the endogenous IAA in the extracts is made by reference to calibration plots derived from a series of standards routinely subjected to the same purification procedure. Standards containing 500 ng of 5-Me-IAA or indole and either

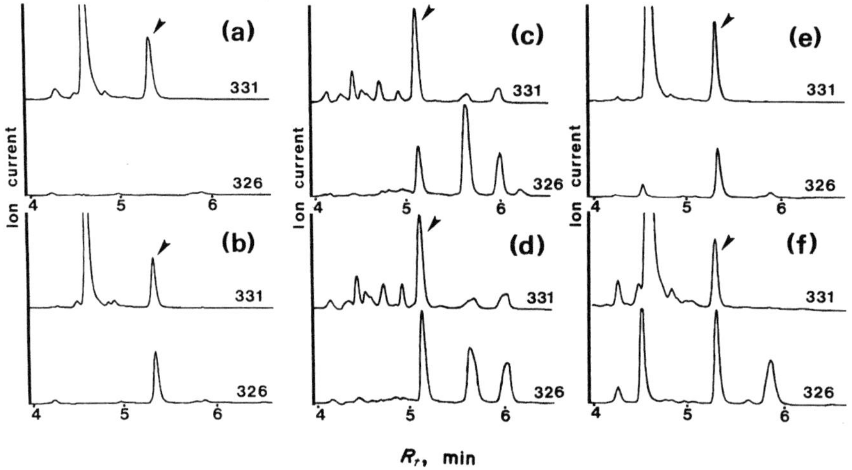

Fig. 21 a–f. Selected ion chromatogram from extracts of the heptafluorobutyryl ethyl ester derivatives of: **a** standard of 250 pmol ^{2}H$_5$-IAA only; **b** standard of 250 pmol IAA and 250 pmol ^{2}H$_5$-IAA; **c** *Zea* coleoptile; **d** *Zea* mesocotyle; **e** dark-grown *Pisum* epicotyl; **f** light-grown *Pisum* epicotyl. Arrows indicate the peaks from ^{2}H$_5$-IAA-HFB-Et

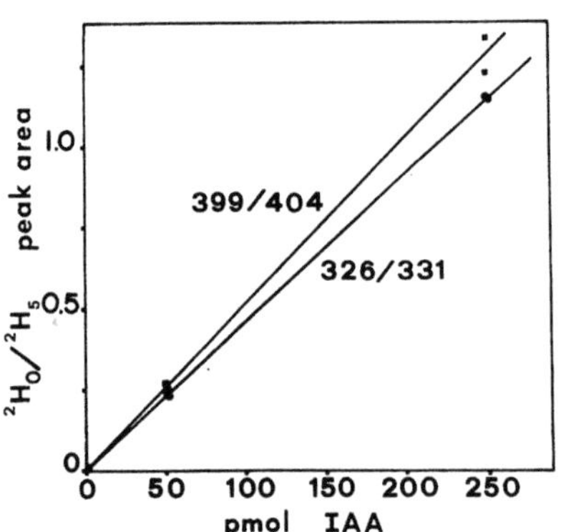

Fig. 22. Typical calibration plot of IAA against the peak area ratios derived from the M$^+$ and β-cleavage fragment ions, using 250 pmol ^{2}H$_5$-IAA as an internal standard

10 or 50 ng of $^2H_5$-IAA together with known amounts ranging from 0 to 50 ng of IAA exhibit a linear relationship between the amount of IAA present and the calculated ^1H and ^2H ratio from either channel couple selected (Fig. 22).

Remark

1. Methyl ester could also be used. However, when using methanol as solvent at any stage of the procedure, some methyl esterification takes place. This happen also by simply reducing to dryness an IAA methanolic solution (Allen et al. 1982a). When using ethylester of IAA, it is possible to differentiate from the artifactually obtained IAA-Me and the true endogenous IAA measured as IAA-Et derivative.

2. When significant amounts of lipids are present in the plant tissue, it is advisable to remove them first by partitioning the Na-acetate buffer with 5 ml of n-hexane. The IAA does not move into the hexane layer at low pH.

3. If very small quantities of starting material have to be measured, it is sometimes possible to avoid the TLC purification and/or the HFBI derivatization (Rivier and Pilet 1983).

17 Conclusions

The GC-MS analyses of auxins are well documented in only very few cases. IAA detection and quantification can be now considered as routine in the same plant materials. Much of work has still to be done to promote the accuracy and precision of the analysis of other auxins. Appropriate IS are still not easily available and have to be synthesized. However, by using GC-MS methods where the potential techniques already exist, it is now a matter of adaptation of standard procedures to meet particular cases.

Acknowledgments. I am particularly grateful to Prof. P.E. Pilet for his continuing interest in and support of my research on endogenous growth regulators and for sharing with me his broad knowledge of plant growth hormonology. I would also like to thank Professors B. Bandurski, J. MacMillan, R.P. Pharis, and Dr. J. Cohen for their critical comments and advice on reading the final version of the manuscript and for providing me with technical details and unpublished results. To Drs. E. Jensen, J.-E. Rebeaud, and M. Saugy I express my gratitude for their continuous support during the preparation and completion of this text.

References

Allen JRF, Baker DA (1980) Free tryptophan and indole-3-acetic acid levels and vascular pathways of *Ricinus communis* L. Planta 148:69–74

Allen JRF, Holmstedt B (1980) The simple β-carbolines. Phytochemistry 19:1573–1582

Allen JRF, Rebeaud JER, Rivier L, Pilet PE (1982a) Quantification of indole-3-acetic acid
    by GC-MS using deuterated labeled internal standard. In: Schmidt HL, Forstel HL,
    Heinziger K (eds) Stable isotopes. Elsevier, Amsterdam, pp 529–534
Allen JRF, Rivier L, Pilet PE (1982b) Quantification of indol-3-yl-acetic acid in pea and
    maize seedlings by gas chromatography-mass spectrometry. Phytochemistry 21:523–
    534
Arteca RN, Poovaiah BW, Smith OE (1980) Use of high performance liquid chromatog-
    raphy for the determination of endogenous hormone levels in *Solanum tuberosum* L.
    subjected to carbon dioxide enrichment of the root zone. Plant Physiol 65:1216–1219
Atsumi S, Kuraishi S, Hayashi T (1976) An improvement of auxin extraction procedure
    and its application to cultured plant cells. Planta 129:245–247
Bandurski RS (1979) Chemistry and physiology of conjugates of indole-3-acetic acid. In:
    Mandava N (ed) Plant growth substances. Am Chem Soc Symp Ser 111:1–17
Bandurski RS (1984) Metabolism of indole-3-acetic acid. In: Crozier A, Hillman JR (eds)
    The biosynthesis and metabolism of plant hormones. Cambridge, Soc of Exp Biol, Sem
    Ser 23:183–200
Bearder J (1980) Plant hormones and other growth substances – their background, struc-
    ture and occurrence. In: MacMillan J (ed) Hormonal regulation of development I. En-
    cyclopedia of plant physiology new serie, vol 9. Springer, Berlin Heidelberg New York,
    pp 9–112
Bialek K, Meudt WJ, Cohen JD (1983) Indole-3-acetic acid (IAA) and IAA conjugates ap-
    plied to been stem sections. Plant Physiol 73:130–134
Budzikiewicz H, Djerassi C, Williams DH (1964) Interpretation of mass spectra of organic
    compounds. Holdan-Day, San Fransisco, pp 251–253
Bush ED, Trager WF (1981) Analysis of linear approaches to quantitative stable isotope
    methodology in mass spectrometry. Biomed Mass Spectrometry 8:211–218
Caruzo JL, Zeisler CS (1983) Indole-3-acetic acid in douglas fir seedlings: a reappraisal.
    Phytochemistry 22:589–590
Chapman JR (1978) Computers in mass spectrometry. Academic Press, London New York
    San Fransisco, pp 265
Cohen JD (1984) Convenient apparatus for the generation of small amounts of diazo-
    methane. J Chromatogr 303:193–196
Cohen JD, Bandurski RS (1982) Chemistry and physiology of the bound auxins. Ann Rev
    Plant Physiol 33:403–430
Cohen JD, Bialek K (1984) The biosynthesis of indole-3-acetic acid in higher plants. In:
    Crozier A, Hillman JR (eds) The biosynthesis and metabolism of plant hormones.
    Cambridge, Soc Exp Biol, Sem Ser 23:165–181
Cohen JD, Schulze A (1981) Double-standard isotope dilution assay. I. Quantitative assay
    of indole-3-acetic acid. Anal Biochem 112:249–257
Dawson PH (ed) (1976) Quadrupole mass spectrometry and its applications. Elsevier, Am-
    sterdam Oxford New York, pp 349
Elliott MC, Stowe BB (1971) Indole compounds related to auxins and goitrogens of wood
    (*Isatis tinctoria* L.). Plant Physiol 47:366–372
Engvild KC, Egsgaard H, Larsen E (1981) Determination of 4-chloroindoleacetic acid
    methyl ester in *Viciae* species by gas chromatography-mass spectrometry. Physiol Plant
    53:79–81
Epstein E, Cohen JD (1981) Microscale preparation of pentafluorobenzyl ester: electron-
    capture GC detection of IAA from plants. J Chromatogr 209:413–420
Epstein E, Cohen JD, Bandurski RS (1980) Concentration and metabolic turnover of in-
    doles in germinating kernels of *Zea mays* L. Plant Physiol 65:415–421
Freeman RR (ed) (1979) High resolution gas chromatography. Hewlett-Packard, Avon-
    dale, pp 162
Harrison AG (1983) Chemical ionization mass spectrometry. Chemical Rubber Company
    Press, Boca Raton, pp 630
Hofinger M (1980) A method of quantification of indole auxins in the picogram range by
    high performance GC of their heptafluorobutyryl derivative. Phytochemistry 19:219–
    221

Iino M (1982) Action of red light on IAA status and growth in coleoptiles of etiolated maize seedlings. Planta 156:21–32

Iino M, Yu ST, Carr DJ (1980) Improved procedure for the estimation of nanogram quantities of indole-3-acetic acid in plant extracts using the indolo-pyrone fluorescence method. Plant Physiol 66:1099–1105

Jamieson WD, Hutzinger O (1970) Identification of simple naturally occuring indoles by mass spectrometry. Phytochemistry 9:2029–2036

Jennings W (1980) Gas chromatography with capillary columns 2nd ed. Academic Press, New York, pp 184

Knox JP, Wareing PF (1984) Apical dominance in *Phaseolus vulgaris* L.: the possible roles of abscisic acid and indole-3-acetic acid. J Exp Bot 35:239–244

Letham DS, Goodwin PB, Higgins TJV (ed) Phytohormones and related compounds, a treatise, vol 1: the biochemistry of phytohormones and related compounds, and vol 2: phytohormones and the development of higher plants. Elsevier, Amsterdam

MacMillan J (ed) (1980) Hormonal regulation of development I. Encyclopedia of plant physiology new serie, vol 9. Springer, Berlin Heidelberg New York, pp 681

Magnus V, Bandurski RS, Schultze A (1980) Synthesis of 4,5,6,7 deuterium-labeled IAA for use in mass spectrometric assays. Plant Physiol 66:775–781

Mann JD, Jaworski EG (1970) Minimizing loss of indoleacetic acid during purification of plant extracts. Planta 92:285–291

Martin GC, Scott IM, Neill SJ, Horgan R (1982) Identification of abscisic acid glucose ester, indole-3-acetic acid, zeatin and zeatin riboside in receptacles of pear. Phytochemistry 21:1079–1082

McDougall J, Hillman JR (1978) Analysis of indole-3-acetic acid using GC-MS techniques. In: Hillman JR (ed) Isolation of plant growth substances. Cambridge University Press, London, pp 1–25

McDougall J, Hillman JR (1980) Derivatives of indole-3-acetic acid for SIM GC-MS studies. Z Pflanzenphysiology 98:89–93

McFadden WH (ed) (1973) Techniques of combined gas chromatography-mass spectrometry: application in organic analysis. Wiley, New York London Sydney Toronto, pp 463

Millard BJ (1978) Quantitative mass spectrometry. Heyden, London Philadelphia Rheine, pp 171

Munson B (1977) Chemical ionization mass spectrometry: ten years after. Analyt Chem 49:772 A–778 A

Noma M, Koike N, Sano M, Kawashima N (1984) Endogenous indole-3-acetic acid in the stem of tobacco in relation to flower neoformation as measured by mass spectrometric assay. Plant Physiol 75:257–260

Nonhebel HM, Bandurski RS (1984) Oxidation of indole-3-acetic acid and oxindole-3-acetic acid to 2,3-dihydro-7-hydro-2-oxo-1 *H* indole-3-acetic acid-7'-*O*-β-glucopyranoside in *Zea mays* seedlings. Plant Physiol 76:979–983

Nonhebel HM, Crozier A, Hillman JR (1983) Analysis of [^{14}C]indole-3-acetic acid metabolites from the primary roots of *Zea mays* seedlings using reverse-phase high-performance liquid chromatography. Physiol Plant 57:129–134

Pengelly WL, Hall PJ, Schulze A, Bandurski RS (1982) Distribution of free and bound indole-3-acetic acid in the cortex and stele of *Zea mays* mesocotyl. Plant Physiol 69:1304–1307

Pilet PE, Rebeaud JER (1983) Effect of abscisic acid on growth and indolyl-3-acetic acid levels in maize roots. Plant Sci Lett 31:117–122

Pilet PE, Saugy M (1985) Effect of applied and endogenous IAA on maize root growth. Planta 164:254–258

Plett T, Böttger M, Hedden P, Graebe J (1984) Occurrence of 4-Cl-indoleacetic acid in broad bean and correlation of its levels with seed development. Plant Physiol 74:320–323

Powers JC (1968) The mass spectrometry of simple indoles. J Org Chem 5:2044–2050

Purcell JE (1982) Quantitative capillary gas chromatographic analysis. Chromatographia 15:546–558

Rayles DL, Purves WK (1967) Isolation and identification of indole-3-ethanol (tryptophol) from cucumber seedlings. Plant Physiol 42:520–524

Reeve DR, Crozier A (1980) A quantitative analysis of plant hormones. In: MacMillan J (ed) Hormonal regulation of development I. Encyclopedia of plant physiology new serie vol 9. Springer, Berlin Heidelberg New York, pp 203–280

Reinecke DM, Bandurski RS (1983) Oxindole-3-acetic acid, an IAA catabolite in *Zea mays* L. Plant Physiol 71:211–213

Rivier L, Pilet PE (1971) Composés hallucinogènes indoliques naturels. Ann Biol 10:129–149

Rivier L, Pilet PE (1974) Indolyl-3-acetic acid in cap and apex of maize roots: identification and quantification by mass fragmentography. Planta 120:107–112

Rivier L, Pilet PE (1983) Simultaneous gas chromatographic-mass spectrometric determination of abscisic acid and indol-3yl-acetic acid in the same plant tissue using ^{2}H-labelled internal standard. In: Frigerio A (ed) Recent developments in mass spectrometry in biochemistry, medicine and environmental research, vol 8. Elsevier, Amsterdam, pp 219–231

Rivier L, Saugy M (1986) Chemical ionization mass spectrometry of IAA and ABA: evaluation of negative ion detection and quantification of ABA in growing maize roots. J Plant Growth Regul, in press

Salvidge RA, Wareing PL (1983) Seasonal variation in endogenous indole-3-acetic acid and abscisic acid levels in *Pinus conforta* Dougl. Can J Forest Res 23:123–134

Sandberg G (1984) Biosynthesis and metabolism of indole-3-ethanol and indole-3-acetic acid by *Pinus sylvestris* L. needles. Planta 161:1–6

Sandberg G, Jensen E, Crozier A (1984) Analysis of 3-indole carboxylic acid in *Pinus sylvestris* needles. Phytochemistry 23:99–102

Sandberg G, Crozier A, Ernstsen A (1986) Indole-3-acetic acid and related compounds. In: Rivier L, Crozier A (eds) Principles and practice of plant hormone analysis. Academic Press, London Oxford, in press

Saugy M, Pilet PE (1984) Enodgenous indol-3yl-acetic acid in stele and cortex of gravistimulated maize roots. Plant Sci Lett 37:93–99

Schneider EA, Wightman F (1978) Auxins. Phytohormones and related compounds – a comprehensive treatise. In: Letham DS, Goodwin PB, Higgins TJV (eds) The biochemistry of phytohormones and related compounds, vol 1. Elsevier/North-Holland, Amsterdam, pp 29–105

Schneider EA, Kazakoff CW, Wightman F (1985) Gas chromatography-mass spectrometry evidence for several endogenous auxins in pea seedlings organs. Planta 165:232–241

Scott TK (ed) (1984) Hormonal regulation of development II: the functions of hormones from the level of the cell to the whole plant. Encyclopedia of plant physiology new serie, vol 10. Springer, Berlin Heidelberg New York Tokyo, pp 309

Seeley SD, Powell LE (1973) Gas chromatography and detection of microquantities of gibberellins and IAA as their fluorinated derivatives. Anal Biochem 58:39–46

Segal LM, Wightman F (1982) Gas chromatographic and GC-MS evidence for the occurrence of 3-indolylpropionic acid and 3-indolylacetic acid in seedlings of *Cucurbita pepo*. Physiol Plant 56:367–370

Smith TA (1977) Tryptamine and related compounds in plants. Phytochemistry 16:171–175

Sundberg B, Sandberg G, Jensen E (1985) Identification and quantification of 3-indole methanol in etiolated seedlings of Scots pine (*Pinus sylvestris*). Plant Physiol 77:952–955

Wightman F, Lichty DL (1982) Identification of phenylacetic acid as a natural auxin in the shoots of higher plants. Pyhsiol Plant 55:17–24

Yokota T, Murofushi N, Takahashi N (1980) Extraction, purification and identification. In: MacMillan J (ed) Hormonal regulation of development I. Encyclopedia of plant physiology, new serie, vol 9. Springer, Berlin Heidelberg New York, pp 113–201

GC-MS Methods for the Quantitative Determination and Structural Characterization of Esters of Indole-3-Acetic Acid and myo-Inositol

R. S. BANDURSKI and A. EHMANN

1 Introduction

No previous reviews have been devoted to gas chromatographic-mass spectrometric (GC-MS) methods for the analysis and characterization of indole-3-acetic acid (IAA) esters of myo-inositol. Many reviews have dealt with the metabolism and function of the "bound auxins" (Bandurski 1978–1980, 1982–1984; Bandurski and Nonhebel 1984; Bentley 1958, 1961; Cohen and Bandurski 1982; Rivier and Crozier 1985). Other reviews have appeared concerning conjugates of IAA and conjugates of the other plant hormones (Sembdner 1974; Sembdner et al. 1981; Schliemann and Liebisch 1984).

The terms "bound auxin" or "bound hormone" have historical precedence (e.g., Went and Thimann 1937) over such terms as conjugated hormone. However, it has been suggested (Cohen and Bandurski 1982; Lang 1970; Schliemann and Liebisch 1984) that the phrase "conjugated hormone" be used to indicate a covalently linked hormone and that the term "bound hormone" be restricted to hormones adsorbed to a protein binding site. This review deals with covalently conjugated IAA esters, primarily esters with myo-inositol, and includes literature through 1984.

1.1 Discovery of IAA-Inositols

Conjugated auxins were first demonstrated by Cholodny (1935), who observed that a moistened piece of seed endosperm induced growth when applied to an oat coleoptile. He measured the conjugated auxin quantitatively by means of the growth induced and could arrange the cereal kernels according to their content of conjugated auxin. Cholodny (1935) may also be credited with the first observation of hydrolysis of a conjugate to yield an active "free" auxin, since he recognized that the seed endosperm must be moistened with water, but not alcohol, to elicit the coleoptile growth response.

Other workers, including Pohl (1935), Laibach and Meyer (1935), Hatcher and Gregory (1941), Avery et al. (1940), devised methods for the extraction and hydrolysis of the conjugated auxin (e.g., Cohen and Bandurski 1982). In particular, van Overbeek (1941) realized that only 5% of the auxin in the plant was in "available" form and Kögl et al. (1934), Avery et al. (1941), Haagen-Smit et al. (1942), proved that the auxin was, in fact, IAA, and most probably was conjugated in ester linkage.

The "chemical era" of auxin conjugate research began 20 years after Cholodny's work with the discovery by Andreae and Good (1955) that exogenously applied IAA was conjugated by the plant to form IAA aspartate. Zenk (1961) found that applied IAA could also be esterified to form IAA glucose. Then Labarca et al. (1965) separated the "bound auxins" of *Zea mays* into a high and a low molecular weight fraction and began a series of works which showed the low molecular weight fractions to be mainly IAA inositols (Cohen and Bandurski 1982) and the high molecular weight fraction to be IAA esterified to beta-1,4-linked glucans (Piskornik and Bandurski 1972). Mass spectrometry was first used on an IAA-inositol in 1969 (Ueda and Bandurski 1969a) with a later report by Ueda et al. in 1970.

1.2 Occurrence of IAA Conjugates

To the best of current knowledge, conjugates of IAA occur in every plant and plant tissue examined (Bandurski and Schulze 1977). As a general rule, IAA esters predominate in the cereal plants, whereas IAA amide linked to an amino acid is more prominent in dicotyledonous plants (Bandurski and Schulze 1977; Cohen 1982).

The IAA inositol esters (Fig. 1) have been studied primarily in *Zea mays* sweet corn since sweet corn provides the richest known source of IAA inositols. IAA inositol esters also occur in field corn, pop corn and even in teosinte (Ehmann, personal communication). IAA inositol has also been found in rice (*Oryza sativa*) (Hall 1980) and most recently has been observed in the aborted ovules of the horse chestnut (Bottle-brush Buckeye, *Aesculus parviflora*, Walt) (Domagalski et al. 1985). This observation may portend a much wider distribution for IAA inositol, since the chestnut is very distantly related to the *Zea* tribe.

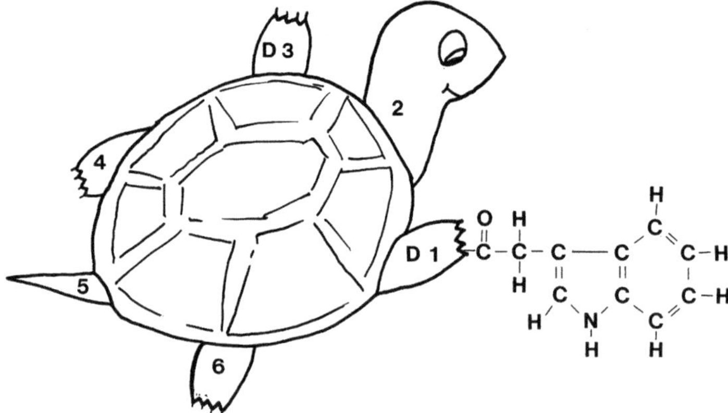

Fig. 1. The structure of IAA-D-1-O-myo-inositol. The use of a turtle as a mnemonic is patterned after Agranoff (1978). Myo-inositol is not optically active but substitution on hydroxyls 1,3,4, and 6 yield enantiomers. Looking down on the shell of the turtle, the numbering is counterclockwise, starting with the front paw as D-1. Numbering clockwise, with the left front paw as L-1, shows the L positions

1.3 Importance of Measuring and Identifying Hormone Conjugates

Although the existence of IAA esters has been known for 50 years, it has only recently become apparent that IAA conjugates are (a) ubiquitous in nature, and (b) occur in larger amounts than the free hormone, and thus must play an important role in plant hormone regulation. We know the following functions and importance: (1) The IAA esters function as storage forms of IAA in the seed, being synthesized during kernel maturation (e.g., Cohen and Bandurski 1982; Corcuera 1967; Corcuera and Bandurski 1982) and hydrolyzed during kernel germination (Chisnell 1984b; Epstein et al. 1980; Ueda and Bandurski 1969b). (2) IAA-myo-inositol (Nowacki and Bandurski 1980) and IAA-myo-inositol-galactoside (Komoszynski and Bandurski 1984) serve as seed auxin precursors, moving from the endosperm of the seed to the shoot, there to be in part, hydrolyzed to free IAA. This concept of "bound auxin" serving as a seed-auxin-precursor was first studied by Skoog (1937). (3) IAA conjugates are immune to destruction by peroxidase (Cohen and Bandurski 1978). Protection against oxidative attack may be important in view of the rapidity with which applied IAA is catabolized (e.g., Bandurski and Nonhebel 1984). (4) IAA conjugates serve as "slow-release" forms of IAA (Hangarter and Good 1981) making them particularly valuable for tissue culture studies. (5) IAA conjugates are involved in homeostatic regulation of IAA amounts in the tissue, for example being formed at the expense of free IAA during photoinhibition of growth (Bandurski et al. 1977) and asymetrically transported during geotropic stimulation (Momonoki and Bandurski 1984). (6) Conjugation of IAA oxidation products (Nonhebel and Bandurski 1984), may serve to increase the hydrophilicity of the conjugate, thus aiding in their vacuolar localization and subsequent metabolism.

The above listing is probably not complete. This is suggested by the extreme diversity of the IAA conjugates. There is the question of why there should be both low and high molecular weight forms. Why should there be IAA-inositols and IAA-inositol-arabinosides and galactosides? We believe the different conjugating moieties may be "zip" codes to designate the site in the plant and in the cell to which the conjugate should be transported.

2 Quantitative Analysis and Identification of the IAA-Inositols

2.1 Analysis After Hydrolysis

As will be discussed below, little use has been made of GC-MS for the quantitative analysis of the IAA conjugates, including IAA inositol and IAA amino acid conjugates. The major use has been in the structural characterization of the conjugates, and owing to the minute amounts of conjugates usually available, GC-MS methods have been essential to the development of the field. Table 1 illustrates the amounts of IAA inositols found in the richest source known, the kernels of *Zea mays* sweet corn.

Table 1. Amounts of IAA and IAA conjugates in kernels of *Zea mays*. (Cohen and Bandurski 1982

Compound	Amount in dry seed mg/kg^{-1}	Percent of total
Indole-3-acetic acid	0.5	0.8
Indole-3-acetyl-myo-inositol	10.1	15.2
Indole-3-acetyl-myo-inositol arabinoside	15.4	23.1
Indole-3-acetyl-myo-inositol galactoside	5.4	8.1
Trace compounds	0.2	0.3
Indole-3-acetyl-glucan	35.0	52.5

2.1.1 Methods for Hydrolysis of IAA Conjugates

The simplest and easiest method for the quantitative analysis of the IAA conjugates is to hydrolyze the conjugate and quantitatively assay the resultant free IAA by the methods reviewed by Rivier in this Volume. This method does not give information as to the nature of the conjugating moiety, but it does give information concerning the nature of the chemical linkage between IAA and the conjugating moiety.

Alkaline hydrolysis is the most frequently used method for hydrolysis of IAA conjugates. In general NaOH is used under the following conditions: 0.1 to 1.0 M NaOH at room temperature, about 22 ° to 25 °C, for 15 to 60 min. To the best of current knowledge, 1 M NaOH at 25 °C will hydrolyze all IAA esters, without hydrolysis of IAA amide conjugates. Cohen et al. (1985) have made the most careful study of the conditions for alkaline hydrolysis of IAA conjugates. They find that 1 h at room temperature hydrolyzed all the esters tested, including the ethanol, t-butanol, pentafluorobenzyl, and p-nitrophenyl esters of IAA. No hydrolysis of the amide conjugates of IAA with alanine, aspartate, glycine, phenylalanine, leucine, or glutamate occurred under these conditions. Thus, there is a clear separation, in ease of hydrolysis, of ester and amide-linked IAA, and there should be no cross-contamination of the ester and amide fractions.

When IAA is linked to the aldehyde carbon of a reducing sugar, the linkage is not an ester but is essentially an anhydride. Such anhydride conjugates are even more readily hydrolyzed in alkali than esters, and so would be included as ester IAA. Similarly, IAA linked in an acyl-sulfur linkage, as in IAA-S-CoA would also be hydrolyzed. Thus, in all the data reported to date, the category "ester" would include true esters and the anhydrides. It seems possible that weaker alkali or lower temperatures might be used to separately estimate the anhydride linked IAA.

The use of ammonium hydroxide for alkaline hydrolysis of IAA esters is advantageous in that the sample may be evaporated to dryness, following hydrolysis, without residue from the alkali, and thus be ready for derivitization and chromatography. However, concentrated ammonium hydroxide yields approximately 50% IAA and 50% of the amide of indole-3-acetic acid owing to ammonolysis of the ester. This is a useful reaction in proving that the IAA carboxyl was linked in ester linkage, but it renders the use of ammonium hydroxide for quantitative estimation of IAA unsatisfactory.

The amounts of ester and amide conjugates of IAA in various sources are illustrated by the data of Table 2. Again, as can be seen, even kilogram amounts of plant material will yield less than milligram amounts of conjugate.

A further point, established by the data of Tables 1 and 2, is that our knowledge of IAA conjugates is restricted to seeds and seedling plants. There are, for example, no data on the total amount of IAA conjugates found in mature plants. Thus, we do not know whether any of the conjugates accumulates as the plant ages. This knowledge is important in determining whether some of the conjugates are end-products or whether all are further metabolized, ultimately to return to the general metabolic pool.

Table 2. Amounts of free and conjugated IAA in various plant tissues. (Bandurski and Schulze 1977)

Species	Tissue	IAA Content		
		Free IAA[a] $\mu g/kg^{-1}$	Ester-IAA[b] $\mu g/kg^{-1}$	Amide-IAA[c] $\mu g/kg^{-1}$
Cereals				
Avena sativa	Vegetative tissue	16	5	69
Avena sativa	Seed	440	7,620	n.d.[f]
Hordeum vulgare	Seed (milled)	40[e]	329	–[g]
Oryza sativa	Seed	1,703	2,739	–
Panicum miliaceum	Seed	633	3,198	–
Triticum aestivum	Seed	123	511	–
Zea mays	Vegetative tissue	24	328	60
Zea mays	Seed	500 to 1,000	71,600 to 78,500	–
Legumes				
Glycine max	Seed	4	50[e]	524
Phaseolus vulgaris	Seed	20[e]	30[e]	136
Pisum sativum	Vegetative tissue	35	5	43
Pisum sativum	Seed	93	n.d.	202
Others				
Cocos nucifera	Liquid endosperm	0	905	–
Fagopyrum esculentum	Seed	40	127	25
Helianthus annuus	Seed	30[f]	110[f]	–
Lycopersicum esculentum	Fruit	Trace	Trace	–
Saccharomyces cerevisea	Packed cells	290	n.d.	–

[a] No alkaline hydrolysis.

[b] IAA after hydrolysis with 1 N alkali minus the free IAA.

[c] IAA after hydrolysis with 7 N alkali minus the free and ester IAA.

[d] Seedlings and fruits are fresh weight, seeds are air dry and yeast cells contain 30% dry matter.

[e] A visual estimate of IAA was made on a TLC plate since colorimetry was precluded by contaminants.

[f] n.d. (not detectible). Where the ester content is high, small amounts of IAA escape detection.

[g] A dash (–) indicates the assay was not done.

2.1.2 Use of Internal Standards

Quantitative analytical methods for the determination of IAA are given in the chapter by Rivier. It is appropriate to indicate here that IAA labeled with carbon-14, or tritium, or with deuterium (Magnus et al. 1980), or carbon-13 (Cohen et al. 1985) has been, almost exclusively, used as the internal standard to measure the IAA liberated by hydrolysis of conjugates of IAA. The ideal internal standard would be a high specific activity conjugate identical in chemical properties to the conjugate being measured. In practice this is not often possible, since the appropriate internal standard may not be available or the identity of the conjugates may be unknown. Thus, when using free IAA as an internal standard during conjugate hydrolysis, there is the possibility that the conjugated IAA may be more or less stable to the conditions of hydrolysis than the free IAA, leading to an inaccurate estimate of losses during hydrolysis. Two careful studies have been made of this potential problem. Cohen (1982) quantitatively estimated the amount of IAA aspartate in soybean seeds using either labeled IAA or IAA aspartate as internal standard. The results for total IAA in the seed following hydrolysis with 7 N alkali were 18 μmol kg^{-1} using IAA-aspartate as internal standard and 19 μmol kg^{-1} using IAA as internal standard. Thus no error was introduced by using free IAA as internal standard. Since the conditions of hydrolysis for amide-linked IAA (7 N NaOH, 100 °C, 3 h) are more drastic than for ester hydrolysis (1 N NaOH, 25 °C, 1 h) it would seem that IAA is an acceptable internal standard for assays involving hydrolysis of both amide and ester IAA. Chisnell (1984a) has used either [^{14}C]-IAA or [^{3}H]-IAA-inositol as internal standard for measuring ester IAA in *Zea mays*. Agreement was within 10 to 20%, so again it may be concluded that either free IAA or ester IAA is an acceptable internal standard for IAA-ester determination. The generality of this conclusion should be further tested.

Deuterium and Carbon-13 Labeled Internal Standards. Conjugates are usually assayed by hydrolyzing the conjugate and measuring the resultant free IAA by means of an internal standard, as discussed elsewhere in this Volume by Rivier. The necessity for hydrolysis in strong alkali imposes restraints on the kinds of internal standards that may be employed. For example, IAA-2,2-d$_2$ has proven unsatisfactory, owing to losses of deuterium from the side chain under alkaline conditions (Caruso et al. 1978; Magnus et al. 1980). IAA labeled in the 4,5,6, and 7 positions is most satisfactory although 2,4,5,6,7 pentadeutero IAA may also be used (Magnus et al. 1980). Even the polydeuterated compounds lead to some problems owing to GC peak shapes and retention times, which differ from those of the unlabeled compound (Cohen et al. 1985). The recent synthesis of IAA labeled with 6 C-13 in the benzene ring should provide an ideal internal standard (Cohen et al. 1985).

2.2 Analysis Before Hydrolysis

A Quantitative Estimation of IAA-Inositol
Using [^{3}H]-IAA-myo-Inositol as an Internal Standard

There appears to be only one example in the literature where IAA-inositol was quantitatively estimated using labeled IAA-inositol as the internal standard. Chisnell (1984a, b) found an average of 74 nmol kg^{-1} fresh weight of IAA-inositol in *Zea mays* seedling shoot tissue. Thus, he could conclude that IAA-inositol constituted 19% of the ester IAA of the *Zea mays* shoot tissue.

3 Qualitative Analysis of IAA-Inositols

By far the greatest use of GC-MS has been in the identification and structural characterization of the IAA conjugates including the IAA inositols. A dramatic illustration of the gains in sensitivity provided by GC-MS is obtained by comparing the original studies of Labarca et al. (1965) with those of Ehmann (1974) and Ehmann and Bandurski (1974). The earlier studies utilized 50 mg of IAA-inositol to yield sufficient inositol for identification by chromatography and by melting point, whereas the later structural studies were accomplished with 0.2 mg of IAA-inositols, and actually using only 50 µg.

3.1 The Inositol Moiety

IAA-aspartate (Andreae and Good 1955) and IAA-glucose (Zenk 1961) were the first IAA conjugates to be chemically characterized. Both had been formed following the application of IAA to the plant, and in that sense, these compounds were not endogenous compounds. Thus, the IAA-inositols (Labarca et al. 1965) were the first endogenous IAA esters to be fully characterized, including, the use of GC-MS (Ueda et al. 1970). It remains uncertain as to why myo-inositol is the esterifying moiety. Myo-inositol is ubiquitous in nature, having been found in every living species examined, except for a few bacteria (Anderson 1972; Posternak 1965). Inositol is not abundant, as compared to other carbohydrates. Thus, the requirement of myo-inositol as a growth factor for many tissue cultures (e.g., Posternak 1965) and the occurrence of myo-inositol in membrane liquids suggests that the esterification of the growth hormone, IAA, to myo-inositol may not be a fortuitous association.

Myo-inositol was isolated over 100 years ago by Scherer (e.g., Posternak 1965) from muscle and given the name inosit. It has subsequently been called meso-inositol, or i-inositol, although the name myo-inositol is now widely accepted. The structure of myo-inositol may best be envisaged as a turtle (Agranoff 1978) as described below: The 2 OH is axial and projects above the plane of the ring as would the head of a turtle (Fig. 1). Hydroxyls 1,3, and 5 are equatorial, but project above the plane of the ring. Thus, the D pair, 1 and 3, may be envisaged respectively as

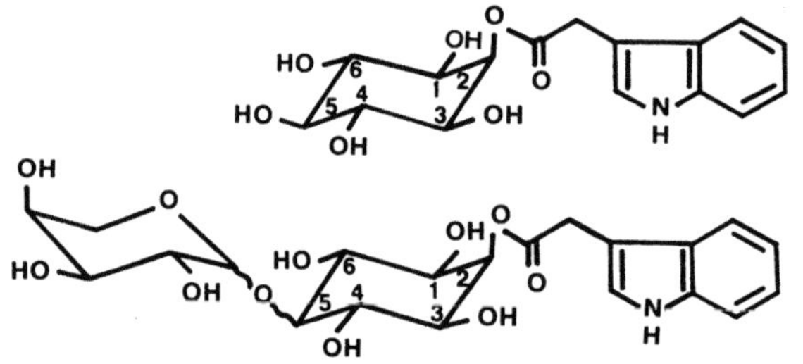

Fig. 2. A The structure of indole-3-acetyl-2-O-myo-inositol. Note the axial 2-O and the plane of symmetry about a line through the 2 and 5. The inositol is numbered as though it were the L series. If the IAA were on the 1-OH it would be a 1-L-1-O compound. If the IAA were on (here numbered as 3), the compound would be 1-D-1-O. Cyclitol nomenclature has been discussed by Agranoff (1978). **B** The structure of 5-O-L-arabinopyranosyl-2-O-indole-3-acetyl-myo-inositol. (Ueda and Bandurski 1974)

the right and left front paws of the turtle. The D pair 4 and 6 project down from the plane of the ring and may be visualized respectively as the left and right hind paws with the 5 OH projecting slightly up and to one side as the tail. With the mnemonic of Fig. 1, patterned after Agranoff (1978), the reader will remember the stereochemistry of myo-inositol.

The structure of a cylitol, such as myo-inositol, permits the acyl substituent, IAA, to migrate freely about the ring, particularly between cis-hydroxyls. Thus, there are six isomeric IAA inositols, the D,L-1 pair, the DL 4 pair, and the nonoptically active 2-O and 5-O esters (Fig. 2). Four of the six isomers are separable by chromatographic methods and four distinct IAA inositols are, in fact, observed by gas chromatography. The D,L pairs could be resolved by use of chiral columns although this has not yet been accomplished. An enzyme could recognize all six isomers or, if only one of the isomers is biologically active, this could be a subtle and difficult to study form of biological control.

The phenomenon of acyl migration makes the assignment of an exact structure to an equatorially substituted myo-inositol difficult. It is only the axial 2 OH which is recognizable by NMR (Nicholls 1967; Nicholls et al. 1971) and thus it is only the IAA-2-O-myo-inositol and the related axial IAA-2-O-myo-inositol glycosides which are uniquely characterized. The facile migration to the 1 or 3 (D,L-1) hydroxyls makes possible a tentative identification of the IAA-inositols that are esterified to these equatorial hydroxyls. Both M. Ueda and A. Ehmann (Ehmann and Bandurski 1974; Ueda and Bandurski 1974) have made careful attempts to recognize the isomers. It must be recognized that the differences in mass-spectral fragmentation patterns are subtle and subject to influences such as that of source pressure. Thus, comparisons of the mass spectra for purposes of identifying the isomers must be done under conditions of identical sample size.

3.2 Derivitization of IAA-Inositols for GC-MS

The most commonly used derivative of the IAA-inositols has been the trimethylsilyl (TMS) ethers. Most commonly 30 to 300 nmoles of IAA-inositol is treated for 45 min at 45 °C with 10 to 50 µl of N,O-bis(trimethylsilyl)trifluoroacetamide (BSTFA) containing 1% trimethylchlorosilane (TMCS). Derivitization improves if the samples are stored in a desiccator containing anhydrous calcium sulfate for 12 or more additional h at 25 °C. Under these conditions five TMS groups are added to the inositol and one TMS group is added to the nitrogen of the IAA. The derivatized IAA inositols may usually be stored for weeks, or even months, at -20 °C for later analysis. In earlier studies, derivitization was accomplished using N-(trimethylsilyl)-imidazole, 1:1 (v/v) pyridine. This reagent does not fully derivatize the indole nitrogen so that a mixture of one-half fully derivatized and one-half partially derivatized IAA-inositol is obtained. Rigorously anhydrous conditions are necessary, and the TMS derivatives have an uncertain storage life, owing, in part to hydrolysis and, in part, to decompostion.

Chisnell (1984 a, b) acetylated the IAA-inositols using a 1:1 (v/v) mixture of pyridine and acetic anhydride. Both reagents were distilled and stored over anhydrous $CaSO_4$. For derivitization, the samples were first evaporated to dryness, then using a N_2 flushed dry box, 50 µl of pyridine containing 0.5% (w/v) 4-dimethyl-aminopyridine was added to the sample followed by 50 µl of acetic anhydride. The mixture was reacted at 22 °C for 10 min to complete acetylation. Under these conditions, all five free hydroxyls of the inositol are acetylated and an acetyl group may, or may not add to the nitrogen. The acetylated derivatives are more reproducibly stable during storage.

Acetylation is essentially instantaneous although rigorously dry conditions, a reasonably clean sample, and a large excess of reagent to sample are required. Failure to observe the above conditions results in failure to acetylate the indole nitrogen or even failure to obtain a GC-volatile product.

Chromatography of the acetylated derivatives has been on 3% OV-17 at 300 °C and using a 2 mm × 0.6 m column temperature programmed from 200 (2 min hold) to 300 at 20 °C per minute and a He gas flow rate of 26 ml min^{-1}. Under these conditions the retention times for the isomers ranged from 7.9 to 9.4 min.

For TMS derivatives, a 2% OV-1 column was used with helium as carrier gas and a flow rate of 25 ml min^{-1}. Ionizing energy was 70 eV, with the flash heater at 270 °C and the molecular separator at 260 °C. Using a 0.3 m column retention times ranging between 5 and 12 min were observed for the four isomeric peaks resolved.

3.3 Mass Spectral Fragmentation Pattern

A complete mass spectral fragmentation pattern for an IAA-myo-inositol has not been published. Tables 3 and 4 show the complete 70 eV spectra for the 6 TMS derivative of an equatorial IAA-inositol, presumably the 1-O-ester, and for the axial, 2-O-ester (Fig. 3). These data were presented in monograph form by Ehmann et al. (1975) and the interpretation is that of Ehmann, as published in Ehmann and Bandurski (1974).

Table 3. The mass spectral fragmentation pattern of 1-DL-1-O-(indole-3-acetyl)-myo-inositol (6 TMS, MW = 769, Ehmann et al. 1975)

m/e	Relative intensity (%)	m/e	Relative intensity (%)	m/e	Relative intensity (%)	m/e	Relative intensity (%)
31	0.4	129	4.5	204	6.8	307	0.7
39	0.1	130	7.0	205	1.8	317	0.8
41	0.4	131	1.4	207	0.5	318	7.5
42	0.1	132	0.1	215	0.7	319	3.5
43	0.5	133	2.2	216	0.5	320	1.6
44	0.3	134	0.5	217	10.4	321	0.7
45	2.6	135	0.1	218	2.8	331	0.7
47	0.5	142	0.3	219	1.5	332	0.8
50	0.5	143	1.2	220	0.3	333	0.3
51	0.5	144	0.4	221	1.3	342	0.4
53	0.5	145	0.6	222	0.2	343	2.0
54	0.1	146	0.5	223	0.1	344	1.0
55	0.3	147	16.3	227	0.5	345	1.0
57	0.5	148	2.2	228	3.0	361	0.5
58	0.4	149	1.4	229	100.0	362	0.3
59	0.7	153	0.3	230	21.6	390	0.4
61	0.5	155	0.5	231	6.5	417	1.0
67	0.1	156	0.5	232	1.3	418	0.5
69	0.3	157	15.1	233	0.5	419	0.5
72	0.9	158	2.0	239	0.4	432	0.5
73	74.3	159	0.6	242	0.3	433	1.7
74	6.4	160	0.4	243	1.3	434	0.9
75	6.0	161	0.5	244	0.5	435	0.5
76	0.3	163	0.3	245	0.7	462	0.4
77	0.3	169	0.5	246	1.1	507	1.4
81	0.3	172	0.4	247	3.5	508	0.7
84	0.1	173	0.4	248	0.8	509	0.3
89	0.4	175	0.7	255	0.5	597	0.3
91	0.1	176	0.3	265	0.1	607	0.3
93	0.1	177	0.5	266	0.7	678	0.4
95	0.1	186	0.9	267	0.6	679	0.9
101	0.7	187	0.5	271	0.7	680	0.5
102	0.5	188	0.3	276	0.5	681	0.3
103	3.3	189	0.9	290	0.4	697	0.7
104	0.5	190	0.5	291	1.0	698	0.4
105	0.1	191	8.9	292	0.4	699	0.3
111	0.1	192	2.0	293	1.1	767	0.1
113	0.3	193	1.0	295	0.1	768	0.9
115	0.4	199	0.4	296	0.4	769(M^+)	11.2
116	0.4	200	2.4	303	0.4	770	8.2
117	0.7	201	2.3	304	1.6	771	3.3
119	0.3	202	33.7	305	3.6	772	1.3
127	0.3	203	7.2	306	1.3	773	0.2

Table 4. The mass spectral fragmentation pattern of 2-O-(indole-3-acetyl)-myo-inositol (6 TMS, MW = 769, Ehmann et al. 1975)

m/e	Relative intensity (%)	m/e	Relative intensity (%)	m/e	Relative intensity (%)	m/e	Relative intensity (%)
31	0.6	103	3.6	187	0.7	291	1.3
39	0.3	104	0.6	189	1.2	292	0.6
41	2.3	105	0.3	190	0.6	293	1.2
42	0.9	106	0.6	191	8.8	294	0.6
43	2.1	109	0.2	192	1.8	295	0.2
44	1.1	111	0.6	193	1.2	303	0.6
45	3.3	113	0.6	200	3.1	304	4.4
47	0.5	116	0.6	201	3.1	305	4.7
51	0.6	117	1.4	202	48.4	306	2.0
53	0.7	118	0.8	203	10.0	307	0.8
54	0.3	119	0.5	204	5.3	317	1.0
55	0.2	120	0.6	205	2.7	318	8.3
56	0.1	121	0.7	207	0.6	319	4.1
57	0.6	125	0.6	208	0.1	320	2.1
58	0.6	127	0.6	215	0.9	321	0.7
59	0.9	129	5.1	216	1.0	331	1.0
61	0.7	130	2.8	217	9.8	332	1.2
67	0.5	131	1.5	218	2.9	333	0.6
68	0.5	132	0.3	219	1.4	343	3.3
69	2.3	133	2.7	221	1.8	344	1.3
70	0.1	134	0.6	222	0.2	345	1.1
71	0.8	135	0.1	223	0.2	417	1.4
72	1.2	137	0.7	227	0.6	418	0.6
73	85.1	143	0.9	228	3.9	432	0.6
74	7.0	145	0.6	229	100.0	433	2.3
75	7.0	147	17.5	230	22.0	434	1.0
76	0.4	148	2.5	231	5.8	435	0.6
77	0.6	149	2.1	232	1.4	462	0.8
78	0.6	150	0.6	233	0.6	507	2.1
79	0.1	151	0.5	243	1.4	508	1.0
81	1.4	153	0.5	244	0.6	509	0.6
82	0.3	155	0.6	245	0.7	574	0.6
83	0.7	156	0.6	246	1.8	697	0.8
84	0.2	157	3.1	247	2.6	698	0.5
85	0.2	158	1.0	248	0.8	767	0.1
91	0.1	161	0.8	255	0.6	768	0.7
93	0.2	169	0.5	265	1.8	769(M.$^+$)	15.0
95	0.7	172	0.6	266	0.7	770	10.5
96	0.3	173	0.6	267	0.8	771	4.6
97	0.6	175	0.7	271	0.7	772	1.3
98	0.5	177	0.6	276	1.2	773	0.1
101	0.8	186	1.5	290	0.7		

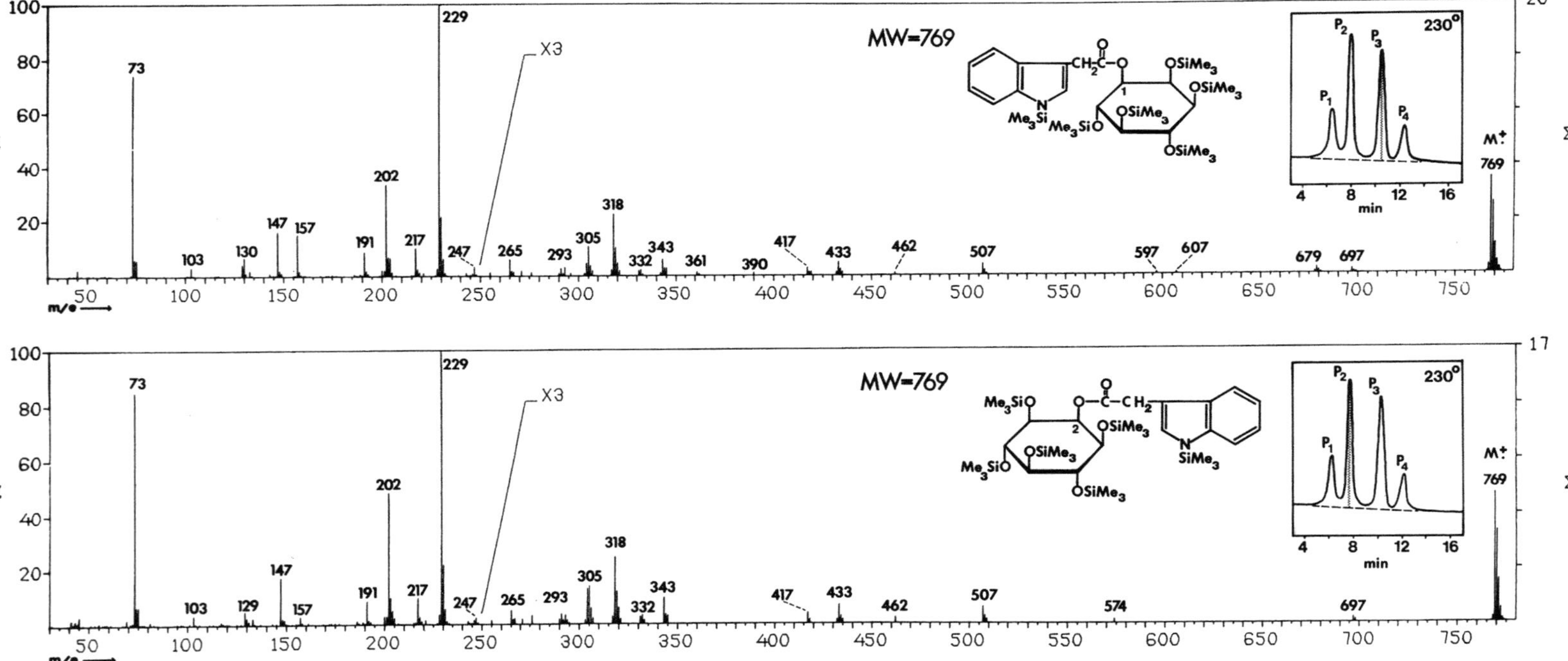

Fig. 3. *Upper* the 70 eV mass spectrum of the trimethylsilyl derivative (6 TMS) of L-1-O-(indole-3-acetyl)-myo-inositol. *Lower* The 70 eV mass spectrum of the trimethylsilyl derivative (6 TMS) of 2-O-(indole-3-acetyl)-myo-inositol. The *Inserts* show the GC profiles and *shaded areas under the peaks* indicate the portion of the peak scanned

3.3.1 1-DL-1-O-(Indole-3-Acetyl)-myo-Inositol (6 TMS, MW 769)

The molecular ion (M^+ at m/e 769) of the fully derivatized IAA inositol with a TMS group on the imino nitrogen of the indole is more stable than the 5 TMS IAA-inositol, lacking a TMS on the indole (Ueda and Bandurski 1974). The pentaacetyl derivative described by Chisnell (1984 a, b) shows an even more stable molecular ion. The molecular ion (at m/e 769) loses 72 daltons to yield 697, since a hydrogen from the departing TMS transfers to the indole nucleus (Ehmann and Bandurski 1974). The loss of TMSOH yields 679, which is supported by a metastable ion at m/e 600 (calc. 599.5). Loss of $CH_2(CH_3)_2Si$ as a radical yields 607. Base peak at 229 represents the TMS-IAA cleaved from the molecular ion at the ester bond.

3.3.2 2-O-(Indole-3-Acetyl)-myo-inositol (6 TMS, MW 769)

An important question is whether the equatorial esters can be differentiated from the 2-O (axial) ester. The answer is a qualified, yes, subject to the conditions discussed above. The 2-O ester lacks ions at 679, 607, 597, and 361 but does have 574, which is lacking in the equatorial ester. These differences, plus the GC retention time differences, permit identification. Ueda earlier addressed this question (Ueda and Bandurski 1974), using the penta TMS esters. In that case, the molecular ion (m/e = 697) of the 2-O ester fragmented with loss of the entire indolyl moiety (m/e = 174) to yield a 523, whereas the equatorial ester lost 90 TMSOH to yield a 607.

These characteristic fragmentation patterns plus the marked differences in retention time on GC or even thin layer chromatography make the IAA inositols easily differentiated from IAA glucose. A full description of the GC-MS characteristics of the known IAA glucose anhydride (1-O) plus the IAA glucose esters 2,4, and 6-O has been published (Ehmann 1974; Ehmann et al. 1975).

3.3.3 Di-O-[N-(Trimethylsilyl)Indole-3-Acetyl]-O-Tetra-O-Trimethylsilyl-myo-Inositol

Both $(IAA)_2$-myo-inositol and $(IAA)_3$-myo-inositol occur in extracts of the kernels of *Zea mays* sweet corn. The occurrence of these unusual compound indicates that the enzyme which transfers IAA from IAA-glucose to inositol (indole-3-acetyl-glucose : myo-inositol indol-3-acetyl transferase; Michcalczuk and Bandurski 1982) will utilize IAA-inositol, or even IAA_2-inositol as an acceptor for the IAA transferred from IAA-glucose to the recipient inositol. Why the enzyme should do this is not clear, since there are ample reserves of free inositol in the tissue.

The 70 eV mass spectral fragmentation patterns have been discussed by Ehmann (Ehmann and Bandurski 1974), but the complete tabular mass spectra have hitherto been available only in monograph form (Ehmann et al. 1975).

GC of the $(IAA)_2$-inositol shows three peaks with 80% of the total ester in the peaks described in Table 5 and Fig. 4A. The ion at 926 indicates the compound to be di-O-[N-(trimethylsilyl)indole-3-acetyl]-O-tetra-O-trimethylsilyl-myo-inositol. M^+ loses CH_3 to yield 911 and also loses 72 daltons to yield m/e

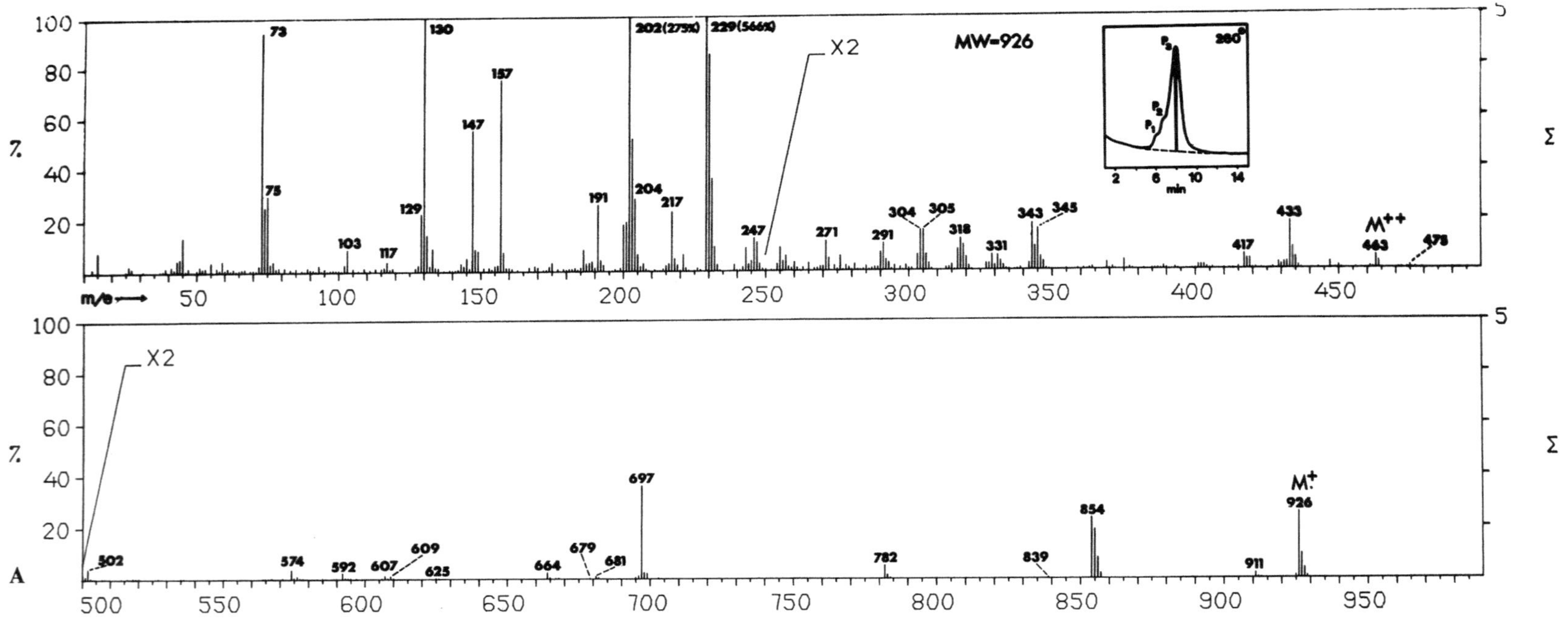

MW=926
73
130
202(275X)
229(566X)
X2
157
147
75
129
103
117
191
204
217
247
271
291
304
305
318
331
343
345
417
433
M++
463
478
280°
P3
P2
P1
2 6 10 14 min
m/e
X2
502
574
592
607
609
625
664
679
681
697
782
839
854
911
M.+
926
%
Σ

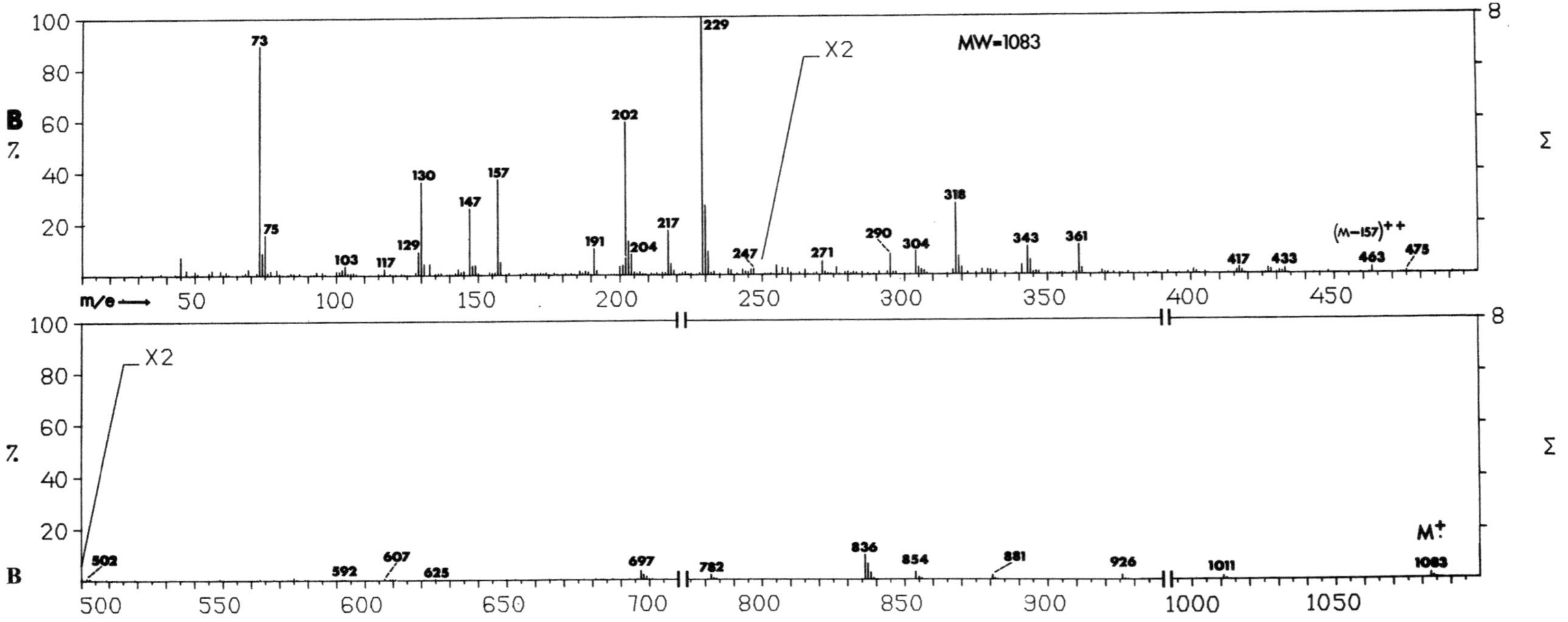

Fig. 4. A The 70 eV mass spectral fragmentation pattern of Di-O-(N-trimethylsilyl-indole-3-acetyl)-O-tetra-O-trimethylsilyl-myo-inositol (MW = 926). The portion of the GC peak scanned is shown as P-3 in the *inset*. (Ehmann et al. 1975) **B** The 70 eV mass spectral fragmentation pattern of tri-O-(N-trimethylsilyl-indole-3-acetyl)-O-tri-O-trimethylsilyl-myo-inositol (MW = 1083). The sample was introduced using a solid probe

Table 5. The mass spectral fragmentation pattern of Di-O-(N-trimethylsilyl-indole-3-acetyl)-O-tetra-O-trimethylsilyl-myo-inositol (MW = 926, Ehmann et al. 1975)

m/e	Relative intensity (%)	m/e	Relative intensity (%)	m/e	Relative intensity (%)	m/e	Relative intensity (%)
31	0.7	93	2.6	151	1.5	209	0.4
37	0.5	94	0.5	153	1.5	211	0.7
38	0.8	95	1.3	154	1.0	212	0.2
39	1.8	96	0.6	155	2.3	213	0.4
41	2.3	97	0.7	156	2.7	214	1.0
42	1.3	98	0.9	157	75.0	215	2.4
43	4.7	99	0.8	158	7.6	216	3.1
44	5.3	100	0.7	159	1.5	217	23.5
45	13.6	101	0.5	160	1.5	218	5.1
46	0.7	102	2.8	161	1.2	219	2.6
47	1.9	103	8.7	163	0.7	220	0.6
48	0.4	105	1.4	164	0.1	221	6.7
49	0.4	106	0.3	166	0.2	222	1.6
50	1.0	107	0.5	167	1.7	223	0.6
51	2.4	108	0.1	168	0.3	225	0.4
52	1.6	109	1.6	169	2.1	226	1.4
53	1.9	111	0.9	170	1.7	227	0.7
55	4.0	113	1.3	172	0.5	229	566.0
56	0.1	115	1.4	173	0.8	230	85.0
57	1.9	116	1.9	174	2.0	231	36.1
58	1.1	117	4.0	175	3.4	232	9.6
59	4.4	118	1.0	177	1.0	233	2.7
60	1.0	119	1.5	179	1.0	234	0.5
61	1.7	120	0.3	180	0.8	237	0.3
62	0.7	121	0.1	181	1.1	239	2.7
63	1.4	123	0.3	182	1.0	240	0.6
64	0.6	124	0.1	183	1.4	241	0.6
65	0.9	127	1.5	184	0.8	242	1.7
66	1.2	128	2.7	185	1.5	243	9.0
67	1.3	129	22.7	186	8.5	244	2.8
69	0.9	130	100.0	187	3.1	245	4.1
70	0.9	131	14.3	188	3.4	246	12.9
71	0.7	132	2.3	189	4.0	247	11.3
72	2.6	133	9.1	190	1.5	248	2.9
73	94.4	134	1.8	191	26.2	249	1.1
74	25.4	135	1.5	192	4.4	250	0.4
75	29.8	136	0.4	193	2.7	252	0.3
76	3.3	137	0.6	195	0.4	253	0.2
77	4.3	138	0.2	196	0.4	254	1.4
78	1.6	139	0.8	197	1.1	255	4.6
79	1.8	140	0.5	198	0.4	256	2.0
81	1.8	141	0.7	199	1.4	257	3.0
83	1.2	142	1.0	200	18.4	258	0.9
84	0.6	143	3.3	201	19.4	259	0.6
85	1.6	144	2.3	202	275.0	260	1.7
87	0.8	145	5.3	203	51.7	261	0.7
88	0.4	146	1.4	204	28.4	262	0.2
89	1.8	147	55.1	205	6.6	263	0.7
90	1.1	148	8.8	206	2.0	265	1.6
91	0.9	149	8.1	207	.3.1	267	0.5
92	0.5	150	0.3	208	1.0	268	0.6

Table 5 (continued)

m/e	Relative intensity (%)	m/e	Relative intensity (%)	m/e	Relative intensity (%)	m/e	Relative intensity (%)
269	0.9	332	1.9	412	0.2	518	0.3
270	0.9	333	1.0	413	0.2	519	0.2
271	5.8	334	0.4	415	0.5	520	0.2
272	2.5	335	0.3	417	3.0	526	0.1
274	1.1	337	0.2	418	2.1	527	0.2
275	0.4	338	0.3	419	2.1	535	0.2
276	2.9	339	0.5	420	0.3	536	0.2
277	0.3	342	1.4	421	0.4	556	0.1
278	1.1	343	9.2	423	0.2	564	0.2
279	0.7	344	4.7	426	0.2	565	0.2
280	0.3	345	8.1	427	0.4	566	0.2
281	1.3	346	2.7	429	1.3	567	0.3
282	0.3	347	1.7	430	0.9	571	0.3
283	0.2	348	0.4	431	1.3	574	2.0
284	0.4	349	0.2	432	1.5	575	0.4
285	0.9	350	0.1	433	9.3	576	0.7
286	0.3	355	2.5	434	4.3	577	0.3
287	0.5	358	0.1	435	2.4	578	0.2
288	0.7	359	1.6	436	0.7	579	0.1
289	0.7	361	1.7	437	0.3	586	0.1
290	3.6	362	0.8	443	0.3	589	0.2
291	5.3	363	0.4	445	0.1	590	0.2
292	2.1	364	0.6	447	1.4	592	1.3
293	1.6	369	1.5	448	0.1	593	0.4
294	0.4	370	0.3	449	0.2	594	0.3
295	1.0	371	0.6	450	0.7	595	0.3
297	0.7	372	0.2	451	0.2	597	0.2
298	0.2	375	2.0	456	0.2	599	0.1
299	1.1	376	0.2	461	0.5	602	0.1
300	0.5	377	0.5	462	0.3	604	0.1
301	0.5	378	0.3	463(M^{2+})	2.7	606	0.2
303	3.1	382	0.2	464	1.6	607	0.8
304	8.2	385	0.3	465	0.1	608	0.3
305	7.8	386	0.3	471	0.2	609	0.7
306	3.2	387	0.4	472	0.3	610	0.3
307	1.4	388	0.2	474	0.3	618	0.1
308	0.4	389	0.7	475	0.7	625	0.3
313	0.6	390	0.4	476	0.1	626	0.2
314	0.7	391	0.1	477	0.3	646	0.1
315	1.2	392	0.1	479	0.1	663	0.1
316	0.2	393	0.5	490	0.1	664	1.3
317	4.1	394	0.2	492	0.1	665	0.4
318	6.2	396	0.1	500	1.0	667	0.2
319	5.0	399	0.2	501	0.7	668	0.3
320	2.6	400	0.2	502	2.1	679	0.2
321	0.9	401	0.9	503	1.8	681	0.6
322	0.2	402	0.9	504	0.5	682	0.4
327	1.4	403	0.9	507	0.2	689	0.1
328	1.3	404	0.5	508	0.1	695	0.4
329	3.0	405	0.2	514	0.2	696	0.7
330	1.0	406	0.4	516	0.2	697	18.1
331	2.9	410	0.3	517	0.1	698	1.3

Table 5 (continued)

m/e	Relative intensity (%)	m/e	Relative intensity (%)	m/e	Relative intensity (%)	m/e	Relative intensity (%)
699	1.2	784	0.2	857	1.0	926 (M$^+$)	12.6
703	0.2	839	0.2	911	1.0	927	4.6
708	0.1	854	11.6	912	0.2	928	2.0
782	2.6	855	9.4	913	0.1	929	0.4
783	0.8	856	4.0	925	0.4		

854 which in turn yields 697 by loss of the indole-3-ketene radical. The 697 ion plus the fact that the (IAA)$_2$-inositol can be selectively hydrolyzed to yield IAA-inositol is diagnostic.

3.3.4 Tri-O-[N-(Trimethylsilyl)Indole-3-Acetyl]-O-Tri-O-Trimethylsilyl-myo-Inositol

This compound, (IAA)$_3$-inositol, decomposes under GC conditions to yield (IAA)$_2$-inositol. A mass spectrum has been obtained by means of the solid probe and shows a molecular ion at m/e 1083 which corresponds to (IAA)$_3$-inositol, the complete 70 eV mass spectrum of this compound is shown in Table 6 and Fig. 4 B. Again, the ion at 1083 plus the yield of (IAA)$_2$-inositol under GC conditions are diagnostic (Ehmann and Bandurski 1974; Ehmann et al. 1975).

3.3.5 IAA-myo-Inositol-Arabinoside and IAA-myo-Inositol-Galactoside

The 70 eV mass spectra of the 7 TMS IAA-myo-inositol-arabinoside (Fig. 2) and the 8 TMS IAA-myo-inositol-galactoside have been published (Ueda et al. 1970; Ueda and Bandurski 1974). The major ions are shown in Table 7. Under the derivitization conditions used (hexamethyldisilazane + trimethylchlorosilane in pyridine or trimethylsilylimidazole in pyridine), the indole nitrogen is not substituted. A later publication (Corcuera et al. 1982) in which the more vigorous silylation reagent, NO-bis(trimethylsilyl)trifluoroacetamide containing 1% trimethylchlorosilane was used, results in the 9 TMS IAA-myo-inositol-galactoside. The 8 TMS IAA-myo-inositol-galactoside shows a molecular ion at 1075 which loses CH$_3$ to form m/e = 1060 or M$^+$ minus TMSOH to form 985. An identical fragmentation occurs for the pentoside. Both the galactoside and pentoside yield an IAA-inositol at 769 or 697, depending on the substitution of the indole nitrogen. These ions plus the characteristic sugar ion at 361 are diagnostic.

3.4 Uses of GC-MS to Identify and Characterize IAA-Esters

The first published report of the use of GC-MS to characterize an IAA-inositol is that of Udea and Bandurski (1969 a). This and two subsequent reports (Ueda et al. 1970; Ueda and Bandurski 1974) were concerned with the structural char-

Table 6. The mass spectral fragmentation pattern of tri-O-[N-(trimethylsilyl-indole-3-acetyl)-O-tri-O-trimethylsilyl]-myo-inositol (MW = 1083, Ehmann et al. 1975)

m/e	Relative intensity (%)	m/e	Relative intensity (%)	m/e	Relative intensity (%)	m/e	Relative intensity (%)
31	0.7	105	1.0	171	0.8	223	1.4
37	0.3	106	1.2	172	0.9	224	0.5
38	0.9	107	0.7	173	0.7	225	0.9
43	0.7	110	0.3	174	1.1	227	0.4
45	7.5	112	0.3	175	0.4	228	1.0
46	0.1	115	0.9	177	0.9	229	100.0
47	2.4	116	0.6	178	0.4	230	27.4
49	0.4	117	2.8	179	0.2	231	9.3
50	1.8	118	0.3	180	0.4	232	1.0
51	0.9	119	0.7	181	0.8	233	1.6
54	0.2	120	0.3	182	0.4	234	0.3
55	1.2	121	0.6	183	0.8	235	0.7
56	2.1	123	0.7	184	0.4	237	0.2
58	0.2	124	0.5	185	0.1	238	2.5
59	2.0	125	0.4	186	1.8	239	2.0
60	0.5	126	0.3	187	1.1	241	0.8
61	1.5	128	1.4	188	1.8	243	2.4
62	0.5	129	9.3	189	1.3	244	1.5
65	0.4	130	36.4	190	0.4	245	1.3
67	0.5	131	4.8	191	10.7	246	2.3
68	1.0	133	4.8	192	2.0	247	2.6
69	2.6	134	0.2	193	0.2	249	0.2
72	1.1	135	0.7	194	0.3	251	0.4
73	89.9	136	0.9	195	0.1	252	0.3
74	8.9	137	1.0	196	0.4	253	0.4
75	16.0	139	0.6	197	0.6	254	0.2
76	1.2	140	0.5	198	0.6	255	2.0
77	2.0	141	0.4	199	0.4	256	0.3
78	0.4	142	1.1	200	3.6	257	1.4
79	2.5	143	2.7	201	4.1	259	1.4
80	0.9	144	1.3	202	59.6	260	0.5
83	0.5	145	1.8	203	13.6	261	0.2
84	1.0	146	0.4	204	8.4	262	0.2
85	0.9	147	26.3	205	1.4	263	0.5
86	0.2	148	4.0	206	0.9	265	1.1
87	0.9	149	4.2	207	1.3	266	0.3
88	0.3	150	0.9	208	0.8	268	0.3
89	0.3	151	0.2	209	0.8	269	0.5
90	0.3	154	1.4	210	0.2	271	2.8
91	0.5	155	1.0	211	0.9	272	0.7
92	0.6	156	1.2	212	0.6	273	0.4
93	1.5	157	37.6	213	1.1	274	0.4
94	0.2	158	5.4	214	0.4	275	0.2
95	1.3	160	0.6	215	1.2	276	1.6
96	0.5	161	1.0	216	1.0	277	0.3
97	0.1	164	0.2	217	17.7	279	0.6
100	1.9	165	0.6	218	4.7	280	0.7
101	1.7	166	0.7	219	2.0	281	0.2
102	2.5	167	1.0	220	0.3	283	0.5
103	4.0	169	0.7	221	6.8	284	0.1
104	1.1	170	0.7	222	1.0	285	0.6

Table 6 (continued)

m/e	Relative intensity (%)	m/e	Relative intensity (%)	m/e	Relative intensity (%)	m/e	Relative intensity (%)
287	0.4	352	0.3	445	0.2	610	0.3
289	0.3	353	0.2	448	0.5	611	0.3
291	0.7	354	0.3	451	0.1	621	0.2
294	0.7	355	0.4	456	0.1	625	0.4
295	4.1	356	0.2	459	0.3	636	0.1
296	0.6	359	0.5	462	0.3	637	0.2
297	0.8	360	0.5	463	1.3	643	0.1
300	0.3	361	5.7	$(M-157)^{2+}$		650	0.2
302	0.2	362	0.8	472	0.2	653	0.2
304	4.7	369	0.3	473	0.3	665	0.2
305	1.6	370	0.5	474	0.3	667	0.1
306	1.1	371	0.5	475	0.6	669	0.2
307	0.9	373	0.5	476	0.2	677	0.2
308	0.4	375	0.4	477	0.2	681	0.3
310	0.1	376	0.3	479	0.2	683	0.2
312	0.1	378	0.5	491	0.4	684	0.2
315	0.3	382	0.2	493	0.1	687	0.3
317	1.0	384	0.2	499	0.3	688	0.1
318	14.0	387	0.4	500	0.4	693	0.3
319	3.7	388	0.4	501	0.2	695	0.2
320	1.6	392	0.6	502	0.5	696	0.3
321	0.2	396	0.3	509	0.3	697	1.2
322	0.6	399	0.5	516	0.1	698	0.6
325	0.5	401	0.9	519	0.3	699	0.4
326	0.2	402	0.6	531	0.1	782	1.0
327	1.1	403	0.2	548	0.2	783	0.4
328	0.3	405	0.5	549	0.4	784	0.2
329	1.1	406	0.2	562	0.2	836	4.8
330	1.0	413	0.3	563	0.5	837	3.2
331	0.4	414	0.2	565	0.1	838	1.4
332	0.8	416	0.8	568	0.4	839	0.4
335	0.4	417	1.3	571	0.3	854	1.4
338	0.2	418	0.8	573	0.2	855	0.4
339	0.4	419	0.2	574	0.2	856	0.2
340	0.2	425	0.2	575	0.6	881	1.0
341	2.0	426	0.3	576	0.2	882	0.4
342	0.2	427	1.2	579	0.3	883	0.2
343	5.6	428	0.9	580	0.2	926	1.0
344	2.9	430	0.4	590	0.3	927	0.2
345	0.6	431	0.7	591	0.1	1011	0.8
346	0.8	432	0.6	592	0.3	1012	0.4
347	0.7	433	1.1	593	0.2	$1083(M^{+})$	1.0
350	0.3	434	0.2	607	0.5	1084	0.4
351	0.3	443	0.2	609	0.3		

Table 7. The 70 eV mass spectral fragmentation pattern of the trimethylsilyl ester of IAA-myo-inositol-arabinoside and IAA-myo-inositol-galactoside. (Ueda and Bandurski 1974)

m/e	Compound			
	IAA-inositol pentose	IAA-inositol hexose	Origin	Ion species
73	100.0	100.0	P&H[a]	
103	10.1	9.9	P&H	
129	11.1	11.9	P&H	
130	37.8	23.6	P&H, ROH	
133	7.5	7.7	P&H	
147	35.8	39.8	P&H	
157	38.8	29.2	P&H, ROH	
169	2.8	2.3	P	
189	3.5	3.4	P&H	
191	15.6	12.0	P&H	
204	44.3	52.8	P&H	
217	30.2	21.2	P&H	
221	15.6	21.2	P&H	
259	23.2	2.3	P	
265	1.6	2.4	P&H	
271	3.7	3.2	H	
291	1.5	0.8	P&H	
293	–	0.6	P&H	
305	2.8	2.5	P&H	
318	3.5	2.6	P&H	
343	4.9	2.8	P&H	
349	4.6	0.7	P	
361	+	8.4	H	
390	+	–	P&H	
417	+	–	P&H	
433	2.0	1.8	P&H	
479	0.2	–	P&H	m/e 653-IAA
563	0.4	0.3	P&H	m/e 653-TMSOH
607	0.4	0.3	P&H	m/e 697-TMSOH
609	+	0.2		
653	0.6	0.9	P&H	
682	+	0.1	P&H	m/e 697-Me
697	2.1	1.9	P&H	IAA-inositols
739	+	+	P&H	
868	0.2	–	P	M-Me-TMSOH
880	–	0.07	H	M-Me-2TMSOH
883	+	–	P	M-TMSOH
970	–	0.2	H	M-Me-TMSOH
973	3.8	–	P	M
985	–	0.03	H	M-TMSOH
1,060	–	0.05	H	M-Me
1,075	–	1.4	H	M

[a] The three isomeric IAA-inositol-pentose peaks co-elute during gas chromatography with the three isomeric IAA-inositol-hexose peaks. Thus the designation P (pentose) or H (hexose) indicates whether the ion fragment, most probably, originated from a pentoside or hexoside.

acterization of IAA-myo-inositol, IAA-myo-inositol-arabinoside and IAA-myo-inositol-galactoside. Owing to the small amounts of these compounds isolated from corn kernels, GC-MS was essential to the work. Ehmann and Bandurski (1974) used GC-MS to structurally characterize the isomers of IAA-myo-inositol and to identify and characterize $(IAA)_2$-myo-inositol and $(IAA)_3$-myo-inositol. Ehmann (1974) also identified the 2-O, 4-O, and 6-O, IAA-glucose esters. This work, and the later chromatographic studies of Michalczuk and Bandurski (1982) indicated that acyl migration from the 1-O position to the 2-O, 4-O, and 6-O of glucose was so facile as to make isolation of the 1-O isomer from a natural source, a very difficult procedure (Zenk 1961). Other useful applications of GC-MS to IAA esters has been to prove the identity of the product of an enzyme-catalyzed reaction, as for examples, IAA-glucose, IAA-myo-inositol (Michalczuk and Bandurski 1982), or IAA-myo-inositol-galactoside (Corcuera et al. 1982). In the case of high specific activity labeled compounds, synthesized by enzymatic (Michalczuk and Chisnell 1982) or chemical means (Nowacki et al. 1978), it has proven useful to use GC-MS to characterize unlabeled samples, and then compare the labeled compounds by other methods with the GC-MS authenticated compounds, thus avoiding contamination of the source of the mass spectrometer.

4 Conclusions

Further study of the conjugates of IAA is mandated, if for no other reason, than by their widespread distribution in nature and by their presence in large amounts relative to the free hormone. The very occurrence of IAA conjugates is sufficient to perturb the pool size of the free and active hormone. Unfortunately, the conjugates of only two plant species are under active study, *Zea mays*, sweet corn, and *Glycine max*, soybean. Thus, much remains to be learned. For example, we know little concerning the occurrence of conjugates in mature plants, since most studies have dealt with seeds or seedling plants. Similarly, we know little of the structures of the conjugates that occur in vegetative tissue. We know almost nothing of the localization of the conjugates within the cell. It would be helpful to know which conjugates, if any, are in the vacuole, and which might be in an active cytoplasmic pool. We know little about where the conjugates are in the intact plant, for example, the coleoptile tip and mesocotyl node, or root cap and root tip.

The methodologies for studying the conjugates are available since separation and purification techniques, such as HPLC, capillary GC, and mass spectrometry have developed so well. It seems likely that the techniques here described together with further extensions, such as the use of antibody techniques, will lead to a better concept of how the conjugates are involved in the overall hormonal metabolism of the plant.

5 Abbreviations

BSTFA   N,O-bis(trimethylsilyl)trifluoroacetamide
GC/MS   gas chromatographic/mass spectrometric
IAA   indole-3-acetic acid
Inositol myoinositol
TMCS   trimethylchlorosilane
TMS   trimethyl silyl, $(CH_3)_3Si$
TMSOH $(CH_3)_3SiOH$.

Acknowledgment. We are indebted to Ms. Wendy Whitford and Mr. Mark D. Desrosiers for aid in manuscript preparation. Supported by the Metabolic Biology Section of the U.S. National Science Foundation (PCM 8204017, ORD 30668) and the National Aeronautic and Space Administration (NASA-NAGW 97, ORD 33355).

References

Agranoff BW (1978) Cyclitol confusion. Trends Biochem Sci 3:N283–285
Anderson L (1972) The Cyclitols. In: Pigman W, Horton D (eds) The carbohydrates. Academic, New York, pp 520–580
Andreae WA, Good NE (1955) The formation of indoleacetylaspartic acid in pea seedlings. Plant Physiol 30:380–382
Avery GS, Berger J, Shalucha B (1941) The total extraction of free auxin and auxin-precursor from plant tissues. Am J Bot 28:596–607
Avery GS, Creighton HB, Shalucha B (1940) Extraction methods in relation to hormone content of maize endosperms. Am J Bot 27:289–300
Bandurski RS (1978) Chemistry and physiology of *myo*-inositol esters of indole-3-acetic acid. In: Wells WW, Eisenberg Jr F (eds) Cyclitols and phosphoinositides. Academic, New York, pp 35–54
Bandurski RS (1979) Chemistry and physiology of conjugates of indole-3-acetic acid. In: Mandava NB (ed) Plant growth substances. Am Chem Soc, Washington DC, pp 1–17
Bandurski RS (1980) Homeostatic control of concentrations of indole-3-acetic acid. In: (1979) Skoog F (ed) Plant growth substances. Springer, Berlin Heidelberg New York, pp 37–49
Bandurski RS (1982) Auxin biosynthesis and metabolism. In: Wareing P (ed) Plant growth substances. Academic, London, pp 3–11
Bandurski RS (1983) Mobilization of seed indole-3-acetic acid reserves during germination. In: Nozzolillo C, Lea PJ, Loewus F (eds) Mobilization of reserves in germination. Plenum, New York, pp 213–228
Bandurski RS (1984) Metabolism of indole-3-acetic acid. In: Crozier A, Hillman JR (eds) The biosynthesis and metabolism of plant hormones. Cambridge University Press, Cambridge, pp 183–200
Bandurski RS, Nonhebel HM (1984) Auxin. In: Wilkins MB (ed) Advanced plant physiology. Pitman, London, pp 1–20
Bandurski RS, Schulze A (1977) The concentration of indole-3-acetic acid and its derivatives in plants. Plant Physiol 60:211–213
Bandurski RS, Schulze A, Cohen JD (1977) Photo-regulation of the ratio of ester to free indole-3-acetic acid. Biochem Biophys Res Commun 79:1219–1223
Bentley JA (1958) The naturally-occurring auxins and inhibitors. Annu Rev Plant Physiol 9:47–80

Bentley JA (1961) The states of auxin in the plant. In: Ruhland W (ed) Handbuch der Pflanzenphysiologie. Springer, Berlin Heidelberg New York, 14:609–619

Caruso JL, Smith RG, Smith LM, Cheng TY, Daves GD (1978) Analysis of indole-3-acetic acid in douglas fir using a deuterium analog and combined gas chromatography-mass spectrometry. Plant Physiol 62:841–845

Chisnell JR (1984a) *Myo*-inositol esters of indole-3-acetic acid are endogenous components of *Zea mays* L. shoot tissue. Plant Physiol 74:278–283

Chisnell JR (1984b) The presence and translocation of indole-3-acetyl-*myo*-inositol in shoots of *Zea mays* L. PhD Thesis, Michigan State University

Cholodny NG (1935) Über das Keimungshormon von Gramineen. Planta 23:289–312

Cohen JD (1982) Identification and quantitative analysis of indole-3-acetyl-L-aspartate from seeds of Glycine max L. Plant Physiol 70:749–753

Cohen JD, Bandurski RS (1978) The bound auxins: protection of indole-3-acetic acid from peroxidase-catalyzed oxidation. Planta 139:203–208

Cohen JD, Bandurski RS (1982) Chemistry and physiology of the bound auxins. Annu Rev Plant Physiol 33:403–430

Cohen JD, Baldi BG, Slovin JP (1986) $^{13}C_6$-[Benzene ring]-indole-3-acetic acid: a new internal standard for quantitative mass spectral analysis of indole-3-acetic acid in plants. Plant Physiol 80:14–19

Corcuera L (1967) Estudio de indolacetilinositoles en granos de maiz. BS Thesis Univ Catolica de Chile, Santiago

Corcuera LJ, Bandurski RS (1982) Biosynthesis of indole-3-yl-acetyl-*myo*-inositol arabinoside in kernels of *Zea mays*. Plant Physiol 70:1664–1666

Corcuera LJ, Michalczuk L, Bandurski RS (1982) Enzymatic synthesis of indol-3-yl-acetyl-*myo*-inositol galactoside. Biochem J 207:283–290

Domagalski W, Schulze A, Bandurski RS (1985) Isolation and characterization of indole-3-acetyl-*myo*-inositol from the chestnut *Aesculus parviflora*. Plant Physiol 77 (Supplement):3

Ehmann A (1974) Identification of 2-O-(indole-3-acetyl)-D-glucopyranose, 4-O-(indole-3-acetyl)-D-glucopyranose and 6-O-(indole-3-acetyl)-D-glucopyranose from kernels of *Zea mays* by gas-liquid chromatography-mass spectrometry. Carbohydr Res 34:99–114

Ehmann A, Bandurski RS (1974) The isolation of di-O-(indole-3-acetyl)-*myo*-inositol and tri-O-(indole-3-acetyl)-*myo*-inositol from mature kernels of *Zea mays*. Carbohydr Res 36:1–12

Ehmann A, Bandurski RS, Harten J, Young N, Sweeley CC (1975) Mass spectrometry of indoles and trimethylsilyl-indole derivatives. Michigan State University (private printing)

Epstein E, Cohen JD, Bandurski RS (1980) Concentration and metabolic turnover of indoles in germinating kernels of *Zea mays* L. Plant Physiol 65:415–421

Haagen-Smit AJ, Leech WD, Bergren WR (1942) The estimation, isolation and identification of auxins in plant material. Am J Bot 29:500–506

Hall PJ (1980) The occurrence of indole-3-acetyl-*myo*-inositol in kernels of *Oryza* sativa. Phytochemistry 19:2121–2123

Hangarter RP, Good NE (1981) Evidence that IAA conjugates are slow-release sources of free IAA in plant tissues. Plant Physiol 68:1424–1427

Hatcher ESJ, Gregory FG (1941) Auxin production during the development of grains of cereals. Nature 148:626

Kögl F, Erxleben H, Haagen-Smit AJ (1934) Über die Isolierung der Auxine a und b aus pflanzlichen Materialien. IX. Mitteilung über pflanzliche Wachstumsstoffe. Z Physiol Chem 225:215–229

Komoszynski M, Bandurski RS (1984) Metabolism of [^{3}H]-5-indole-3-acetyl-*myo*-inositol-[^{14}C]galactoside by seedlings of *Zea mays*. Plant Physiol 75(S):108

Labarca C, Nicholls PJB, Bandurski RS (1965) A partial characterization of indoleacetylinositols from *Zea mays*. Biochem Biophys Res Commun 20:641–646

Laibach F, Meyer F (1935) Über die Schwankungen des Auxingehaltes bei *Zea mays* und Helianthus annus im Verlauf der Ontogenese. Senckenbergiana 17:73–86

Lang A (1970) Gibberellins: structure and metabolism. Annu Rev Plant Physiol 21:537–570

Magnus V, Bandurski RS, Schulze A (1980) Synthesis of 4,5,6,7 and 2,4,5,6,7 deuterium-labeled indole-3-acetic acid for use in mass spectrometric assays. Plant Physiol 66:775–781

Michalczuk L, Bandurski RS (1982) Enzymatic synthesis of indol-3-yl-acetyl-1-O-B-D-glucose and indol-3-yl-acetyl-*myo*-inositol. Biochem J 207:273–291

Michalczuk L, Chisnell JR (1982) Enzymatic synthesis of 5-[^{3}H]-indole-3-acetic acid and 5-[^{3}H]-indole-3-acetyl-*myo*-inositol from 5-[^{3}H]-L-tryptophan. J Labeled Compd Radiopharm 19:121–128

Momonoki YS, Bandurski RS (1984) Induction by gravity of an asymmetric distribution of [^{14}C]-glucose and [^{3}H]-IAA-*myo*-inositol in the mesocotyl of *Zea mays*. Plant Physiol 75(S):178

Nicholls PB (1967) The isolation of indole-3-acetyl-2-O-*myo*-inositol from *Zea mays*. Planta 72:258–264

Nicholls PB, Ong BL, Tate ME (1971) Assignment of the ester linkage of 2-O-indoleacetyl-*myo*-inositol isolated from *Zea mays*. Phytochemistry 10:2207–2209

Nonhebel HM, Bandurski RS (1984) Oxidation of indole-3-acetic acid and oxindole-3-acetic acid to 2,3-dihydro-7-hydroxy-2-oxo-1*H* indole-3-acetic acid-7'-O-B-D-glucopyranoside in *Zea mays* seedlings [1]. Plant Physiol 76:979–983

Nowacki J, Bandurski RS (1980) *Myo*-inositol esters of indole-3-acetic acid as seed auxin precursors of *Zea mays* J. Plant Physiol 65:422–427

Nowacki J, Cohen JD, Bandurski RS (1978) Synthesis of ^{14}C-indole-3-acetyl-*myo*-inositols. J Labeled Compd Radiopharm 15:325–329

Piskornik Z, Bandurski RS (1972) Purification and partial characterization of a glucan containing indole-3-acetic acid. Plant Physiol 50:176–182

Pohl R (1935) Über den Endosperm-Wuchsstoff und die Wuchsstoffproduktion der Koleoptilspitze. Planta 24:523–526

Posternak T (1965) The cyclitols. Holden-Day, Publishers, San Francisco, California

Rivier L, Crozier A (1985) Auxins and related compounds. In: Rivier L, Crozier A (eds) Methods in plant hormone analysis. Academic (in press)

Schliemann W, Liebisch HW (1984) Enzymic and chemical hydrolysis of plant hormone conjugates. Biochem Physiol Pflanz 179:533–552

Sembdner G (1974) Conjugation of plant hormones. In: Schreiber K, Schutte HR, Sembdner G (eds) Biochemistry and chemistry of plant growth regulators. Inst Plant Biochem Acad Sci, Halle, GDR, pp 283–302

Sembdner G, Gross D, Liebisch HW, Schneider G (1981) Biosynthesis and metabolism of plant hormones. In: MacMillan J (ed) Hormonal regulation of development. I. Molecular aspects of plant hormones. Springer, Berlin Heidelberg New York (Encyclopedia of plant physiology ser, vol 9) pp 281–444

Skoog F (1937) A deseeded Avena test method for small amounts of auxin and auxin precursors. J Gen Physiol 20:311–334

Ueda M, Bandurski RS (1969a) Gas chromatographic and mass spectrometric analysis of indole-3-acetic acid-*myo*-inositol esters. Plant Physiol 44 (Supplement):27

Ueda M, Bandurski RS (1969b) A quantitative estimation of alkali-labile indole-3-acetic acid compounds in dormant and germinating maize kernels. Plant Physiol 44:1175–1181

Ueda M, Bandurski RS (1974) Structure of indole-3-acetic acid *myo*-inositol esters and pentamethyl-*myo*-inositols. Phytochemistry 13:243–253

Ueda M, Ehmann A, Bandurski RS (1970) Gas-liquid chromatographic analysis of indole-3-acetic acid *myo*-inositol esters in maize kernels. Plant Physiol 46:715–719

van Overbeek J (1941) A quantitative study of auxin and its precursor in coleoptiles. Am J Bot 28:1–10

Went FW, Thimann KV (1937) Phytohormones. MacMillan, New York

Zenk MH (1961) 1-(indole-3-acetyl)-B-D-glucose, a new compound in the metabolism of indole-3-acetic acid in plants. Nature 191:493–494

GC-MS Methods for Cytokinins and Metabolites

L. M. S. Palni, S. A. B. Tay, and J. K. MacLeod

1 Introduction

Cytokinin is a generic term proposed to define both naturally occurring and synthetic compounds that induce cell division in plant tissue cultures (e.g., tobacco stem pith, soybean cotyledonary callus and carrot secondary phloem etc.) grown on defined medium in the presence of an optimal concentration of auxin. Although the idea that specific chemical substances may control cell division in plants dates back to the nineteenth century (Wiesner 1892), it first received experimental support only in the early 1920's following the observation that wounding induced cell division in many plant tissues, e.g., potato parenchyma (Haberlandt 1921). Kinetin (Fig. 1) was the first cytokinin to be identified, and its isolation in 1956 from autoclaved herring sperm DNA was a direct consequence of studies on growth requirements of plant tissue cultures (Miller et al. 1956). It does not occur per se in living tissues, and had resulted as an artefact of the autoclaving process in the original work. Synthetic kinetin was found to be a very potent promoter of cell division, and induced cell division activity in the tobacco pith assay

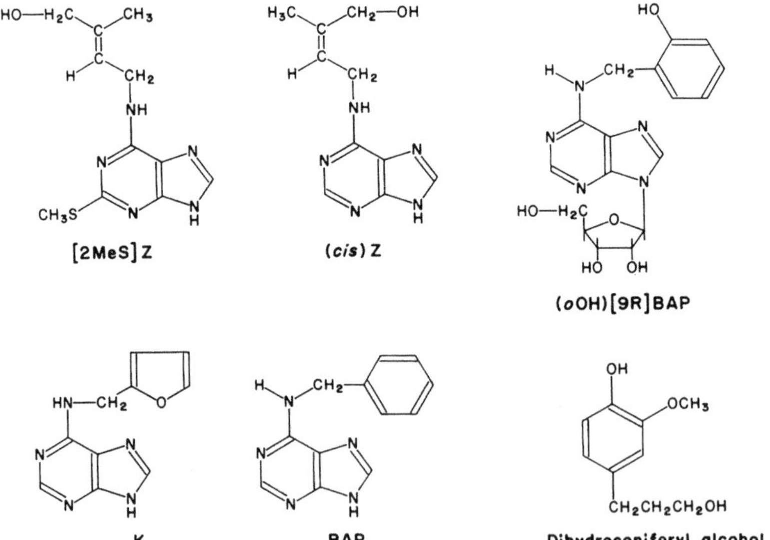

Fig. 1. Chemical structures for some naturally occurring and synthetic cytokinins. Although dihydroconiferyl alcohol is also active in cell division tests, it is not classified as a cytokinin. See Fig. 2 for commonly occurring natural cytokinins

at concentrations as low as 1 µg l^{-1}. The realization that the kinetin effect could be replaced by extracts from many plant tissues triggered the search for endogenous cytokinins, resulting in the isolation of the first natural cytokinin, zeatin [6-(4-hydroxy-3-methylbut-*trans*-2-enylamino) purine] from immature *Zea mays* kernels (Letham 1963; Letham et al. 1964, 1967). The number of cytokinins identified so far from diverse natural sources is nearing 30. In addition to their occurrence as free compounds in higher plants, cytokinins have also been isolated from transfer ribonucleic acid (tRNA) hydrolysates of plants, animals and microorganisms (Letham and Wettenhall 1977; Letham and Palni 1983).

Almost all natural cytokinins identified to date, like zeatin, are 6-substituted aminopurines, and generally possess an isoprenoid side chain (Figs. 1 and 2). The biological activity of these compounds ranges from highly active to totally inactive. Thus, in contrast to their physiological definition, from a chemical viewpoint cytokinins also include compounds that are inactive in cell-division tests, e.g., zeatin-7-glucoside (Letham et al. 1983). On the other hand dihydroconiferyl alcohol, a nonadenine derivative, although active in cell division tests (Lee et al. 1981) is not classified as a cytokinin. In the context of this chapter the consideration will be limited to cytokinins which are substituted 6-aminopurines. The system used for numbering of the purine nucleus is also used in cytokinin nomenclature to define the position of substituents (see Z in Fig. 2). Many systems of abbreviations for cytokinins, including those based on the established system for nucleic acid derivatives, have been used in the literature. The nomenclature for cytokinins proposed by Letham (1978) has been used in this chapter. As shown in Table 1 zeatin (Z), isopentenyladenine (iP), benzyladenine (BAP), and kinetin (K) are regarded as basic compounds; substitutions on the purine nucleus are denoted in square brackets, e.g., [9R]iP for isopentenyladenosine, and substitutions on the sidechain are denoted in round brackets, e.g., (diH OG)Z for O-glucosyl dihydrozeatin.

Although originally recognised by their cell division inducing activity in plant tissue cultures, cytokinins also evoke a diversity of other physiological responses in plants and excised plant parts. Like tissue culture assays, many of the other responses have been utilized to develop alternative cytokinin bioassays, e.g., the radish cotyledon, the oat/barley leaf senescence, the cucumber cotyledon greening and the *Amaranthus* betacyanin test, etc. (Letham 1978; Horgan 1984). Bioassays are essential for the detection of new growth regulating substances, and are a means of monitoring biological activity of such compounds (e.g., cytokinins) during isolation and purification. The ease with which "cytokinin-like activity" can be detected in a plant extract has led to the widespread use of bioassays for the purpose of identification and quantification of cytokinins. This approach, however, is unsatisfactory for a variety of reasons (Letham 1978; Horgan 1984). Consequently attempts of plant physiologists in the past to correlate growth with obviously "faulty determinations" of the endogenous levels of plant growth substances have at times, if anything, brought some disrepute to the growth hormone field in general. Like other plant growth substances, cytokinins affect a range of developmental processes, and careful consideration of many factors is necessary before the importance of cytokinins (or any other growth substance) in regulating particular aspects of plant growth can be assessed. A minimum requirement is the

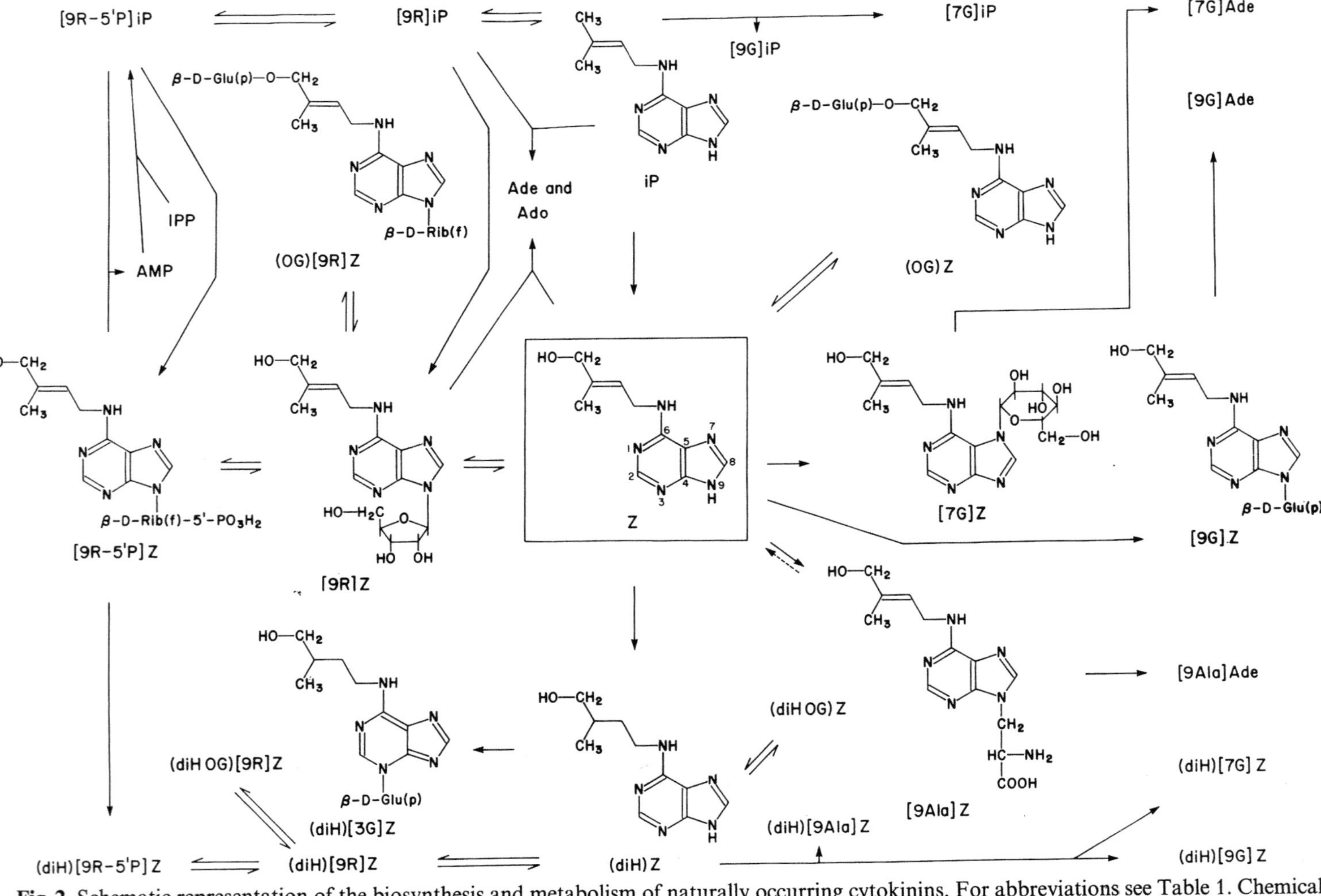

Fig. 2. Schematic representation of the biosynthesis and metabolism of naturally occurring cytokinins. For abbreviations see Table 1. Chemical structures for some cytokinins are also given. IPP = Δ^2-Isopentenyl pyrophosphate

Table 1. Naturally occurring and synthetic cytokinins

Cytokinins, abbreviations	Mol wt.	Other common abbreviations	References to selected syntheses and mass spectra
iP = Isopentenyladenine, and its derivatives			
iP: N^6-(Δ^2-isopentenyl)adenine	203	i^6Ade, 2iP, IP	Robins et al. (1967) (A), Young (1977) (H), Wang et al. (1980) (D), (1982) (M), Cole et al. (1975) (O)
[9R]iP: 9-β-D-ribofuranosyl-iP	335	i^6A, 2iPA, IPA	Robins et al. (1967) (A,D), Hashizume et al. (1976) (A,D), (1979) (B), Ernst et al. (1983a) (B,G), Tsui et al. (1983) (G,I,K), Young (1977) (H), Cole et al. (1975) (O), Scott and Horgan (1984) (H)
[9R-5′P]: 5′-monophosphate of [9R]iP	415	i^6AMP, IPMP	Grimm and Leonard (1967) (A), Fig. 8 (P), Hong et al. (1975) (A)
[7G]iP: 7-glucopyranosyl-iP	365	i^6Ade7G, IP7G	McGaw et al. (1985) (G)
[9G]iP: 9-glucopyranosyl-iP	365	i^6Ade9G, IP9G, i^6AG	Hashizume et al. (1982b) (A,D,G), Sugiyama et al. (1983) (B)
[2MeS]iP: 2-methylthio-iP	249	ms^2i^6Ade, 2msIP	Burrows et al. (1969) (A)
[2MeS 9R]iP: 2-methylthio-[9R]iP	381	ms^2i^6A, 2msIPA	Burrows et al. (1969) (A,D), (1970) (D), Hashizume et al. (1979) (B,G), Sugiyama and Hashizume (1980) (B), Hecht et al. (1970a) (D), Murai et al. (1980) (G)
Z = Zeatin, and its derivatives			
Z: 6-(4-hydroxy-3-methylbut-*trans*-2-enylamino)purine	219	io^6Ade	Corse and Kuhnle (1972) (A), Shaw et al. (1966) (A), Kadir et al. (1984) (A), Leonard et al. (1971) (*A), Summons et al. (1979a) (B), Scott and Horgan (1980) (C), (1984) (H), Shannon and Letham (1966) (D), Fig 3a (D), Dauphin et al. (1979) (G), Einset and Skoog (1977) (*G), Ludewig et al. (1982) (I), Palni et al. (1985) (J), Cole et al. (1975) ($\triangle$O), Young (1977) (H), Wang et al. (1981) (M)
[9R]Z: 9-β-D-ribofuranosyl-Z	351	io^6A, ZR, RZ	Shaw et al. (1966) (A), Scott and Horgan (1980) (C), 1984 (H), Summons et al. (1979a) (B), Hashizume et al. (1979) ($\triangle$B), Playtis and Leonard (1971) (*A), Fig. 3b (D), Hecht et al. (1970a) (D), Dauphin et al. (1979) (G), Morris et al. (1981) (*H), Ludewig et al. (1982) (I), Babcock and Morris (1970) (*D), Cole et al. (1975) (*O), Murai et al. (1980) (*G), Young (1977) (H)
[9R-5′P]Z: 5′monophosphate of [9R]Z	431	io^6AMP, ZMP	Shaw et al. (1967), (1968) (A), Summons et al. (1983) (B,E,N), Palni et al. (1983a) ($\square$A)
[7G]Z: 7-glucopyranosyl-Z	381	Z7G, io^6Ade7G	Cowley et al. (1978) (A,D), Summons et al. (1977) (B), Entsch et al. (1980) (B), Parker and Letham (1973) (D), Fig. 3c (D), MacLeod et al. (1976) (G), Scott and Horgan (1984) (C,G), Palni et al. (1983b) (G)

Table 1 (continued)

Cytokinins, abbreviations	Mol wt.	Other common abbreviations	References to selected syntheses and mass spectra
[9G]Z: 9-glucopyranosyl-Z	381	Z9G, io^{6}Ade9G	Cowley et al. (1978) (A,D), Summons et al. (1977) (B), Scott et al. (1980) (F,G), (1982) (C), Entsch et al. (1980) (B), Palni and Horgan (1982) (F), MacLeod et al. (1976) (G), Parker and Letham (1974) (D)
[9Ala]Z: L-β-[6-(4-hydroxy-3-methylbut-*trans*-2-enylamino)purin-9-yl]alanine, Lupinic acid	306	LA, Ala^{9}io^{6}Ade	Duke et al. (1978) (A,D,E), Entsch et al. (1980) (B), Parker et al. (1978) (D), Fig. 3e (D), Murakoshi et al. (1977) ($\square$A)
[2MeS]Z: 2-methylthio-Z	265	ms^{2}io^{6}Ade, 2msZ	Hecht et al. (1970b) (A), Guerin et al. (1984) (G), Cole et al. (1975) (*O)
[2MeS 9R]Z: 2-methylthio-[9R]Z	397	ms^{2}io^{6}A, 2msZR	Burrows et al. (1970) ($\triangle$A,D), Hashizume et al. (1979) ($\triangle$B,G), Sugiyama and Hashizume (1980) (B), Vreman et al. (1974) ($\triangle$A, D), (1978) (G), Morris et al. (1981) (*H), Murai et al. (1980) (*G)
(OG)Z: O-β-D-glucopyranosyl-Z	381	ZOG	Duke et al. (1978) (A,D), Scott et al. (1982) (C), Summons et al. (1979a) (B), (1980) (G,L), Morris (1977) (F), Parker et al. (1978) (D), Fig. 3d (D), Scott and Horgan (1984) (H)
(OG)[9R]Z: 9-β-D-ribofuranosyl-(OG)Z	513	ZROG	Duke et al. (1979) (A,D), Scott et al. (1982) (C), Summons et al. (1979a) (B), 1980 (D,E,L), Morris (1977) (F)
(diH)Z = Dihydrozeatin, and its derivatives			
(diH)Z: 6-(4-hydroxy-3-methylbutyl-amino)purine	221	H$_2$io^{6}Ade, DHZ	Koshimizu et al. (1967) (A), Leonard et al. (1969) (A), Summons et al. (1979a) (B), Sugiyama and Hashizume (1980) (B), McGaw et al. (1984a) (G), Scott and Horgan (1984) (C,H)
(diH)[9R]Z: 9-β-D-ribofuranosyl-(diH)Z	353	H$_2$io^{6}A, DHZR	Leonard et al. (1969) (A), Summons et al. (1979a) (B), Sugiyama and Hashizume (1980) (B), Hecht et al. (1970a) (D), Palni and Horgan (1982) (G), Scott and Horgan (1984) (C,H)
(diH)[9R-5'P]Z: 5'-monophosphate of (diH)[9R]Z	433	H$_2$io^{6}AMP, DHZMP	Scott and Horgan (1984) (B)
(diH)[3G]Z: 3-glucopyranosyl-(diH)Z	383	DHZ3G, H$_2$io^{6}Ade3G	McGaw et al. (1984a, b) (G)
(diH)[7G]Z: 7-glucopyranosyl-(diH)Z	383	DHZ7G, H$_2$io^{6}Ade7G	McGaw et al. (1984a, b) (G), Scott and Horgan (1984) (C,G)
(diH)[9G]Z: 9-glucopyranosyl-(diH)Z	383	DHZ9G, H$_2$io^{6}Ade9G	Summons et al. (1980) (A,D,G), McGaw et al. (1984a, b) (G), Palni et al. (1983b) (G)
(diH)[9Ala]Z: dihydrolupinic acid	308	DLA, Ala^{9}H$_2$io^{6}Ade	Parker et al. (1978) (A,D,E), Summons et al. (1979a) (B)

(diH OG)Z: 0-β-D-glucopyranosyl-dihydrozeatin	383	DHZOG	Duke et al. (1979) (A,D), Summons et al. (1979a) (B)
(diH OG)[9R]Z: 9-β-D-ribofuranosyl-(diH OG)Z	515	DHZROG	Duke et al. (1979) (A,E), Fig. 6 (D,E,N), Summons et al. (1979a) (B)
BAP = Benzyladenine, and its derivatives			
BAP: 6-benzylaminopurine	225	bzl[6]Ade, BA	Bullock et al. (1956) (A), Daly and Christensen (1956) (A), Skinner and Shive (1955) (A), Shannon and Letham (1966) (D), Fox et al. (1971) (D)
[9R]BAP: 9-β-D-ribofuranosyl-BAP	357	bzl[6]A	Ernst et al. (1983b), (B,G), Dyson et al. (1972) (D), Fox and Chen (1967) (A), Fleysher et al. (1969) (A)
[9R-5'P]BAP: 5'-monophosphate of [9R]BAP	437	bzl[6]AMP, BAMP	Tao et al. (1983) (A,E,L), Hong et al. (1975) (A)
[3G]BAP: 3-β-D-glucopyranosyl-BAP	387	BA3G, 3gp-BAP	Letham et al. (1975) (A,D,G), Tao et al. (1983) (E)
[7G]BAP: 7-β-D-glucopyranosyl-BAP	387	BA7G, 7gp-BAP	Cowley et al. (1978) (A,D), Wilson et al. (1974) (D), MacLeod et al. (1976) (G)
[9G]BAP: 9-β-D-glucopyranosyl-BAP	387	BA9G, 9gp-BAP	Cowley et al. (1978 (A,D), Wilson et al. (1974) (D), MacLeod et al. (1976) (G), Tao et al. (1983) (E)
[9Ala]BAP: L-β-(6-benzylaminopurin-9-yl)alanine	312	Ala[9]bzl[6]Ade	Letham et al. (1979) (A,D), Murakoshi et al. (1983) ($\square$A)
(o OH)[9R]BAP: 6-(o-hydroxybenzylamino)purine riboside	373	(OH[2]bzl)[6]A	Horgan et al. (1975) (A,D,G), Sugiyama et al. (1980) (B)
(o OH)[2MeS 9G]BAP: 6-(o OH benzylamino)-2-methylthio-9-β-D-glucosylpurine	449	ms[2](OH[2]bzl)[6]Ade9G	Chaves das Neves and Pais (1980) (A,D,G)
K = Kinetin, and its derivatives			
K: 6-furfurylaminopurine	205	fr[6]Ade, Kin	Miller et al. (1956) (A), Bullock et al. (1956) (A), Daly and Christensen (1956) (A), Shannon and Letham (1966) (D)
[9R]K: 9-β-D-ribofuranosyl-K	337	fr[6]A	Doree and Guern (1967) (A)
[9R-5'P]K: 5'-monophosphate of [9R]K	417	fr[6]AMP	Hong et al. (1975) (A)

Synthesis of unlabelled (A), [2]H (B) and [15]N (C) labelled compounds. El-MS probe spectrum of underivatized compound (D), TMSi (E) and permethyl (F) derivatives. GC-El-MS spectrum of TMSi (G), permethyl (H), TFA (I) and tBuDMSi (J) derivatives. Cl-MS probe spectrum of underivatized compound (K) and TMSi derivative (L). GC-CI-MS spectrum (M), DCl spectrum (N), FD spectrum (O) and FAB spectrum (P). *=*cis*-isomer, $\triangle$=*cis*- and *trans*-isomers, $\square$=enzymatic synthesis.
The di- and triphosphate derivatives of ribosides are denoted by [9R-5'PP] and [9R-5'PPP] respectively; eg [9R-5'PPP]Z for 5'-triphosphate of [9R]Z; (*cis*)[2MeS 9R]Z denotes *cis*-isomer of [2MeS 9R]Z etc.

unambiguous identification of cytokinin(s) in the plant tissue (or compartment) under consideration; additional quantitative information would also be required before a correlative role for cytokinins could be proposed. Hence the need for techniques to accurately identify and precisely quantify cytokinins.

The isolation and identification of trace compounds like cytokinins in plant extracts poses some difficulties, mainly due to technical problems associated with the purification of these compounds. This is reflected in the fact that although there are numerous reports of the detection of cytokinin like activity in plant extracts, there are relatively few reports of unambiguous identification of cytokinins from plant sources. Indeed proper identification and quantification of cytokinins has been carried out in less than half a dozen plant species so far. However, as a result of the continuing development of techniques in high performance liquid chromatography (HPLC), gas chromatography (GC) and mass spectrometry (MS), particularly in the last 5 years or so, most technical problems have largely been overcome. It is hoped that due to the availability of improved techniques for purification, identification and quantification of cytokinins, future progress will be rapid and provide a sound basis for an understanding of the physiological role of cytokinins in the integrated development of plants.

There are a number of reviews (Brenner 1981; Horgan 1981; Yokota et al. 1980) which describe general procedures used in the analysis of plant growth substances, and those dealing specifically with the extraction, purification, identification and quantification of cytokinins (Horgan 1978; Horgan and Scott 1985). Although a number of modern analytical techniques are available, and are required for successful and rapid cytokinin analysis, GC, MS, and/or combined GC-MS have played a major role in the identification of all cytokinins isolated so far. Gas chromatography with its resolving power combined with the sensitive and selective detection qualities of the mass spectrometer have made combined GC-MS singularly the most powerful technique for cytokinin analysis. In this chapter these techniques will be discussed in some detail.

2 Gas Chromatography (GC)

2.1 Instrumentation

The ability to form volatile cytokinin derivatives has led to the exploitation of GC as a powerful and rapid separation technique for cytokinins often as the last step in a multistep purification scheme. In GC or more correctly, gas-liquid chromatography (GLC) the sample constituents are separated due to selective partitioning between the carrier gas phase (mostly N_2; He for GC-MS) and the liquid stationary phase evenly distributed on a solid support or on the walls of the column.

The solid support for the GC column should be of uniform particle size, have a large surface area to volume ratio, and show stability towards both sample and the liquid phase coated on to its surface. Its mechanical strength should be such that handling and vibration of the column does not cause particle fracture. Diatomaceous silica is commonly used as solid support and is silanized for increased inertness. The stationary liquid phase should have low volatility (bleed) and be thermally stable at the maximum temperatures employed for the analysis. Stationary phases can be classified as either selective or non-selective. Selective phases separate on the basis of chemical differences, e.g., polarity, while non-selective phase separation depends on differences in boiling points.

The retention time (R_t) of a substance on a non-selective phase is therefore a function of its boiling point. High column temperatures are necessary for the analysis of low volatility high molecular weight substances if short analysis times are required. As a general rule acceptable analysis times can be obtained by operating column temperatures about 50 °C below the boiling point of the sample. Temperature programming is commonly used to overcome the problems of stationary phase "bleed", baseline instability and column deterioration that are often associated with high temperature operation. However, certain compounds undergo thermal rearrangement [e.g. [3G]BAP (both as $(TMSi)_3$ and permethylated derivative; MacLeod et al. (1976)] or decomposition [e.g., TMSi derivatives of sidechain O-glucosides of [9R]Z and (diH)[9R]Z; Summons et al. (1979 b)], and are not suitable for gas chromatography.

2.1.1 Liquid Stationary Phases and Columns

For GC analysis of cytokinins, liquid stationary phases of high temperature stability and low volatility are required because the cytokinin derivatives, particularly of more polar cytokinins, elute at high temperatures. Liquid phase loadings of less than 5% on a very inert support are preferred to minimise column bleed interference. Non-polar polydimethylsiloxane (e.g., OV-1, SE-30, DB-X, BP-1, CP-Sils etc.) or polyphenylmethylsiloxane (e.g., OV-17, SE-54, BP-10 etc.) stationary phases have been used for cytokinin analysis.

Packed Columns. Glass or vitreous silica columns are used exclusively in GC analysis of cytokinins. Although packed columns are inexpensive and robust, they generally give poor resolution. This would possibly limit their use in trace analysis where "active site" decomposition may be pronounced. However, quantitative measurements of cytokinins in the ng (10^{-9} g) range using a 2 m × 2 mm ID glass column (3% OV-1 on 100–200 mesh Gas Chrom Q) have been done (S.A.B. Tay et al., unpublished work). Most et al. (1968) have reported separation of TMSi derivatives of Z, (diH)Z, [9R]Z, iP and [9R]iP on a glass column packed with 3% SE-52 on 80–100 mesh Diataport S (a silanized diatomaceous earth). A common problem with most packed columns is peak tailing due to chemical interaction between sample components and the support. This can be minimised by increasing the stationary phase load or by modification of the support surface by silanization. The latter is preferable as high stationary phase loadings would lead to excessively long retention times and intolerably high column bleed. Improved performance has been reported with the use of bonded GC packings (e.g., Permabond). In these a mono-molecular layer of the liquid phase is chemically and uniformly bonded onto an inert support. The resulting low concentration of liquid phase gives high efficiency, rapid elution times, low column bleed and better temperature stability.

Capillary Columns. Although capillary columns have been in use for many years they have only recently been used in the GC analysis of cytokinins. Jennings (1980) has discussed capillary column GC in detail. The advent of flexible vitreous silica or fused silica columns has revolutionised capillary column analysis. There are two main types of capillary columns. (1) Wall coated upon tubular (WCOT), in which the stationary phase film is coated directly onto the deactivated wall of the capillary tubing made of vitreous silica. (2) Support coated open tubular (SCOT) in which fine particles of deactivated diatomaceous earth are coated with the stationary phase and deposited on the glass surface. The flexible vitreous silica columns can be inserted close to the GC detector or connected directly into the ion source of a MS, thus reducing dead volume.

Capillary columns in general provide considerably improved resolution, shorter retention times (or lower elution temperature) with minimal column bleed compared with packed columns. For example WCOT capillary columns typically show efficiencies of up to 10^4–10^5 theoretical plates per metre which represents a 10–10^2 fold increase in net efficiency over packed columns. However, these are expensive, have low load capacity, require relatively skilful manipulation, sample injection is more difficult, and the columns degrade over a shorter period of time especially when relatively impure plant extracts are analysed. The use of shorter, wide bore columns with more stable bonded stationary phases such as BP-1 or BP-5 would obviate some of these problems. The recently introduced wide bore (0.53 mm ID) fused silica open tubular NON-PAKD columns have potential for cytokinin analysis in plant extracts. These can be installed and operated either as ordinary packed columns, or as wide bore capillary columns. Their 1.2 μ thick film of bonded phase allows higher capacity without excessive retention times. Horgan and Scott (1985) have provided a direct comparison of the GC analysis of a range of permethylated cytokinins carried out on packed and capillary columns. The capillary system gave considerably superior resolution, and it is interesting to note that the diasterioisomers of permethyl (diH OG)Z could also be partially resolved. Kemp and Andersen (1981) have reported excellent separation of a range of cytokinin bases and ribosides (as TMSi derivatives) using fused silica capillary column. The column bleed was minimal and did not interfere with GC-MS analyses. Other reported uses of capillary columns for the analysis of cytokinins include Kemp et al. (1982, 1983), Ludewig et al. (1982), Ernst et al. (1983a), Scott and Horgan (1984), Taylor et al. (1984), Palni et al. (1985), Lee et al. (1985), and Palmer and Wong (1985).

2.1.2 Injectors

Sample introduction is particularly critical when capillary columns are used. Several types of injectors (on-column, split/splitless) are available, the choice depending upon such factors as volatility and relative purity of the analysates. As it is impractical to change injectors frequently, a good understanding of the type on hand helps to prolong column life time. McMahon (1985) has discussed results of a collaborative study to evaluate quantification utilizing different injection modes for capillary GC.

In the split/splitless (Grob type) injectors the sample is vaporised in the hot injector (maintained at 220 °–250 °C), and thus introduction of non-volatile impurities into the column is prevented. However, the high operating temperatures needed to vaporise high molecular weight cytokinin derivatives (e.g. glucosides) may cause sample decomposition. Sample decomposition is also increased if active surfaces in the glass line tubing (GLT) of the injector are exposed; if possible quartz liners should be used. However, injectors with GLT should be regularly cleaned (chromic acid) and re-silanised, and kept deactivated by regular injection of silylating reagents, e.g., BSA, BSTFA or MSTFA (see Table 2) during the course of use. Kemp et al. (1982) have evaluated split and splitless injection modes for TMSi derivatives of cytokinins using a cold trapping technique (40 °C initial temperature). In the splitless mode a 52-fold greater FID response for TMSi-[9R]Z was found over the split mode.

The simplest method of sample introduction into a capillary column is by on-column injection. In this method boiling point discrimination is eliminated, and thermal and adsorption effects are minimised. However, build-up of nonvolatile impurities at the point of injection is detrimental to column performance. This is an important consideration in view of the inherently impure nature of most cytokinin samples from plant extracts. Although bonded phase capillary columns can be washed with solvents to remove contaminants without stripping off the stationary phase, this may not necessarily restore column performance. In practice the contaminated part of the column from the inlet end should be cut off. Although the on-column technique provides the most reliable and quantitative injection for capillary columns, users are particularly cautioned against injecting relatively impure samples.

The injection technique for both packed and capillary columns can also have a dramatic influence on the quality of the resultant chromatogram. Ideally, the sample delivered to the column should accurately represent the injected material, and should occupy a minimum length of column. The in-needle injection is a poor technique as pointed out by Grob and Grob (1979). Introduction of solution filled needles into the inlet causes the more volatile sample components to distil off from the needle, thus the remaining material will be richer in the less volatile compounds. For a more representative injection, the needle should be empty when inserted into the inlet. The plunger can be retracted to withdraw the sample into the glass barrel, the needle is then inserted, a few seconds are allowed to warm up the empty needle and the injection is finally accomplished by thrusting the plunger home rapidly. Bad injection techniques can cause peak doublets leading to faulty results. Microsyringes without calibrated needle volumes do not allow representative sampling, therefore the use of microsyringes with no dead volume is recommended for quantitative analysis. A simple graphical method has been suggested by Guha (1984) for obtaining the calibration graph and the true volume injected with microsyringes with a dead volume.

2.1.3 Detectors

Flame ionization detectors (FID) respond to compounds that ionize upon combustion in a hydrogen flame. Although these detectors are sensitive to ng quantities of organic compounds, they exhibit poor selectivity. Thus extensive purification of the sample is necessary before this technique can be used for the analysis of trace compounds like cytokinins. Furthermore, for quantitative use of nonselective detectors such as the FID, it is imperative that the peaks being measured should contain only the compounds of interest. Most et al. (1968) found that for TMSi derivatives of authentic cytokinins the minimum amount detectable with FID was 5 ng. Babcock and Morris (1970) have observed that the relative detector response varied for different cytokinins. Kemp and Andersen (1981) have used capillary GC in combination with FID to resolve a number of closely related cytokinins. Upper et al. (1970) and Hahn (1975) reported identification of cytokinins using this technique. However, cytokinin analysis by GC/FID is not satisfactory due to its lack of selectivity which is based solely on retention times, even when relatively purified plant extracts are examined.

A more realistic approach to GC analysis of cytokinins has been obtained with selective detectors like the nitrogen-phosphorous (NPD), the flame-photometric (FPD) and the electron capture (ECD) detectors. The use of a selective detector for the GC analysis of cytokinins was first demonstrated by Zelleke et al. (1980): using permethylated cytokinin bases they reported detection of as little as 0.1 pg iP, BAP and K, and 10 pg Z with NPD. Cytokinins are nitrogen rich, and some contain sulphur or phosphorous, and therefore lend themselves to the increased sensitivity provided by the selective NPD and FPD. Stafford and Corse (1982) found that cytokinins with large nitrogen/carbon (N/C) ratios gave good NPD responses, although the response decreased with additions on the N-6 sidechain which effectively lowered the N/C ratio. In combination with fused silica capillary columns, pg quantities of permethylated cytokinins were detected. Using the same combination, Kemp et al. (1983) determined cytokinins in wheat germ tRNA. Their estimations, based on GC peak areas, indicated 8–10 fold greater response with NPD over the FID for TMSi-[9R]Z and per TMSi N-6-methyladenosine. Cytokinins in bean leaves, coconut milk and wheat germ tRNA have also been measured by Whenham (1983) using NPD and FID. In comparison with FID, the NPD gave 25–50 times better response for a mixture of permethylated cytokinin standards. Furthermore, an additional advantage due to the selectivity of NPD was that cleaner chromatograms were obtained because of less interference from phase bleed. The selectivity of FPD response between sulphur- and non-sulphur containing compounds was reported to be $10^4 : 1$ (Ehrlich et al. 1981). Whenham (1983) investigated the suitability of FPD for the analysis of permethylated derivatives of methylthio cytokinins ([2 MeS]iP, [2 MeS 9R]iP, and [2 MeS]Z) and estimated the detection limit for these to be approximately 50 ng. Use of the FPD provided greatly enhanced selectivity rather than improved sensitivity for methylthio cytokinins in wheat germ tRNA hydrolysates. In the same paper ECD was reported to have no significant increase in sensitivity over FID for permethylated cytokinins. Derivatization with halogen-containing functional groups would improve the sensitivity for analysis by ECD, in addition such derivatives would also be particularly suitable for negative CI-MS. It should, however, be noted that other compounds present in plant extract could also react with the derivatising reagents and affect selectivity of detection by ECD. Ludewig et al. (1982) have used trifluoroacetyl (TFA) derivatives in combination with ECD to measure cytokinins by GC. By comparison with the pentafluoropropionyl and heptafluorobutyryl derivatives, the TFA compounds were the most stable and reaction by-products were negligible. Although they reported detection of Z, iP and their ribosides in sunflower exudates using this method, further support for the identification was not provided.

Cytokinin analysis in plant extracts by GC using FID and to a lesser extent by NPD, FPD or ECD without some additional criterion of identification (e.g. HPLC, MS etc.) would be of questionable value. The most selective detector which can be coupled to a GC is the mass spectrometer. This will be discussed in more detail in a subsequent section.

2.2 Derivatisation of Cytokinins

Derivatisation as an aid to the analysis of organic compounds is a very old concept and a widely used procedure (Blau and King 1978; Knapp 1979; Drozd 1981). A prerequisite for the analysis of cytokinins by GC or GC-MS is the formation of their volatile derivatives. Derivatives commonly used for GC are usually useful also for GC-MS or MS analysis.

Analysis of plant samples by GC and GC-MS often deals with trace amounts of material. The formation of derivatives is a potential source of both qualitative and quantitative errors. Derivatisation reactions are essentially micro-scale organic syntheses where high demand is placed on both the quality of materials used and analytical precision. With amounts in the picogram to microgram range, manipulation of the sample should be minimal to prevent accidental or systematic losses. Whenever possible, the reaction should be designed to require little or no workup as long as reagents or by-products do not interfere with subsequent analysis. For the preparation of derivatives for cytokinin analysis, anhydrous conditions are generally required. The first and critical step is to dry the sample under a stream of inert gas, commonly nitrogen. Residual moisture can then be removed by azeotropic distillation with dry dichlormethane, or by keeping the sample over P_2O_5 under vacuum, or by lyophilization. Some derivatives (e.g., TMSi) are susceptible to hydrolysis in contact with moisture, and even traces of moisture in the carrier gas during GC analysis may cause decomposition. Thus oxygen and moisture absorbents must be installed in the GC carrier gas line.

A number of derivatives suitable for the GC or GC-MS analysis of cytokinins have been reported in the literature and are described below.

2.2.1 Trimethylsilyl (TMSi) Derivatives

Trimethylsilylation involves the substitution of $Si(CH_3)_3$ group for an active hydrogen. TMSi reagents and derivatives readily undergo hydrolysis, therefore particular care should be taken to ensure completely anhydrous conditions during sample manipulation. Reagents prepared for immediate use are commercially available (Pierce 1984). Some of the reagents frequently used for the preparation of cytokinin derivatives are listed in Table 2.

TMSi derivatives for the GC analysis of cytokinins were first described by Most et al. (1968) and were formed by reaction with BSA in CH_3CN (1:2) for

Table 2. Reagents and conditions commonly used for the preparation of silyl derivatives of cytokinins

Reagents	Reaction conditions	Cytokinin derivative(s)	Reference(s)
C_5H_5N/BSTFA/TMCS (10:99:1)	60 °C, 30–60 min	[f][3G]BAP, [f][7G]BAP, [f][9G]BAP, [f][9R-5'P]BAP, [g][7G]Z, [g][9G]Z, [h](OG)[9R]Z, [h,i](diH OH)[9R]Z	MacLeod et al. (1976) Tao et al. (1983) Duke et al. (1979)
C_5H_5N/MSTFA (1:1)	90 °C, 10 min	[f][9R]Z, [f](diH)[9R]Z, [e][9R]iP	Tay et al. (1985)
C_5H_5N/MTBSTFA (1:1)	90 °C, 10 min	[a]iP	Tay et al. (1985)
C_5H_5N/MTBSTFA** (1:1)	90 °C, 10 min	[b]Z, [b](diH)Z	Palni et al. (1985)
C_5H_5N/BSTFA (1:1)	90 °C, 10 min	[g][7G]Z, [g][9G]Z, [g](diH)[9G]Z, [d]Z, [d](diH)Z, [c]iP, [e][9R]iP, [f](diH)[9R]Z	Palni et al. (1983b)
C_5H_5N/BSTFA/TMCS (10:9.9:0.1)	80 °C, 10 min	[g][9R-5'P]Z	Summons et al. (1983)
BSA	80 °C, 30 min	[f][9R]Z, [g][7G]Z, [g](diH)[7G]Z	Scott and Horgan (1980, 1984)
BSA	90 °C, 1 h	[d](diH)Z	Wang et al. (1977)
C_5H_5N/BSTFA (1:2)	70 °C, 1 h	[e][9R]iP, [d,e]Z, *,[f][9R]Z	Dauphin et al. (1977)
DMF/HMDS (1:10 or 1:9)	90 °C, 1 h	[e][9R]iP, [e][2MeS 9R]iP, *,[f][9R]Z, *,[f][2MeS 9R]Z	Hashizume et al. (1979) McCloskey et al. (1979)
C_5H_5N/HMDS/TMCS (7:2:1)	120 °C, 1 h	[e][9R]iP, [f][9G]iP, *,[f][9R]Z	Hashizume et al. (1982a) Sugiyama et al. (1983)
CH_3CN/BSTFA/TMCS (20:10:1)	70 °C, 10 min	*,[e]Z	Sugiyama et al. (1983)
CH_3CN/BSA/TMCS (20:10:1)	70 °C, 5 min	[e]Z, [f,g][9R]Z	Watanabe et al. (1978a)
CH_3CN/BSA (2:1)	70 °C, 10 min	[f][cis][9R]Z	Watanabe et al. (1978b)
BSA	80 °C, 1 h	[f](o OH)[9R]BAP	Thompson et al. (1975)
C_5H_5N/BSA (1:2)	80 °C, 1 h	[e]Z, [d](diH)Z, [f][9R]Z, [g][9G]Z, [f](diH)[9R]Z	Palni and Horgan (1982)
C_5H_5N/BSA (1:1)	90 °C, 2 h	*,[f,g][9R]Z	Palni and Horgan (1983)
C_5H_5N/BSTFA (1:1)	Room temp. 30 min	[f][9R]Z, [e]Z, [e][9R]iP	Chen (1983)

BSA = N,O-bis-(trimethylsilyl)acetamide,   BSTFA = N,O-bis-(trimethylsilyl)trifluoroacetamide, HMDS = Hexamethyldisilazane, MSTFA = N-methyl-N-(trimethylsilyl)trifluoroacetamide, MTBSTFA = N-methyl-N-t-butyldimethylsilyl trifluoroacetamide (containing 1% t-BuDMCS), TMCS = Trimethylchlorosilane.
a = (tBuDMSi), b = (tBuDMSi)$_2$, c = (TMSi), d = (TMSi)$_2$, e = (TMSi)$_3$, f = (TMSi)$_4$, g = (TMSi)$_5$, h = (TMSi)$_6$, i = (TMSi)$_7$, * = Both cis- and trans-isomers, ** = containing 4-dimethylamino pyridine as catalyst, DMF = Dimethylformamide, C_5H_5N = Pyridine, CH_3CN = Acetonitrile.

5 min at 60 °C. Silylation with more reactive reagents such as BSTFA or MSTFA yields volatile by-products with short retention times which are less likely to interfere with the GC analysis of compounds of interest. Solvents such as pyridine (C_5H_5N), acetonitrile (CH_3CN) or dimethylformamide (DMF) are mostly used in silylation reactions.

Many methods for the preparation of TMSi derivatives of cytokinins have been developed (see Table 2). Control of reaction conditions is critical for the production of derivatives of consistent composition. Trace quantities of moisture tend to result in side peaks due to hydrolysis (Most et al. 1968), as well as in the production of multiple derivatives (Horgan 1978). The use of reagents such as BSTFA can lead to silylation of even the less reactive N-6 position e.g., in addition to $(TMSi)_4$-[9R]Z, formation of appreciable amounts of the penta-trimethyl-silyl derivative has been noted (Palni et al. 1984). Highly substituted derivatives, e.g., $(TMSi)_3$-Z and $(TMSi)_5$-[9R]Z were observed when the derivatisation was carried out with $BSA/CH_3CN/TMCS$ (10:20:1) at 70 °C for 5 min (Watanabe et al. 1978a). On the other hand, milder silylation conditions, e.g., reacting in BSA alone at 80 °C for 1 h gave reproducible yields of the lesser substituted cytokinin derivatives $(TMSi)_2$-Z and $(TMSi)_4$-[9R]Z (Horgan 1978). Similarly reaction with BSTFA/TMCS (99:1) in pyridine at 60 °C for 30–60 min, resulted in the production of partially substituted derivatives of N-glucosides of Z and BAP; silylation at the N-6 position did not occur (MacLeod et al. 1976). However, under the same conditions formation of multiple trimethylsilyl derivatives has been observed (Letham et al. 1978), e.g., a mixture of $(TMSi)_7$ and $(TMSi)_6$-derivatives of (diH OG) [9R]Z was reported (Duke et al. 1979). Summons et al. (1980) have also noted formation of $(TMSi)_7$ and $(TMSi)_8$ derivatives of (OG)[9R]Z.

Silylation of cytokinin bases in particular is notable for non-quantitative yields, formation of multiple derivatives (e.g., Dauphin et al. 1977; Purse et al. 1976) and instability during GC. Young (1977) reported that (TMSi)-Z could not be detected below 80–100 ng whereas less than 10 ng levels of (TMSi)-[9R]Z were easily detectable. Although the formation of multiple TMSi derivatives of a cytokinin may provide additional information for GC-MS identification, in the context of quantification of trace levels this effectively reduces the overall sensitivity of the method.

2.2.2 Permethyl Derivatives

Permethyl derivatives were first used in cytokinin analysis by Young (1977) and Morris (1977). The formation of permethyl derivatives is accomplished by reacting the cytokinin first with a strong base followed by methyl iodide (MeI). Young (1977) reported that the use of sodium carbonate, sodium ethoxide, or sodium hydride as base in DMF was not successful, and discrete GC peaks were not seen. However, the use of sodium methylsulfinyl carbanion (dimsyl anion, DMSO$^-$) in DMSO base (von Minden et al. 1973; Hakomori 1964), yielded single GC peaks for permethylated derivatives of iP, Z and their ribosides (Young 1977).

DMSO$^-$ solution can be obtained by heating sodium hydride (200 mg) with DMSO (4 ml) at 65 °C for 2 h (Corey and Chaykovsky 1962). However, an improved procedure for making DMSO⁻ in DMSO using potassium t-butoxide (Eagles et al. 1974) has been adopted for cytokinin analysis by recent workers (Morris 1977; Wang et al. 1981; Martin et al. 1981; Palmer et al. 1981; Stafford and Corse 1982; Tay et al. 1985). 0.1–0.2 M DMSO⁻ solution can be prepared by mixing DMSO (freshly distilled from CaH_2) with potassium t-butoxide (ca. 30 mg ml⁻¹) for 1 h at 45 °C, and keeping the reaction mixture under nitrogen. The reaction mixture can be clarified by filtration or centrifugation, and the final anion concentration can be checked by titration against 0.1 M HCl (phenolphthalein) as reported by Zelleke et al. (1980). Although the reagent can be stored frozen under nitrogen for a few weeks, it should be freshly prepared whenever possible.

Horgan and Scott (1985) have noted that the most suitable concentration of DMSO⁻ in DMSO is ca. 0.1 M. Although Hopping et al. (1979) have successfully used 1 M solution for permethylation of Z, and Morris (1977) used 1.5 M DMSO⁻ in DMSO for permethylation of (OG) [9R]Z and (OG)Z, formation of multiple products for cytokinin ribosides has been observed with high anion concentration (Horgan and Scott 1985). In addition DMSO⁻ solution of high molarity is viscous and difficult to pipette.

For preparing permethylated derivatives the cytokinin sample is thoroughly dried in a reaction vial, dissolved in 5–10 fold molar excess of DMSO⁻ solution (appearance of a red colour when a trace of triphenylmethane is added would indicate an excess of DMSO⁻) and then an excess of purified MeI (or tri deuteromethyl iodide when appropriate) is added. The reaction is left at 45 °C for 30 min, and then quenched with H_2O (50 µl–1 ml depending on reaction scale). The peralkylated derivatives are extracted into chloroform (0.1–1 ml), the chloroform layer is back extracted with an equal vol. of H_2O to remove any excess of base and DMSO, and finally evaporated under nitrogen. The residue is taken up in a small vol. of an appropriate solvent (e.g., ethyl acetate, chloroform etc.) for analysis.

Permethylated cytokinins can be further purified by TLC (Young 1977; Hopping et al. 1979) or HPLC (Martin et al. 1981), and have been reported to show enhanced sensitivity for GC with NPD (Stafford and Corse 1982; Zelleke 1980). The stability of the alkylated derivatives was exploited by Morris (1977) in a double permethylation step to establish the position of the glucose moiety of (OG) [9R]Z. Additional advantages are that multiple derivatives are not formed, and the molecular weight increment on permethylation is only 14 amu per H replaced as compared to 72 amu for trimethylsilylation or 114 for t-BuDMSi derivatives.

Although methylation with DMSO⁻ and MeI has been successful with cytokinin bases and glycosides, including [9G]Z (Scott et al. 1980; Palni and Horgan 1982), poor yields and side products were reported with the 7-glucopyranosides and 7-glucofuranosides of Z and BAP (MacLeod et al. 1976).

2.2.3 tert.-Butyldimethylsilyl (t-BuDMSi) Derivatives

The difficulties encountered in the preparation of TMSi derivatives of cytokinin bases and the number of manipulations involved in permethylation has prompted a search for a more stable derivative which is easy to prepare. Sterically crowded trialkylsilyl groups have been used in studies of nucleosides and other biologically important compounds, as selective protecting groups for nucleoside monomers in the synthesis of oligonucleotides (Ogilvie 1973; Quilliam and Westmore 1978; Quilliam et al. 1980). Silyl ethers formed from these groups have greater stability towards hydrolysis than the TMSi ethers. Alcohols can be converted to t-BuDMSi ethers in high yield under mild conditions using t-BuDMSi chloride with a catalytic amount of imidazole in DMF (Corey and Venkateswarlu 1972). For silylation of cytokinin bases, commercially available MTBSTFA (Table 2) containing 1% t-BuDMCSi is the reagent of choice. t-BuDMSi derivatives of cytokinin bases were prepared by reacting them with equal vols. of MTBSTFA and pyridine at 90 °C for 10 min (Tay et al. 1985). With Z and (diH)Z this procedure can result in the formation of multiple derivatives, which can be avoided by using 4-dimethylamino pyridine (dissolved in pyridine) as a catalyst (Scriven 1983; Palni et al. 1985). These derivatives can be further subjected to TLC or HPLC. t-BuDMSi derivatives of cytokinin ribosides have also been prepared by first converting the ribose component to an isopropylidene derivative (C. H. Hocart et al., unpublished work).

2.2.4 Trifluoroacetyl (TFA) Derivatives

The TFA derivatives have a high electron affinity and thus may be detected with greatly enhanced sensitivity by GC equipped with an electron capture detector. TFA derivatives of iP, [9R]iP, K, BAP, Z, (diH)Z, and [9R]Z were prepared by Ludewig et al. (1982). They were able to detect as little as 1 pg of cytokinin by using an "on-column" injector with a fused silica capillary column and ECD. These derivatives were prepared by reacting 10–500 ng of cytokinin with 50 µl trifluoroacetic anhydride in 200 µl dichloromethane in tightly sealed vials for 35 min at 95 °C. The excess reagent was then removed with a stream of nitrogen and the residue taken up in 500 µl cyclohexane for analysis. The mass spectra of TFA derivatives were also examined by Ludewig et al. (1982), who noted that the purine bases were not acylated at N-9; The TFA groups were substituted at N-6 and at all OH groups in the cytokinin molecule. Thus, for example, Z and [9R]Z were substituted by two and five TFA groups, respectively. A mass increment of 96 amu per H replaced is substantially higher for TFA derivatives than that for other derivatives except t-BuDMSi derivatives. Mass spectra of the TFA derivatives of a range of deoxyribonucleosides and ribonucleosides, including [9R]iP have been examined by Koenig et al. (1971). Using GC-MS Tsui et al. (1983) have identified [9R]iP as its tetra-acyl derivative in corms of water chestnut. They indicated that one of the acyl groups was presumably attached either to the exocyclic nitrogen or to position N-1. The TFA derivative was prepared by reacting the HPLC purified cytokinin fraction with equal vols. of pyridine and N-methyl-bis trifluoroacetamide (MBTFA) at 65 °C for 30 min. Although the mass spectra of TFA derivatives of cytokinins do not provide additional structural information in comparison to other derivatives, Tsui et al. (1983) have noted that the acyl derivatives may be very useful for GC-MS quantification based on selected ion current monitoring.

2.3 Preparative GC

Horgan and Scott (1985) have pointed out that although preparative GC is potentially a very powerful technique for the isolation of novel cytokinins, it has received little use. It is usually not possible to identify the peak(s) of interest in the gas chromatogram of a complex mixture containing putative novel compound(s). Even when GC-MS is used, it is a formidable task to recognize the mass spectrum of a novel cytokinin, and to identify it on the basis of the mass spectrum of its TMSi or permethyl derivatives. These derivatives are not particularly rich in structural information. However, the task may be considerably simplified if biological activity can be assigned to a particular GC peak. This has been done for cytokinins by making the TMSi derivative of the extract and subjecting it to semi-preparative GC. Using a suitable device to split the GC effluent between detector and trap, the material is collected into traps by cooling on a peak or time basis, the TMSi groups removed by mild acid hydrolysis [e.g, using 5% acetic acid in ethanol; Wang et al. (1977)] and the product bioassayed. The biologically active fraction can thus be purified to homogeneity and used for direct probe MS or other analyses for structure determination or confirmation. Although HPLC is now the preparative technique of choice, semi-preparative GC has been used in cytokinin research for identification of compounds (Horgan et al. 1975; Wang et al. 1977; Wang and Horgan 1978; Lee et al. 1981) and also for examining the incorporation of radioactivity into particular cytokinin metabolites (Stuchbury et al. 1979).

3 Mass Spectrometry

Since the previous review of mass spectrometric methods in this series (Biemann 1962) enormous advances have been made in instrumentation, particularly in the methods of sample introduction, ionisation and analysis. This has led to a several-fold increase in sensitivity, with routine detection levels for many compounds being in the subnanogram (picomole) range. In addition, the mass range of compounds that can be analysed has been increased to 10,000 amu. It is possible to comment here only briefly on each of these newer developments and more detailed accounts of instrumental methods are contained in standard mass spectrometric texts (Howe et al. 1981; McFadden 1973; McLafferty 1980; Waller and Dermer 1980; Watson 1985).

3.1 Instrumentation

3.1.1 Sample Introduction

Direct probe and gas chromatographic (GC) inlets are the most widely used methods of introducing samples into the ion source of a mass spectrometer. The direct probe is normally the method of choice for purified compounds of low volatility and for those compounds not amenable to GC conditions. The temperature at which sample molecules volatilize on the probe tip in the ion source (pressure $<10^{-6}$ Torr) is substantially lower than that required for the same sample to pass through a gas chromatograph, which operates at about atmospheric pressure. Derivatisation of the sample can overcome this problem but this can often lead to an undesirable large increase in the molecular weight of the compound as well as introducing an extra step in the analysis process. Details of the GC-MS interface and specific applications of this technique to cytokinin analysis will be discussed later in this and the following section.

The direct coupling of a high performance liquid chromatograph (HPLC) to a mass spectrometer has also been successfully carried out and is now available on a number of commercial instruments. Two interface methods have been developed, the first involving a continuous moving belt made of inert material on which the HPLC effluent is deposited and carried into the ion source through a series of pumping chambers designed to remove the solvent. In the second type, the HPLC effluent is introduced directly into the ion source using either a low flow rate microbore column or a split effluent from a standard analytical column. Cryogenic traps in the mass spectrometer source region are used to remove the bulk of the solvent.

3.1.2 Ionisation Methods

Up to the end of the 1960's, electron impact (EI) mass spectrometry was used almost exclusively to obtain mass spectra of organic compounds. Although EI still remains the most common and easy to use method of ionisation, there have been a plethora of additional ionisation techniques developed over the past 15 years. Some of these (e.g., chemical ionisation, CI) are complementary to EI and can provide molecular weight information on compounds when this is not apparent from the EI mass spectrum. Other methods (e.g., field desorption FD, fast atom bombardment FAB) give mass spectral data on highly polar and involatile compounds for which EI gives no information or is of limited applicability. Table 3 lists a selection of ionisation methods which are available on commercial mass spectrometers.

Because of its reproducibility, from day to day and from one instrument to another, together with the large library of data that has been built up using this technique over the past 25 years, EI mass spectrometry is normally used whenever confirmation of the identity of a specific compound is required. For routine analytical applications, particularly at the sub-nanogram ($<10^{-9}$) level, CI and NCI offer distinct advantages for particular classes of compounds including cytokinins (Table 3).

3.1.3 Analysers

Whereas magnetic sector instruments enjoyed an early dominance of the mass spectrometer market for analysis of organic compounds, the development in the 1970's of the quadrupole analyser with its fast scanning capability made it the preferred instrument for GC-MS studies. The recent introduction of laminated magnets with their intrinsically faster scanning capability and greater stability has

Table 3. Common ionisation methods in organic and biological mass spectrometry

Method	Description	Types of compounds analysed	Recent reviews
Electron Impact (EI)	Sample molecules in gas phase bombarded by electron beam from hot wire filament $M+e \rightarrow M^{+\cdot}+2e$ $M^{+\cdot} \rightarrow A^{+}+B^{\cdot}$	Gases, liquids, solids of low to high volatility. Mass spectra usually show abundant fragment ions	McLafferty (1980) Howe et al. (1981)
Chemical Ionisation (CI)	Ions (e.g. CH_5^+, NH_4^+) formed from reagent gases at ~ 1 Torr react with sample molecules in gas phase $M+BH^+ \rightarrow MH^+ + B$ $M+BH^+ \rightarrow MHB^+$ $M+BH^+ \rightarrow M-H^+ + H_2 + B$	As for EI. Mass spectra show fewer or no fragment ions. Especially useful for compounds with high proton affinities e.g. nitrogen bases	Richter and Schwarz (1978) Harrison (1983)
Desorption Chemical Ionisation (DCI)	As for CI but sample molecules are volatilised by rapid vaporisation from a hot wire	Very low volatility polar compounds, e.g. sugars, peptides	Cotter (1980)
Negative Chemical Ionisation (NCI)	Secondary electrons from ionisation of reagent gas (e.g. CH_4, NH_3) attach to sample molecules in gas phase. $M+e^- \rightarrow M^{-\cdot}$	Low to high volatility compounds with high electron affinity e.g. halogen containing compounds	Budzikiewicz (1981) Dougherty (1981) Bowie (1984)
Field Desorption (FD)	Sample parent ions desorbed from activated emitter in high electric field	Involatile polar compounds, salts, high MW biopolymers. Little fragmentation	Reynolds (1979) Wood (1982)
Fast Atom Bombardment (FAB)	Sample +ve and −ve parent ions expelled from surface of liquid matrix (e.g. glycerol) by momentum transfer from beam of fast atoms (Ar, Xe)	Involatile polar compounds, salts, thermally unstable compounds. High MW biopolymers	Barber et al. (1982)

Selected examples of various ionisation methods used in the mass spectrometric analysis of cytokinins: EI: Hashizume et al. (1976, 1979), Horgan et al. (1975), Leonard et al. (1969), Letham et al. (1967), MacLeod et al. (1976), Morris (1977), Shannon and Letham (1966), Summons et al. (1980), Young (1977) etc. CI: Claeys et al. (1978), Summons et al. (1979a, 1980), Palni et al. (1983b, 1985), Wang et al. (1982). DCI: Summons et al. (1983), MacLeod et al. (1986, Fig. 6). FD: Cole et al. (1975). FAB: Fig. 8.

narrowed the performance gap between the quadrupole and magnet sector mass spectrometers for on-line capillary gas chromatographic analysis. The triple sector quadrupole (TSQ) has been developed as a so-called mass spectrometer-mass spectrometer (MS-MS) system, permitting the analysis of individual components in complex mixtures without chromatographic pre-separation of the components (Yost and Fetterolf 1983). Similar MS-MS experiments can be performed on reverse geometry double-focussing mass spectrometers and on conventional double focussing instruments equipped with a third analyser (McLafferty 1983). The Fourier transform mass spectrometer (FTMS), a development of the Ion Cyclotron Resonance spectrometer, shows potential as a multipurpose mass spectrometer, with the capability to combine the best features of the quadrupole (fast scanning, rapid voltage switching) with those of the dual magnetic/electrostatic sector instruments (high resolution, high mass range) (Johlman et al. 1983).

3.1.4 Data Systems

Undoubtedly one of the major reasons for the rapid advances in the application of GC-MS as an analytical technique has been the simultaneous development of computerised data collection and handling systems. Additionally, the advent of microprocessors has brought many of the control functions of the mass spectrometers and its ancillary instrumentation under computer control, to the extent that GC-MS runs can be almost completely automated on some instruments. This means that mass spectrometer operators do not need to be experienced if the instrument is being used for repetitive routine GC-MS analysis.

3.2 Combined Gas Chromatography-Mass Spectrometry (GC-MS)

The combination of the gas chromatograph with the mass spectrometer (GC-MS) provides one of the most sensitive and specific means of analysing and identifying organic compounds in complex mixtures. In effect, the mass spectrometer acts as a detector for the gas chromatograph and can record the characteristic mass spectrum of each individual component eluting from the gas chromatograph. This information, together with the GC retention time of the component, can allow unequivocal identification of the individual compounds in the GC mixture down to the nanogram level, provided that reference data is available for comparison. When capillary columns are used for multicomponent analysis, the rapid elution time for the individual GC peaks (1–5 s) means that it is essential to have a fast scanning mass spectrometer on line to a dedicated data system with a large storage capacity.

Although the gas chromatograph and the mass spectrometer are compatible in most respects, a major difference between the two lies in their operating pressure. The GC operates just above atmospheric pressure due to the carrier gas, while the mass spectrometer must be maintained under high vacuum ($<10^{-6}$ Torr). As a result, the interface between the two instruments has to cope with a pressure drop of several orders of magnitude. The low carrier gas flow rates (<2 ml min^{-1}) used with capillary columns can usually be contained within the operating limits of the mass spectrometer by an efficient source pumping system. With the advent of flexible fused silica capillary columns, the outlet of the column itself can now be introduced via a heated interface directly into the MS ion source. This direct coupling results in no loss of sample between the GC and the MS. An alternative approach is to use an open split capillary between the GC column and the MS. A variable fraction of the GC carrier gas and sample effluent can then be introduced into the MS and the solvent can be partially diverted.

When using packed columns on a GC-MS system with flow rates of 15–30 ml min^{-1}, the interface device must be capable of removing a large fraction of the carrier gas (usually helium) before it enters the mass spectrometer. Numerous systems (called separators) have been developed over the past 20 years to cope with this, but the one most commonly used nowadays is the single stage glass jet. This device selectively pumps away the lighter carrier gas leaving the jet of enriched sample molecules to pass into the MS. In all interface systems surface decomposition of sample molecules must be avoided by using deactivated materials e.g., silanised glass.

Another important consideration when using the GC-MS combination is the stationary phase used for the GC column. It is essential that the "bleed" from the GC is kept as low as possible otherwise this will produce undesirable background ions in the MS scans. Special low bleed stationary phases have been developed for this purpose.

4 Applications of Mass Spectrometry in Cytokinin Analysis

In the preceding section some of the general features of mass spectrometry and instrumentation have been described. This section is intended only to illustrate the use of MS and combined GC-MS in cytokinin analysis for both identification and quantification purposes. Representative examples giving details of fragmentation patterns of selected cytokinins and their derivatives, and comparison of different MS techniques and their application to the analysis of individual cytokinins can be found in reports by Shannon and Letham (1966), Hecht et al. (1970a), Letham (1973), Horgan et al. (1975), Young (1977), Morris (1977), Hashizume et al. (1976, 1982b), Duke et al. (1978, 1979), Cole et al. (1975), Summons et al. (1980, 1983), MacLeod et al. (1976), Palni et al. (1985), and a valuable recent review by Horgan and Scott (1985).

4.1 Structural Studies

Commencing with the elucidation of the structure of the first reported natural cytokinin, zeatin (Letham et al. 1967), mass spectrometry has played a major role in determining the structures of cytokinins and their metabolites. The reason for this is an obvious one. Cytokinins are present in plant samples at such low concentrations that MS is one of the few methods capable of providing structural information on the minute amounts of compounds isolated.

The EI mass spectrum of zeatin run on the direct probe shows a strong molecular ion (M^+) at m/z 219 and fragment ions for loss of –OH and –CH$_2$OH from the sidechain at m/z 202 and 188 respectively (Fig. 3a). The remainder of the spectrum is dominated by ions (m/z 160, 148, 135, 134, 119) containing the adenine ring moiety as confirmed by accurate mass measurements. The presence of all of these ions in the mass spectrum of an unknown cytokinin is firm evidence

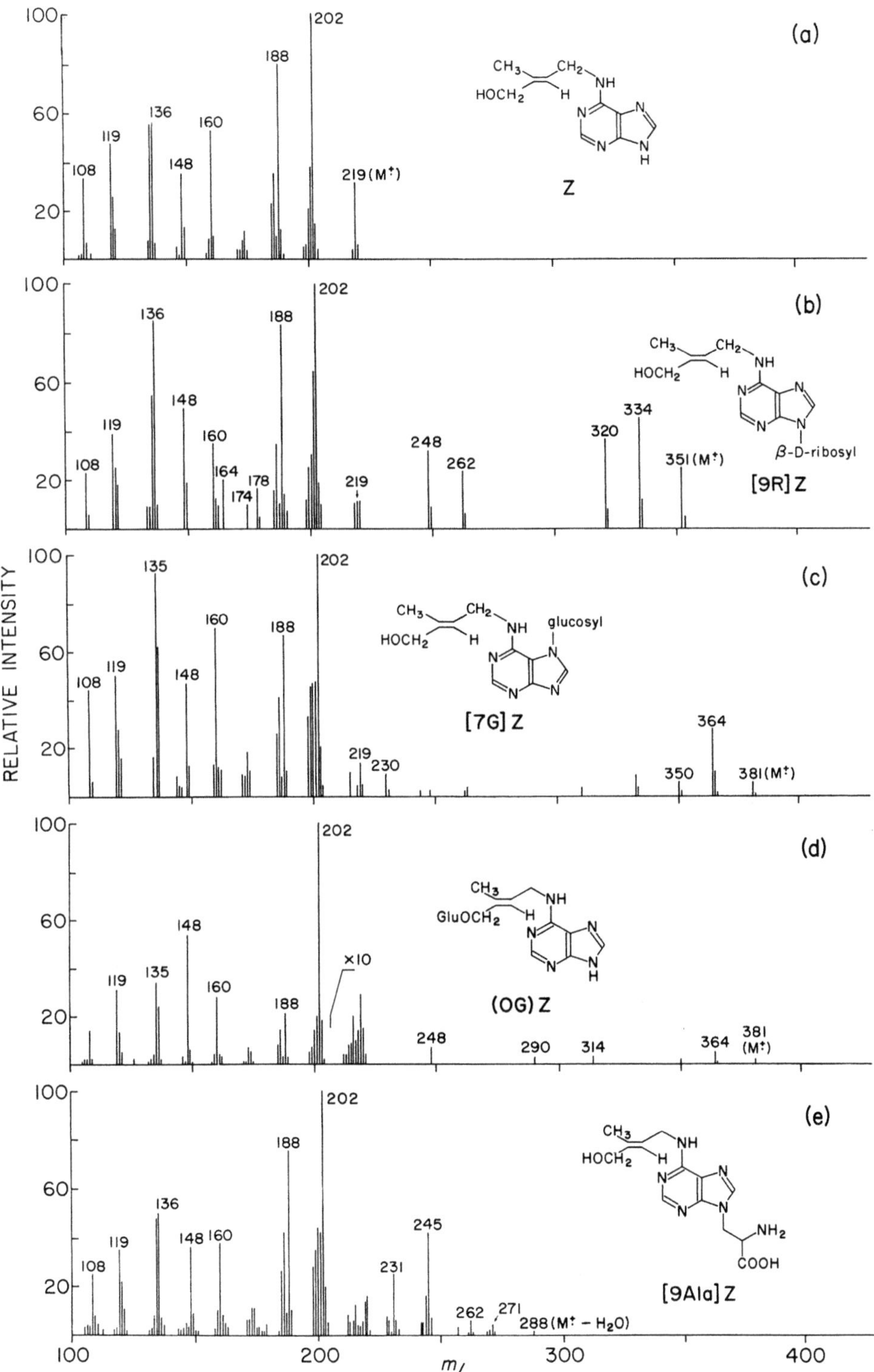

Fig. 3 a–e. Direct probe mass spectra (70 eV, EI) of **a** zeatin **b** zeatin riboside **c** zeatin-7-glucoside (raphanatin) **d** zeatin-O-glucoside, and **e** lupinic acid for the presence of a zeatin derivative. This is illustrated by the direct probe EI mass spectra of zeatin-9-β-D-riboside (Fig. 3 b), zeatin-7-β-D-glucoside (Fig. 3 c), zeatin-0-β-D-glucoside (Fig. 3 d), and lupinic acid (Fig. 3 e). Once the cytokinin has been identified as a zeatin derivative, high resolution accurate mass measurement of the M^+ and other high mass ions is used to establish unequivocally its molecular composition. From this, the substituent(s) on the zeatin nucleus can be determined. Such an approach has been used to identify the gross structures of a large number of zeatin and dihydrozeatin metabolites including those shown in Fig. 3, as well as other cytokinins.

The methodology used in the authors' laboratories is exemplified by the identification of [7G]Z, raphanatin, first as an exogenous and later as a naturally-occurring metabolite of zeatin (Parker et al. 1972; Summons et al. 1977; Cowley et al. 1978). [^{3}H]-Zeatin was supplied exogenously to radish seedlings with their roots excised. After 8 h the seedlings were extracted and chromatographed. The major zone of radioactivity on a paper chromatogram was not identical chromatographically to any of the then known zeatin metabolites. The new metabolite was isolated in μg amounts by a large scale extraction followed by a series of chromatographic purifications using the radiolabelled material as a marker. Its mass spectrum (Fig. 3 c) was typical of that of a zeatin derivative and showed an M^+ at m/z 381, 30 mass units above that of zeatin riboside (Fig. 3 b). An accurate mass measurement on m/z 381 established its composition as $C_{16}H_{23}N_5O_6$, corresponding to a hexose ($C_6H_{11}O_5$) sugar attached to the zeatin ring. The position of substitution of the hexose sugar was determined by UV spectroscopy (Leonard et al. 1965). Identification of the sugar as glucose was obtained by use of the specific glucose oxidase test on the acid hydrolysis product of the new cytokinin, called raphanatin (Parker et al. 1972). Final confirmation of the complete structure of raphanatin was established by synthesis (Cowley et al. 1978). As a follow-up, [^{2}H$_2$] and [^{2}H$_5$]-labelled analogues of raphanatin were synthesised. These stable-isotope labelled compounds were used to confirm the occurrence of raphanatin as an endogenous cytokinin in radish seed using GC-MS (Summons et al. 1977).

4.2 Quantification of Cytokinins

Quantitative mass spectrometry was first described by Sweeley et al. (1966) and has been discussed in detail by Millard (1978). For quantitative analysis by MS or combined GC-MS, maximum sensitivity is obtained by selecting a small number (usually one to four) of characteristic ions in the mass spectrum of the compound to be measured. Monitoring of more ions offers additional selectivity at the cost of a concomitant drop in sensitivity. The output signal due to each of these ions is "continuously" monitored by rapid switching from one ion to the next during the GC elution or probe desorption profile. The technique is commonly known as selected ion monitoring (SIM), multiple ion detection (MID) or multi-peak monitoring (MPM) etc. Detection levels in the picogram (10^{-12} g) range can be readily achieved for many compounds while under favourable circumstances using single ion detection, femtogram (10^{-15} g) amounts of material have been measured. For cytokinins, however, commonly reported detection levels are in the ng range. A data system is a highly desirable though not an essential accessory for MID work, both for control of peak switching and collection of data.

4.2.1 Internal Standards

Substantial losses of cytokinins occur during their purification from plant extracts, and losses vary between extracts (Scott et al. 1982; Tay et al. 1985). Furthermore, losses of individual cytokinins are variable. These problems are best overcome by the use of internal standards. Compounds suitable as internal standards are those which show chemical and physical behaviour similar to the endogenous species. In practice known amounts of one or more compounds (internal standards) are added to the extract at the beginning of purification procedure; the internal standards and the endogenous compounds should suffer identical losses during multistep purification process, but can be distinguished in the final quantitative analysis.

The principle of such an analysis for cytokinins was first demonstrated by Thompson et al. (1975), who quantified (oOH) [9R]BAP from poplar leaves using the unnatural *para*-isomer. The two isomers were resolved during GC, and by monitoring a single common ion (m/z 661) during GC-MS analysis it was possible to quantify the natural (*ortho*-isomer) compound.

As pointed out by Brenner (1981) internal standards should be added at low concentrations, preferably similar to or slightly greater than the expected levels of endogenous compounds in the extract. It is undesirable to add large quantities of internal standards to estimate the recovery of compounds like cytokinins which occur in submicrogram levels in most plant samples. When analysing trace quantities, losses due to irreversible adsorption and decomposition on active sites in the GC column, interfaces and the ion source could be significant. Addition of an excess amount of internal standard different to the compound to be measured is likely to mask all adsorption sites, and is believed to reduce decomposition of endogenous compounds on a statistical basis. This can result in an overestimate of sample recovery. Stable isotopically labelled analogues of the compound to be analysed are therefore preferred as internal standards. Analogues labelled with ^{2}H, ^{13}C, ^{15}N, and ^{18}O are more likely to behave as carriers for small quantities of the endogenous compounds because of the close chemical similarity to the unlabelled compounds, thus reducing the possibility of discriminatory adsorption. Deuterium (^{2}H)-labelled compounds are the easiest to synthesize and least expensive. A slight complication is the possibility of back exchange when the deuterium atom is attached to a labile position of the molecule as has been reported for 2'-^{2}H$_2$-indole acetic acid (McDougall and Hillman 1978). Such exchange cannot occur in the compounds shown in Fig. 4 but should be borne in mind when synthesizing other deuterium labelled analogues.

4.2.2 Stable Isotope Dilution Mass Spectrometry

When stable isotope labelled compounds are use for quantification, the mass spectrometer differentiates between the added labelled internal standard and the

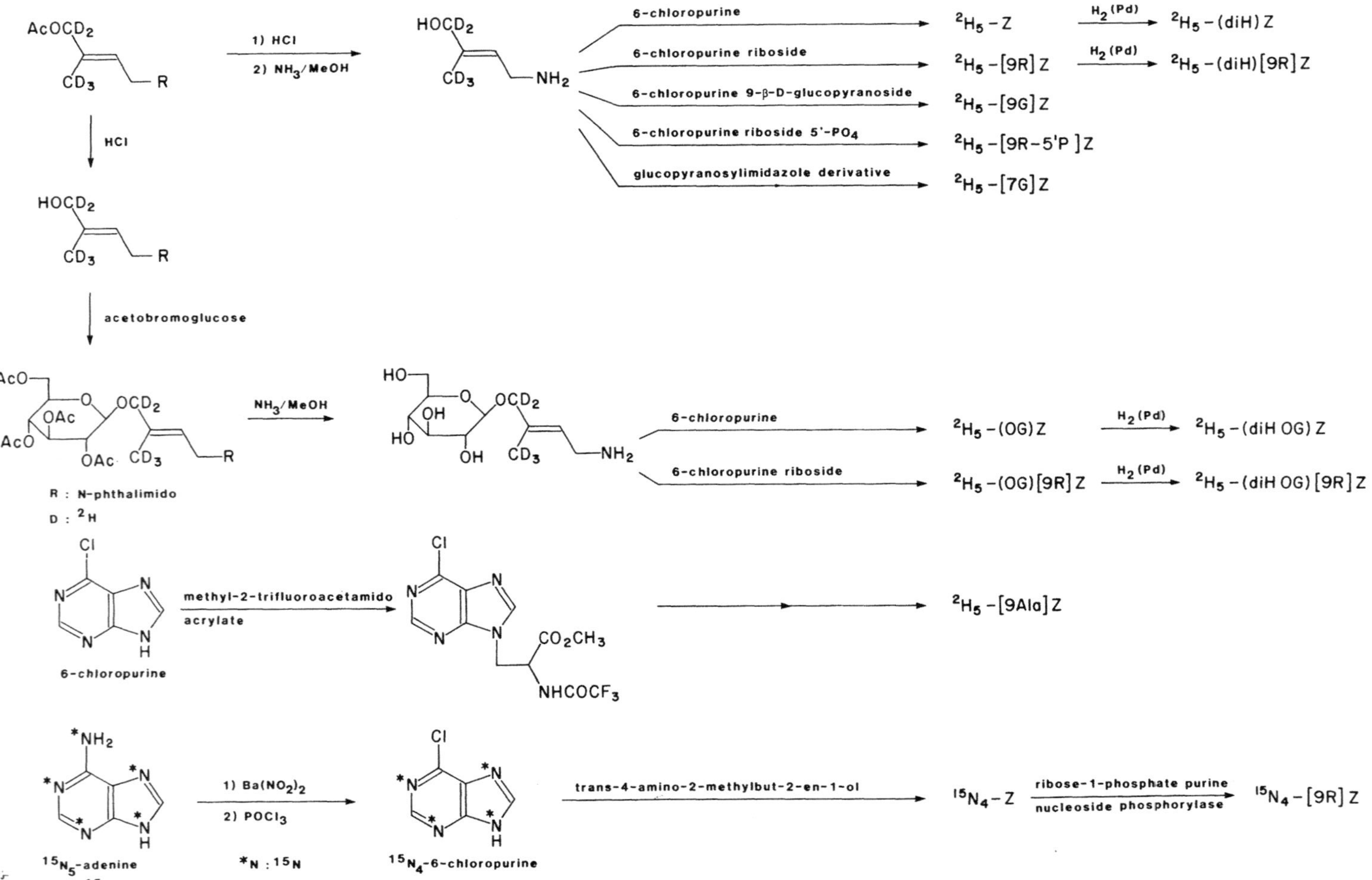

Fig. 4. Outline of synthesis of ²H₅ and ¹⁵N₄ cytokinins, suitable for use as internal standards

endogenous (unlabelled) compound on the basis of the difference in the mass to charge ratio (m/z) or certain ions (e.g.,the molecular ion, M^+) formed from the two otherwise identical compounds. The intensities of selected ions corresponding to both the labelled and unlabelled compounds are measured by MID. The ratio of the unlabelled (endogenous) compound to internal standard (not the absolute amount of endogenous compound) is then determined and used to calculate the quantity of endogenous compound in the original extract, knowing the quantity of added internal standard. This procedure is often referred to as stable isotope dilution mass spectrometry (IDMS). Although details of synthetic procedures for making isotopically labelled compounds will not be discussed here (see Table 1; Horgan and Scott 1985; Entsch et al. 1980; and references therein), key synthetic schemes are given in Fig. 4.

Isotopically labelled internal standards were first used for cytokinin analysis by Summons et al. (1977), who quantified [7G]Z in radish seed using $^2H_2$, i.e., dideuterio labelled internal standard. In later studies $^2H_5$, i.e., D_5, internal standards (Fig. 4) were used for the quantification of cytokinins in *Zea mays* kernels (Summons et al. 1979a), lupin fruits (Summons et al. 1979b), *Datura innoxia* crown gall tissue (Palni et al. 1983b; Summons et al. 1983) and seaweed extract (Tay et al. 1985) etc. Deuterated cytokinins currently used as internal standards have ^2H atoms in the sidechain. One limitation of these deuterium labelled cytokinins is that fragment ions arising from cleavage of the N-6 sidechain are identical for both labelled and endogenous compounds, and thus unsuitable for quantification. With the $^{15}N_4$ labelled cytokinin standards (Fig. 4) cross interference is minimal as the ^{15}N atoms are confined to the purine nucleus, thereby providing a number of fragment ions suitable for detection. A number of these fragment ions in the lower m/z range, however, may not be of much value for quantification due to the lack of selectivity. $^{15}N_4$-labelled cytokinins have been used in the analysis of *Vinca rosea* and *Nicotiana tabacum* crown gall tissues (Scott and Horgan 1980, 1984). For increased sensitivity quantification is normally based on the most intense ion(s), and selecting ion(s) of higher m/z value for this purpose offers enhanced selectivity. For example, under EI-MS the $(M-15)^+$ ion of TMSi cytokinins is usually considerably stronger than the molecular ion and is used for quantification (e.g., see Palni et al. 1983b; Tay et al. 1985). For selection of ions suitable for quantification of various cytokinin derivatives readers are referred to reports by Summons et al. (1979a, b), Hashizume et al. (1979), Dauphin-Guerin et al. (1980), Wang et al. (1982), Martin et al. (1982), Duke et al. (1980), Scott and Horgan (1980, 1984), Scott et al. (1982), and Tay et al. (1985).

For accurate quantification by MID, it is essential that the ions being monitored in both labelled and unlabelled species do not overlap with any other ions (natural isotope abundance ions, column bleed, background ions) during the sample elution. Thus if an internal standard differs from the natural compound by only one or two amu, the intensity of the mass spectral peak characteristic of the internal standard will be inflated due to contributions from natural abundance isotope peaks. On the other hand if the isotopically labelled internal standard contains even a small percentage of unlabelled material, this will contribute towards the "endogenous" mass spectral peak. Therefore to minimise cross talk between ions resulting from the internal standard and its unlabelled analogue the

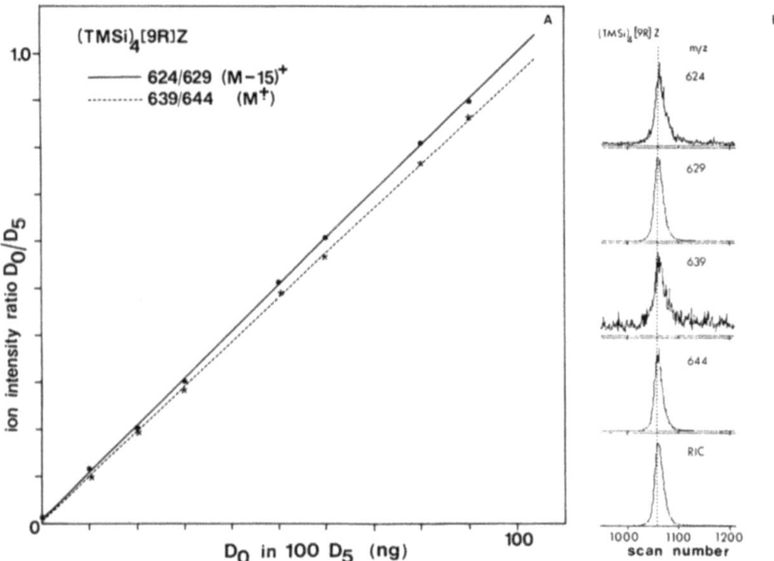

Fig. 5. A Calibration curves for GC-MS-MID quantification of [9R]Z. The m/z 639 (M⁺) and 624 (M-15)⁺ ions of the (TMSi)$_4$-[9R]Z (i.e., D$_0$) were monitored together with the corresponding ions at m/z 644 and 629 due to $^2H_5$ labelled (i.e., D$_5$) internal standard. **B** A multiple-ion-detection tracing on GC-MS of (TMSi)$_4$-[9R]Z purified from tobacco shoots. A packed GC column was used for this analysis. The ions of m/z 644 and 639 (M⁺ for tetra-TMSi-$^2H_5$-[9R]Z and endogenous [9R]Z) were observed at the correct GC R$_t$ for authentic (TMSi)$_4$-[9R]Z. Corresponding (and more intense) ions due to the loss of CH$_3$ could be seen at m/z 629 and 624, and were used to calculate ion intensity ratio (624/629) for quantification using the calibration curve shown in Fig. 5 A. A true ion intensity ratio (for M⁺ –CH$_3$ ions) of 0.040 was obtained by summing scans 1032 to 1100. Less accurate intensity ratios of 0.030, 0.035 and 0.050 were obtained by looking at individual scans at the beginning (1049), top (1058) and tail (1075) of the GC peak, respectively. *RIC* = Reconstituted ion current. A slight GC separation of deuterated from unlabelled (endogenous) molecules can be seen

preferred mass difference between the labelled and unlabelled compounds is 3–5 amu.

In IDMS a calibration plot should be constructed (Fig. 5 A) for individual cytokinins for accurate quantification. Often the straight line plots do not pass through the origin, as in an ideal situation, indicating that the labelled standard is not isotopically pure. Care must be taken when calibrating the system to account for any contribution made by the internal standard to the sample ion(s) being monitored and vice versa. Pickup and McPherson (1976) discuss variations from the standard plot in more detail.

4.2.3 Quantification Using GC-MS

For greater accuracy in GC-MS-MID analysis an average mass spectrum should be derived by summing a series of scans over the ion peak. This aspect has been discussed by Sponsel and MacMillan (1978) in relation to gibberellins. This is necessary as the sample concentration in the MS ion-source changes during the elution of a peak from the GC column, which is reflected in the relative intensities of fragment ions detected during a single MS scan (for example see legend to Fig. 5 B). The other complication encountered in the GC-MS-MID analysis when deuterium labelled internal standards are used is the slight separation between endogenous cytokinins and their labelled analogues (Summons et al. 1977) even with packed columns. As shown in Fig. 5 B the deuterated analogue elutes slightly ahead of the unlabelled $(TMSi)_4$-[9R]Z. For trace analysis a true ion intensity ratio between the internal standard and its natural counterpart can be obtained by determining areas under the GC peaks corresponding to the respective ions (Summons et al. 1977; Hashizume et al. 1979; Scott and Horgan 1980, 1984; Palmer et al. 1981; Tay et al. 1985).

As stated earlier, monitoring of more than one ion for a particular compound and checking their intensity relationship during the run gives additional selectivity, and can provide very satisfactory identification, particularly when GC retention time is also taken into consideration. Any major deviation from the intensity relationship shown in a full mass spectrum of the pure compound run under identical conditions indicates interference from other compounds. However, whenever possible the identity and purity of the compound quantified by multiple- and particularly by single-ion detection should be confirmed by a full or partial MS scan over the GC peak.

4.2.4 Probe Analysis

When a sample is introduced directly into the MS source quantification is usually possible only when the compound of interest can be isolated in a sufficiently pure state. Examples of this type of analysis in cytokinin studies can be found in reports by Summons et al. (1979 b, 1983), who used the direct insertion probe technique to obtain mass spectra. Recently desorption chemical ionisation (DCI) has been successfully used in the scanning mode to quantify underivatized intact cytokinin glucosides (both O-, and N-glucosides) in *Datura* crown gall tissue (MacLeod et al. 1986). Figure 6 illustrates the usefulness of DCI for cytokinin quantification, the notable features in the spectrum of D_5-(diH OG)[9R]Z are the intense protonated molecular ion and an almost total absence of major fragment ions (cf. EI probe spectrum). Satisfactory EI probe spectra have also been obtained in the author's laboratories by desorbing the sample (underivatised or derivatised) from a rapidly heated filament. Identical results were obtained when [9R]Z was quantified in *Datura* crown gall tissue using the desorption electron impact technique (DEI, Fig. 7 A), and as its tetra-TMSi derivative by GC-MS-MID (MacLeod et al. 1986). However, for the analysis of relatively impure samples the method of choice is GC-MS. The development of fast atom bombardment (FAB) as an ionisation technique [particularly in combination with negative ion MS; Horgan and Scott (1985)] holds considerable promise for structural studies of polar cytokinin derivatives e.g., nucleotides and glucosides, by mass spectrometry. The full potential of this technique for the identification and quantification of complex cytokinins as intact molecules is yet to be exploited. Figure 8 shows the positive and negative ion FAB spectra of D_2-[9R-5′P]iP and (OG)[9R-5′P]Z, respectively.

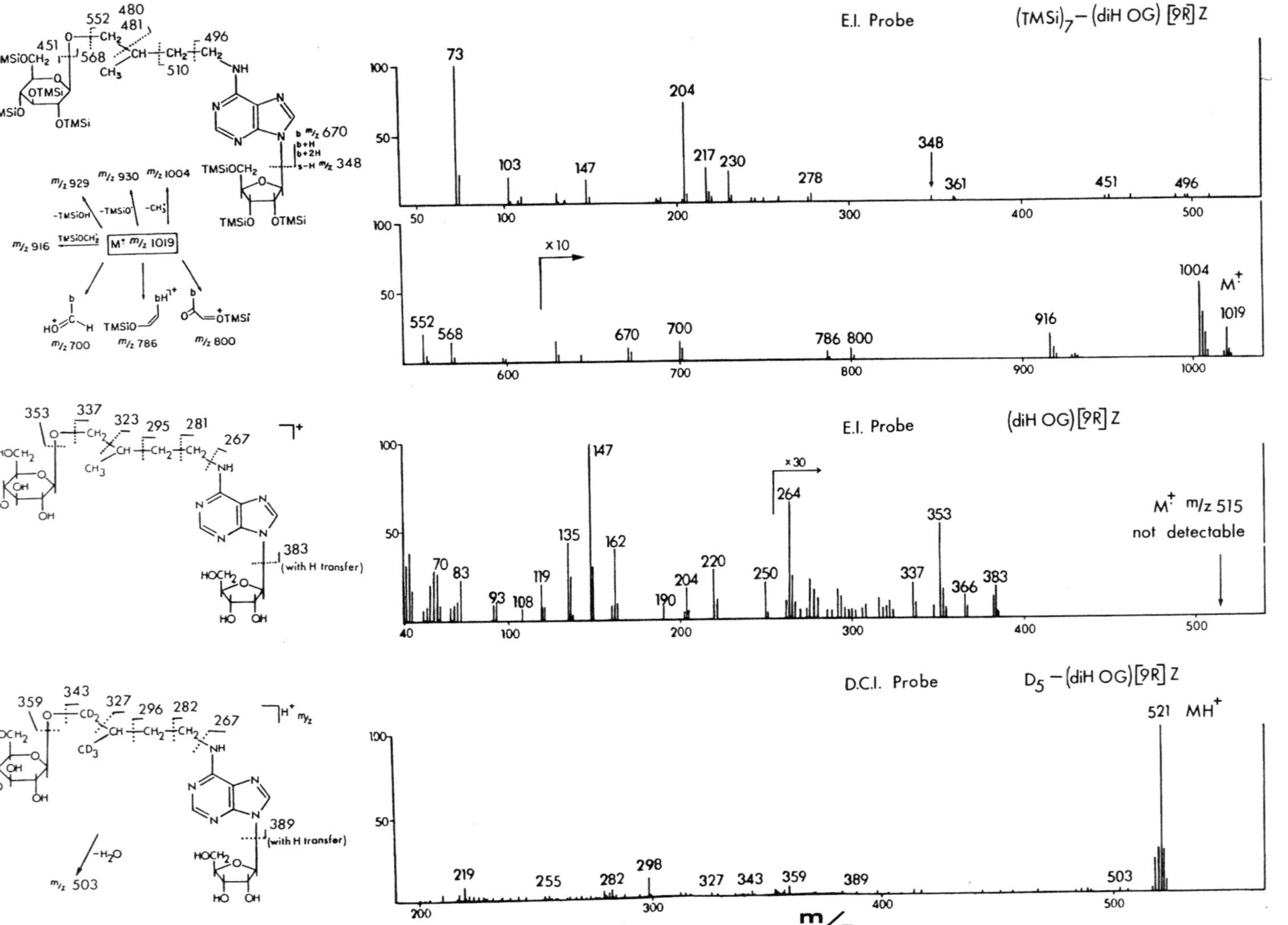

Fig. 6. Probe mass spectra of O-glucoside of dihydrozeatin riboside and its TMSi derivative. Under electron impact ionisation (70 eV) a weak molecular ion was observed for (TMSi)₇-(diH OG)[9R]Z, but not detected for the underivatized compound. In contrast, desorption chemical ionisation (ammonia as reagent gas) gave an intense protonated molecular ion (base peak), and few fragment ions

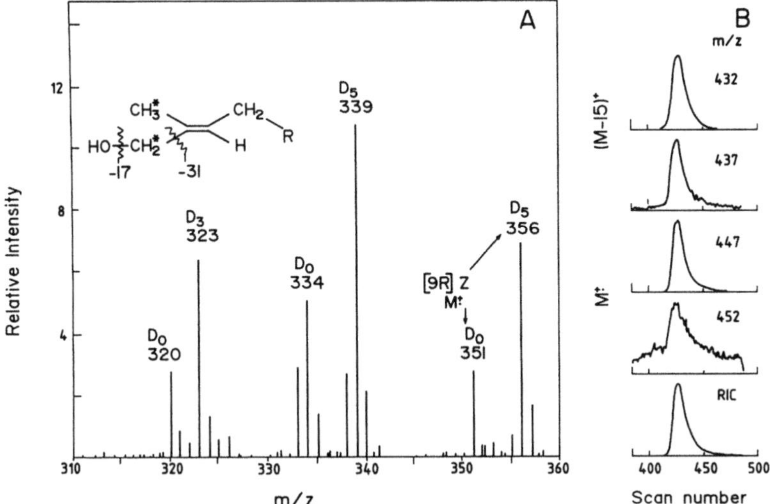

Fig. 7. A Molecular ion region in the DEI mass spectrum of zeatin riboside isolated from *Datura innoxia* crown gall tissue. A known quantity of D_5-standard was added at the time of extraction to permit quantification by MS. The position of deuterium atoms in D_5-standard is indicated by asterisks. **B** A multiple-ion-detection tracing during GC-MS of $(tBuDMSi)_2$-Z purified from *Datura innoxia* crown gall issue. The tissue was incubated with $^{15}N_5$-adenine for 8 h before extraction. The ion profiles at m/z 437 and 452 are due to biosynthesized $^{15}N_5$-Z and the corresponding ion profiles for endogenous Z appear at m/z 432 and 447

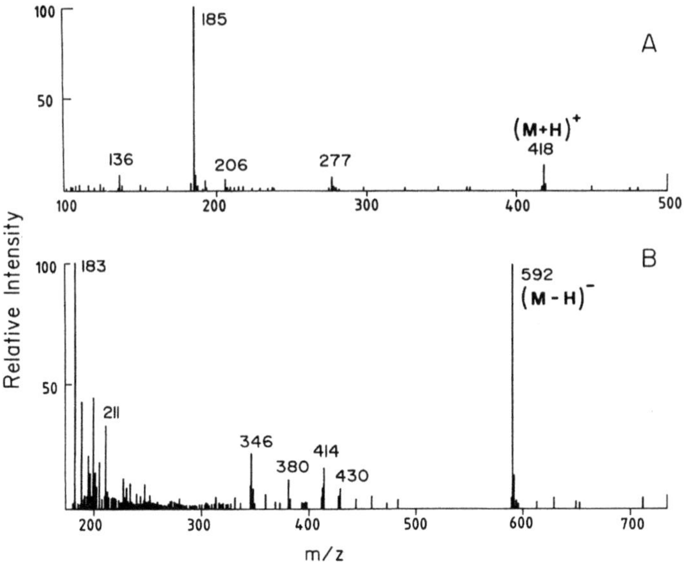

Fig. 8 A, B. Fast atom bombardment (FAB) mass spectra (glycerol matrix) of cytokinins. **A** Positive ion spectrum of D_2-[9R-5′P]iP. **B** Negative ion spectrum of (OG) [9R-5′P]Z; courtesy of Dr. T. L. Wang, John Innes Institute, Norwich, UK

4.3 Metabolic Profiling

Compounds labelled with radioactive isotopes have been used almost exclusively in studies of cytokinin biosynthesis and metabolism. Due to the availability of high specific activity compounds and sensitive detection methods, radioactively labelled cytokinins have been employed very successfully to determine uptake and metabolic stability of these compounds in vivo, and for elucidation of metabolic pathways. Such studies have even facilitated identification of endogenous cytokinins. However, frequently it is not possible to prove co-identity between the radioactive and endogenous metabolites. Even the most reliable identifications are usually based only on co-chromatography of the radioactive metabolite with the authentic compounds. In a few cases, co-crystallisation to constant specific activity has been reported. Horgan (1981) has cited examples where mass spectrometry was used to demonstrate the presence of ^{14}C in gibberellin metabolites to provide direct confirmation of metabolic steps; with ^{3}H labelled compounds this is rarely possible.

Examples where mass spectrometry has provided positive identification of metabolite(s) formed from externally applied cytokinins are listed in a review by Letham and Palni (1983). More recent reports include studies of cytokinin metabolism in oat leaves (Tao et al. 1983), *Phaseolus* embryos (Lee et al. 1985) and de-rooted radish seedlings (McGaw et al. 1984a, 1985).

Cytokinins labelled with both stable and radioactive isotopes would be extremely useful in metabolism studies. The utility of ^{2}H- and ^{3}H-labelled gibberellins has been previously demonstrated (MacMillan 1977). With the use of such compounds it is still possible to conveniently follow the metabolism by monitoring radioactivity. In addition stable isotope labelled metabolite(s) can be distinguished from endogenous compound(s) by mass spectrometry for direct confirmation of identity. Further quantitative measurements are also possible to determine the specific activity of the metabolite(s) identified. Such determinations are necessary in "turnover" studies. Following application of 8-^{3}H and ^{2}H$_5$-(OG)[9R]Z to soybean callus tissue, [9R]Z was shown to be the metabolite responsible for its high biological activity (Palni et al. 1984). In this study MS analysis using probe technique was employed to show unambiguously the hydrolysis of a cytokinin O-glucoside to the corresponding aglycone within a plant tissue. Recently the GC-MS-MID technique has also been used to demonstrate incorporation of ^{15}N$_5$-adenine into Z (Fig. 7B), [9R]Z, and [9R-5'P]Z in *Datura* crown-gall tissue (MacLeod et al. 1986). This confirmed earlier results where radioactively labelled adenine had been used in studies of cytokinin biosynthesis in *Datura* (Palni et al. 1985) and other crown-gall tissues (e.g., Palni et al. 1983a).

It is expected that future studies will make use of the vastly improved sensitivity now available with modern mass spectrometers to provide a sound basis to cytokinin metabolism work.

5 General Remarks and Conclusion

Since cytokinins are present in trace amounts in plant samples, contamination from extraneous sources is a serious problem. It is imperative, therefore, that sample extracts should be prepared and handled very carefully right up to the final analytical step. In practice this means use of immaculately clean glassware, highest quality solvents and reagents. Gaskin and MacMillan (1978) have provided a valuable account of the analytical care required during extraction, purification and final sample preparation for GC-MS analysis of plant growth substances.

An important consideration for successful cytokinin analysis is the inherently low levels of these compounds in plant tissues. Therefore recovery of cytokinins during sample extraction and purification is very critical. Space limitations do not permit discussion of various procedures currently in use for the extraction and purification of these compounds. Horgan and Scott (1985) have recently discussed this aspect in some detail. Choice of a proper extraction procedure is critical as it can influence final results (Tay et al. 1986). It should be emphasised that a clear distinction must be made between qualitative and quantitative analyses. While the former merely demonstrates the presence of a compound in the sample, quantitative analysis determines the actual amount of the specific compound(s) being analysed.

In conclusion combined GC-MS is the most powerful technique available for both qualitative and quantitative analysis of cytokinins and their metabolites. It provides the selectivity and sensitivity required for the analysis of these substances. It is the technique of reference for validation of other analytical methods currently available.

Acknowledgments. We wish to thank colleagues for providing preprints. Particular thanks are due to Dr. M. V. Palmer for his comments on the manuscript and Drs. D. S. Letham, R. Horgan, R. E. Summons, and O. C. Wong for useful discussions.

References

Babcock DF, Morris RO (1970) Quantitative measurement of isoprenoid nucleosides in transfer ribonucleic aicd. Biochemistry 9:3701–3705

Barber M, Bordoli RS, Elliott GJ, Sedgwick RD, Tyler AN (1982) Fast atom bombardment mass spectrometry. Anal Chem 54:645A–657A

Biemann K (1962) Mass spectrometric methods. In: Linskens HF, Tracey MV (eds) Modern methods of plant analysis, vol 5. Springer, Berlin Heidelberg New York, pp 26–50

Blau K, King G (1978) Handbook of derivatives for chromatography. Heyden, London

Bowie JH (1984) The formation and fragmentation of negative ions derived from organic molecules. Mass Spectrom Rev 3:161–207

Brenner ML (1981) Modern methods for plant growth substance analysis. Annu Rev Plant Physiol 32:511–538

Budzikiewicz H (1981) Mass spectrometry of negative ions. Angew Chem Int Ed 20:624–637

Bullock MW, Hand JJ, Stokstad ELR (1956) Syntheses of 6-substituted purines. J Am Chem Soc 78:3693–3696

Burrows WJ, Armstrong DJ, Skoog F, Hecht SM, Boyle JTA, Leonard NJ, Occolowitz J (1969) The isolation and identification of two cytokinins from *Escherichia coli* transfer ribonucleic acids. Biochemistry 8:3071–3076

Burrows WJ, Armstrong DJ, Kaminek M, Skoog F, Bock RM, Hecht SM, Dammann LG, Leonard NJ, Occolowitz J (1970) Isolation and identification of four cytokinins from wheat germ transfer ribonucleic acid. Biochemistry 9:1867–1872

Chaves das Neves HJ, Pais MSS (1980) A new cytokinin from the fruits of *Zantedeschia aethiopica*. Tetrahedron Lett 21:4387–4390

Chen W-S (1983) Cytokinins of the developing mango fruit. Plant Physiol 71:356–361

Claeys M, Messens E, Van Montagu M, Schell J (1978) GC-MS determination of cytokinins in *Agrobacterium tumefaciens* cultures. Fresenius Z Anal Chem 290:125–126

Cole DL, Leonard NJ, Cook JC (1975) System for separation and identification of the naturally occurring cytokinins. High performance liquid chromatography-field desorption mass spectrometry. In: Zdzislaw P (ed) Recent dev oligonucleotide synth chem minor bases tRNA. Int Conf Inst Chem, Poznan, Poland, pp 153–174

Corey EJ, Chaykovsky M (1962) Methylsulfinylcarbanion. J Am Chem Soc 84:866–867

Corey EJ, Venkateswarlu (1972) Protection of hydroxyl groups as tert-butyldimethylsilyl derivatives. J Am Chem Soc 94:6190–6191

Corse J, Kuhnle J (1972) An improved synthesis of *trans*-zeatin. Synthesis 11:618–619

Cotter RJ (1980) Mass spectrometry of non-volatile compounds. Anal Chem 52:1589A–1606A

Cowley DE, Duke CC, Leipa AJ, MacLeod JK, Letham DS (1978) The structure and synthesis of cytokinin metabolites. 1. The 7- and 9-β-D-glucofuranosides and pyranosides of zeatin and 6-benzylaminopurine. Aust J Chem 31:1095–1111

Daly JW, Christensen BE (1956) Purines. VI. The preparation of certain 6-substituted and 6,9-disubstituted purines. J Org Chem 21:177–179

Dauphin B, Teller G, Durand B (1977) A new method for the purification, characterization, and measurement of endogenous cytokinins extracted from *Mercurialis annua* shoots. Physiol Veg 15:747–762

Dauphin B, Teller G, Durand B (1979) Identification and quantitative analysis of cytokinins from shoot apices of *Mercurialis ambigua* by gas chromatography-mass spectrometry computer system. Planta 144:113–119

Dauphin-Guerin B, Teller G, Durand B (1980) Different endogenous cytokinins between male and female *Mercurialis annua* L. Planta 148:124–129

Doree M, Guern J (1967) Sur un mode de degradation possible de quelques cytokinines, par MM. Compt Rend D 265:29–32

Dougherty RC (1981) Negative chemical ionisation mass spectrometry. Anal Chem 53:625A–636A

Drozd J (1981) Chemical derivatization in gas chromatography. Elsevier Scientific Publishing, Amsterdam, pp 1–232 (J Chromatogr Libr vol 19)

Duke CC, MacLeod JK, Summons RE, Letham DS, Parker CW (1978) The structure and synthesis of cytokinin metabolites. II. Lupinic acid and O-β-D-glucopyranosylzeatin from *Lupinus angustifolius*. Aust J Chem 31:1291–1301

Duke CC, Letham DS, Parker CW, MacLeod JK, Summons RE (1979) The structure and synthesis of cytokinin metabolites. IV. The complex of O-glucosylzeatin derivatives formed in *Populus* species. Phytochemistry 18:819–824

Duke CC, Entsch B, Letham DS, MacLeod JK, Parker CW, Summons RE (1980) Mass spectrometric and gas chromatographic-mass spectrometric studies of cytokinin metabolism. Recent Dev Mass Spectrom Biochem Med 6:7–19

Dyson WH, Fox JE, McChesney JD (1972) Short term metabolism of urea and purine cytokinins. Plant Physiol 49:506–513

Eagles J, Laird WM, Self R, Synge RLM (1974) Permethylation for mass spectrometry: Rearrangements for ester linkages and use of potassium t-butoxide. Biomed Mass Spectrom 1:43–48

Ehrlich BJ, Hall RC, Anderson RJ, Cox HG (1981) Sulfur detection in hydrocarbon matrices. A comparison of the flame photometric detector and the 700A Hall electrolytic conductivity detector. J Chromatogr Sci 19:245–249

Einset JW, Skoog FK (1977) Isolation and identification of ribosyl-*cis*-zeatin from transfer RNA of *Corynebacterium fascians*. Biochem Biophys Res Commun 79:1117–1121

Entsch B, Letham DS, Parker CW, Summons RE, Gollnow BI (1980) Metabolites of cytokinins. In: Skoog F (ed) Plant Growth Substances 1979. Springer, Berlin Heidelberg New York, p 527

Ernst D, Schafer W, Oesterhelt D (1983 a) Isolation and quantitation of isopentenyladenosine in an anise cell culture by single-ion monitoring, radioimmunoassay and bioassay. Planta 159:216–221

Ernst D, Schafer W, Oesterhelt D (1983 b) Isolation and identification of a new, naturally occurring cytokinin (6-benzylaminopurine riboside) from an anise cell culture (*Pimpinella anisum* L.). Planta 159:222–225

Fleysher MH, Bloch A, Hakala MT, Nichol CA (1969) Synthesis and biological activity of some new N^6-substituted purine nucleosides. J Med Chem 12:1056–1061

Fox JE, Chen C-M (1967) Characterization of labelled ribonucleic acid from tissue grown on ^{14}C-containing cytokinins. J Biol Chem 242:4490–4494

Fox JE, Sood CK, Buckwalter B, McChesney JD (1971) The metabolism and biological activity of a 9-substituted cytokinin. Plant Physiol 47:275–281

Gaskin P, MacMillan J (1978) GC and GC-MS techniques for gibberellins. In: Hillman JR (ed) Isolation of plant growth substances. Cambridge University Press, Cambridge, p 157

Grimm WAH, Leonard NJ (1967) Synthesis of the "minor nucleotide" N^6-(γ,γ-Dimethylallyl)adenosine 5′-phosphate and relative rates of rearrangement of 1- to N^6-dimethylallyl compounds for base, nucleoside, and nucleotide. Biochemistry 6:3625–3631

Grob K, Grob G (1979) Practical capillary gas chromatography – a systematic approach. J High Resolut Chromatogr/Chromatogr Commun 2:109–117

Guerin B, Kahlem G, Teller G, Durand B (1984) Evidence for host genome involvement in cytokinin metabolism by male and female cells of *Mercurialis annua* transformed by strain 15955 of *Agrobacterium tumefaciens*. Plant Physiol 74:139–145

Guha OK (1984) Effect of injection needle dimensions in gas chromatography. J Chromatogr 292:57–65

Haberlandt G (1921) Wundhormone als Erreger von Zellteilungen. Beitr Allg Bot 2:1–53

Hahn H (1975) Cytokinins: A rapid extraction and purification method. Physiol Plant 34:204–207

Hakomori S (1964) A rapid permethylation of glycolipid, and polysaccharide catalyzed by methylsulfinyl carbanion in dimethylsulfoxide. J Biochem 55:205–208

Harrison AG (1983) Chemical ionization mass spectrometry. CRC Press, Boca Raton, Florida

Hashizume T, McCloskey JA, Liehr JG (1976) Electron impact induced reactions of N^6-(3-methyl-2-butenyl)adenosine and related cytokinins. Biomed Mass Spectrom 3:177–183

Hashizume T, Sugiyama T, Imura M, Cory HT, Scott MF, McCloskey JA (1979) Determination of cytokinins by mass spectrometry based on stable isotope dilution. Anal Biochem 92:111–122

Hashizume T, Suye S, Sugiyama T (1982 a) Isolation and identification of *cis*-zeatin riboside from tubers of sweet potato (*Ipomoea batatas*. L.). Agric Biol Chem 46:663–665

Hashizume T, Suye S, Soedo T, Sugiyama T (1982 b) Isolation and characterization of a new glucopyranosyl derivative of 6-(3-methyl-2-butenylamino)purine from sweet potato tubers. FEBS Lett 144:25–28

Hecht SM, Gupta AS, Leonard NJ (1970 a) Mass spectra of ribonucleoside components of tRNA. II. Anal Biochem 38:230–251

Hecht SM, Leonard NJ, Schmitz RY, Skoog F (1970 b) Cytokinins: Synthesis and growth-promoting activity of 2-substituted compounds in the N^6-isopentenyladenosine and zeatin series. Phytochemistry 9:1173–1180

Hong CI, Tritsch GL, Mittelman A, Hebborn P, Chheda GB (1975) Synthesis and antitumor activity of 5'-phosphates and cyclic 3'-5'-phosphates derived from biologically active nucleosides. J Med Chem 18:465–473
Hopping ME, Young H, Bukovac JM (1979) Endogenous plant growth substances in developing fruit of *Prunus cerasus* L. VI. Cytokinins in relation to initial fruit development. J Am Soc Hortic Sci 104:47–52
Horgan R (1978) Analytical procedures for cytokinins. In: Hillman JR (ed) Isolation of plant growth substances. Cambridge University Press, Cambridge, p 157
Horgan R (1981) Modern methods of plant hormone analysis. In. Reinhold L, Harborne J, Swain T (eds) Progress in phytochemistry, vol 7, pp 137–170
Horgan R (1984) Cytokinins. In: Wilkins MB (ed) Advanced plant physiology. Pitman, Bath, pp 53–75
Horgan R, Scott IM (1985) Cytokinins. In: Crozier A, Rivier L (eds) Principles and practice of plant hormone analysis. Academic Press (in press)
Horgan R, Hewett EW, Horgan JM, Purse J, Wareing PF (1975) A new cytokinin from *Populus robusta*. Phytochemistry 14:1005–1008
Howe I, Williams DH, Bowen RD (1981) Mass spectrometry: Principles and applications, 2nd edn, McGraw-Hill, New York
Jennings W (1980) Gas chromatography with capillary columns. Academic, New York
Johlman CL, White RL, Wilkins CL (1983) Applications of Fourier transform mass spectrometry. Mass Spectrom Rev 2:389–415
Kadir K, Gregson S, Shaw G, Mok MC, Mok DWS (1984) Purines, pyrimidines, and imidazoles. Part 59. A convenient synthesis of (E)-zeatin and E-[8-^{14}C]zeatin. J Chem Res (S)299
Kemp TR, Andersen RA (1981) Separation of modified bases and ribonucleosides with cytokinin activity using fused silica capillary gas chromatography. J Chromatogr 209:467–471
Kemp TR, Andersen RA, Oh J, Vaughn TH (1982) High-resolution gas chromatography of methylated ribonucleosides and hypermodified adenosines. Evaluation of trimethylsilyl derivatization and split and splitless operation modes. J Chromatogr 241:325–332
Kemp TR, Anderson RA, Oh J (1983) Cytokinin determination in tRNA by fused silica capillary gas chromatography and nitrogen-selective detection. J Chromatogr 259:347–349
Knapp DR (1979) Handbook of analytical derivatization reactions. Wiley, New York
Koenig WA, Smith LC, Crain PF, McCloskey JA (1971) Mass spectrometry of trifluoroacetyl derivatives of nucleosides and hydrolysates of deoxyribonucleic acid. Biochemistry 10:3968–3979
Koshimizu K, Matsubara S, Kusaki T, Mitsui T (1967) Isolation of a new cytokinin from immature yellow lupin seeds. Agric Biol Chem 31:795–801
Lee TS, Purse JG, Pryce RJ, Horgan R, Wareing PF (1981) Dihydroconiferyl alcohol: A cell division factor from *Acer* species. Planta 152:571–577
Lee Y-H, Mok MC, Mok DWS, Griffin DA, Shaw G (1985) Cytokinin metabolism in *Phaseolus* embryos. Plant Physiol 77:635–641
Leonard NJ, Carraway K, Helgeson JP (1965) Characterization of N_x, N_y-disubstituted adenines by ultraviolet absorption spectra. J Heterocycl Chem 2:291–297
Leonard NJ, Hecht SM, Skoog F, Schmitz RY (1969) Cytokinins: Synthesis, mass spectra, and biological activity of compounds related to zeatin. Biochemistry 63:175–182
Leonard NJ, Playtis AJ, Skoog F, Schmitz RY (1971) A stereoselective synthesis of *cis*-zeatin. J Am Chem Soc 93:3056–3058
Letham DS (1963) Zeatin, a factor inducing cell division isolated from *Zea mays*. Life Sci 8:569–573
Letham DS (1973) Regulators of cell division in plant tissues. XV. Cytokinins from *Zea mays*. Phytochemistry 12:2445–2455
Letham DS (1978) Cytokinins. In: Letham DS, Goodwin PB, Higgins TJV (eds) Phytohormones and related compounds: A comprehensive treatise, vol 1. Elsevier/North-Holland Biomedical, Amsterdam, pp 205–263

Letham DS, Palni LMS (1983) The biosynthesis and metabolism of cytokinins. Annu Rev Plant Physiol 34:163–197

Letham DS, Wettenhall REH (1977) Transfer RNA and cytokinins. In: Stewart PR, Letham DS (eds) The ribonucleic acids. Springer, Berlin Heidelberg New York, p 374

Letham DS, Shannon JS, McDonald IRC (1964) The structure of zeatin, a factor inducing cell division. Proc Chem Soc 230–231

Letham DS, Shannon JS, McDonald IRC (1967) Regulators of cell division in plant tissues. III. The identity of zeatin. Tetrahedron 23:479–486

Letham DS, Wilson MM, Parker CW, Jenkins ID, MacLeod JK, Summons RE (1975) Regulators of cell division in plant tissues. XXIII. The identity of an unusual metabolite of 6-benzylaminopurine. Biochim Biophys Acta 399:61–70

Letham DS, Summons RE, Entsch B, Gollnow BI, Parker CW, MacLeod JK (1978) Glucosylation of cytokinin analogues. Phytochemistry 17:2053–2057

Letham DS, Summons RE, Parker CW, MacLeod JK (1979) Identification of an amino-acid conjugate of 6-benzylaminopurine formed in *Phaseolus vulgaris* seedlings. Planta 146:71–74

Letham DS, Palni LMS, Tao G-Q, Gollnow BI, Bates CM (1983) Regulators of cell division in plant tissues. XXIX. The activities of cytokinin glucosides and alanine conjugates in cytokinin bioassays. J Plant Growth Regul 2:103–115

Ludewig M, Dorffling K, Konig WA (1982) Electron-capture capillary gas chromatography and mass spectrometry of trifluoroacetylated cytokinins. J Chromatogr 243:93–98

MacLeod JK, Summons RE, Letham DS (1976) Mass spectrometry of cytokinin metabolites. Per(trimethylsilyl) and permethyl derivatives of glucosides of zeatin and 6-benzylaminopurine. J Org Chem 41:3959–3967

MacLeod JK, Tay SAB, Letham DS, Palni LMS (1986) Mass spectrometric studies of cytokinin metabolism; Advances in Mass Spectrometry, vol 10 (in press)

MacMillan J (1977) Some aspects of gibberellin metabolism in higher plants. In: Pilet PE (ed) Plant growth regulation. Springer, Berlin Heidelberg New York, p 305

Martin GC, Horgan R, Scott IM (1981) High-performance liquid chromatographic analysis of permethylated cytokinins. J Chromatogr 219:167–170

Martin GC, Horgan R, Nishijima C (1982) Changes in hormone content of pear receptacles from anthesis to shortly after fertilization as affected by pollination or GA_3 treatment. J Am Soc Hortic Sci 107:479–482

McCloskey JA, Hashizume T, Basile B, Sugiyama T, Sekiguchi S (1979) Determination of cytokinins in bamboo shoots by mass spectrometry using selected ion monitoring. Proc Jpn Acad, Ser B 55:445–450

McDougall J, Hillman JR (1978) Analysis of indole-3-acetic acid using GC-MS techniques. In: Hillman JR (ed) Isolation of plant growth substances. Cambridge University Press, Cambridge, p 157

McFadden W (1973) Techniques of combined gas chromatography-mass spectrometry. Wiley-Interscience, New York

McGaw BA, Heald JK, Horgan R (1984a) Dihydrozeatin metabolism in radish seedlings. Phytochemistry 23:1373–1377

McGaw BA, Scott IM, Horgan R (1984b) Cytokinin biosynthesis and metabolism. In: Crozier A, Hillman JR (eds) The biosynthesis and metabolism of plant hormones. Cambridge University Press, Cambridge, UK, pp 105–133 (SEB Seminar Series no 23)

McGaw BA, Horgan R, Heald JK (1985) Cytokinin metabolism and the modulation of cytokinin activity in radish. Phytochemistry 24:9–13

McLafferty FW (1980) Interpretation of mass spectra. 3rd edn, Univ Sci Books, Mill Valley, California

McLafferty FW (1983) Tandem mass spectrometry. Wiley, New York

McMahon DH (1985) A collaborative study to evaluate quantitation utilizing different injection modes for capillary GC. J Chromatogr Sci 23:137–143

Millard BJ (1978) Quantitative mass spectrometry. Heyden and Son, London

Miller CO, Skoog F, Okumura FS, Von Saltza MH, Strong FM (1956) Isolation, structure and synthesis of kinetin, a substance promoting cell division. J Am Chem Soc 78:1375–1380

Morris RO (1977) Mass spectroscopic identification of cytokinins. Glucosyl zeatin and glucosyl ribosylzeatin from *Vinca rosea* crown gall. Plant Physiol 59:1029–1033

Morris RO, Regier DA, Olson RM Jr, Struxness LA, Armstrong DJ (1981) Distribution of cytokinin active nucleosides in isoaccepting transfer ribonucleic acids from *Agrobacterium tumefaciens*. Biochemistry 21:6012–6017

Most BH, Williams JC, Parker KJ (1968) Gas chromatography of cytokinins. J Chromatogr 38:136–138

Murai N, Skoog F, Doyle ME, Hanson RS (1980) Relationships between cytokinin production, presence of plasmids, and fasciation caused by strains of *Corynebacterium fascians*. Proc Natl Acad Sci USA 77:619–623

Murakoshi I, Ikegami F, Ookawa N, Haginiwa J, Letham DS (1977) Enzymatic synthesis of lupinic acid, a novel metabolite of zeatin in higher plants. Chem Pharm Bull 25:520–522

Murakoshi I, Koide C, Ikegami F, Nasu K (1983) Biosynthesis of β-(6-benzylaminopurin-9-yl)alanine, a metabolite of cytokinin 6-benzylaminopurine in higher plants. Chem Pharm Bull 31:1777–1779

Ogilvie KK (1973) The *tert*-butyldimethylsilyl group as a protecting group in deoxynucleosides. Can J Chem 51:3799–3807

Palmer MV, Wong OC (1985) Identification of cytokinins from xylem exudate of *Phaseolus vulgaris* L. Plant Physiol 79:296–298

Palmer MV, Horgan R, Wareing PF (1981) Cytokinin metabolism in *Phaseolus vulgaris* L. I. Variations in cytokinin levels in leaves of decapitated plants in relation to lateral bud outgrowth. J Exp Bot 32:1231–1241

Palni LMS, Horgan R (1982) Cytokinins from the culture medium of *Vinca rosea* crown gall tissue. Plant Sci Lett 24:327–334

Palni LMS, Horgan R (1983) Cytokinins in transfer RNA of normal and crown gall tissue of *Vinca rosea*. Planta 159:178–187

Palni LMS, Horgan R, Darrall NM, Stuchbury T, Wareing PF (1983a) Cytokinin biosynthesis in crown gall tissue of *Vinca rosea*: The significance of nucleotides. Planta 159:50–59

Palni LMS, Summons RE, Letham DS (1983b) Mass spectrometric analysis of cytokinins in plant tissues. V. Identification of the cytokinin complex of *Datura innoxia* crown gall tissue. Plant Physiol 72:858–863

Palni LMS, Palmer MV, Letham DS (1984) The stability and biological activity of cytokinin metabolites in soybean callus tissue. Planta 160:242–249

Palni LMS, Tay SAB, Nandi SK, Pianca DJ, deKlerk GJM, Wong C, Letham DS, MacLeod JK (1985) Cytokinin biosynthesis in plant tumour tissues. Biol Plant 27:195–203

Parker CW, Letham DS (1973) Regulators of cell division in plant tissues. XVI. Metabolism of zeatin by radish cotyledons and hypocotyls. Planta 114:199–218

Parker CW, Letham DS (1974) Regulators of cell division in plant tissues. XVIII. Metabolism of zeatin in *Zea mays* seedlings. Planta 115:337–344

Parker CW, Letham DS, Cowley DE, MacLeod JK (1972) Raphanatin, an unusual purine derivative and a metabolite of zeatin. Biochem Biophys Res Commun 49:460–466

Parker CW, Letham DS, Gollnow BI, Summons RE, Duke CC, MacLeod JK (1978) Metabolism of zeatin by lupin seedlings. Planta 142:239–251

Pickup JF, McPherson K (1976) Theoretical considerations in stable isotope dilution mass spectrometry for organic analysis. Anal Chem 48:1885–1890

Pierce AE (1984) Handbook of silylation. Pierce Chemical Company, Rockford, Illinois, USA

Playtis AJ, Leonard NJ (1971) The synthesis of ribosyl-*cis*-zeatin and thin layer chromatographic separation of the *cis*- and *trans*-isomers of ribosylzeatin. Biochem Biophys Res Commun 45:1–5

Purse JG, Horgan R, Horgan JM, Wareing PF (1976) Cytokinins of sycamore spring sap. Planta 132:1–8

Quilliam MA, Westmore JB (1978) Sterically crowded trialkylsilyl derivatives for chromatography and mass spectrometry of biologically important compounds. Anal Chem 50:59–68

Quilliam MA, Ogilvie KK, Sadana KL, Westmore JB (1980) Mass spectra of sterically crowded trialkylsilyl derivatives of nucleosides. Org Mass Spectrom 15:207–219

Reynolds WD (1979) Field desorption mass spectrometry. Anal Chem 51:283A–293A

Richter WJ, Schwarz T (1978) Chemical ionisation – a mass spectrometric analytical procedure of increasing importance. Angew Chem Int Ed 17:424–439

Robins MJ, Hall RH, Thedford R (1967) N^6-(Δ^2-Isopentenyl)adenosine. A component of the transfer ribonucleic acid of yeast and of mammalian tissue; methods of isolation and characterization. Biochemistry 6:1837–1848

Scott IM, Horgan R (1980) Quantification of cytokinins by selected ion monitoring using ^{15}N-labelled internal standards. Biomed Mass Spectrom 7:446–449

Scott IM, Horgan R (1984) Mass-spectrometric quantification of cytokinin nucleotides and glycosides in tobacco crown gall tissue. Planta 161:345–354

Scott IM, Horgan R, McGaw BA (1980) Zeatin-9-glucoside, a major endogenous cytokinin of *Vinca rosea* crown gall tissue. Planta 149:472–475

Scott IM, Martin GC, Horgan R, Heald JK (1982) Mass spectrometric measurement of zeatin glycoside levels in *Vinca rosea* L. crown gall tissue. Planta 154:273–276

Scriven EFV (1983) 4-Dialkylaminopyridines: Superacylation and alkylation catalysts. Chem Soc Rev 12:129–161

Shannon JS, Letham DS (1966) Regulators of cell division in plant tissues. IV. The mass spectra of cytokinins and other 6-amino purines. NZ J Sci 9:833–842

Shaw G, Smallwood BM, Wilson DV (1966) Purines, pyrimidines, and imidazoles. XXIV. Syntheses of zeatin, a naturally occurring adenine derivative with plant cell division promoting activity, and its 9-β-D-ribofuranoside. J Chem Soc (C):921–924

Shaw G, Smallwood BM, Wilson DV (1967) Phosphorylated derivatives of the cytokinins, zeatin and its 9-β-D-ribofuranoside; naturally occurring adenine and adenosine derivatives with plant cell division promoting activity. Experientia 23:515–518

Shaw G, Smallwood BM, Wilson DV (1968) Purines, pyrimidines, and imidazoles. XXVII. Synthesis of 9-β-D-ribofuranosylzeatin 5′-phosphate, a naturally occurring adenylic acid derivative with plant cell division promoting activity, and a new synthesis of 6-chloro-9-β-D-ribofuranosylpurine 5′-phosphate. J Chem Soc (C):1516–1519

Skinner CG, Shive W (1955) Synthesis of some 6-(substituted)-aminopurines. J Am Chem Soc 77:6692–6693

Sponsel UM, MacMillan J (1978) Metabolism of gibberellin A_{29} in seeds of *Pisum sativum* cv. progress no. 9; use of (^{2}H) and (^{3}H)-GAs, and the identity of a new GA catabolite. Planta 144:69–78

Stafford AE, Corse J (1982) Fused-silica capillary gas chromatography of permethylated cytokinins with flame-ionisation and nitrogen-phosphorus detection. J Chromatogr 247:176–179

Stuchbury T, Palni LMS, Horgan R, Wareing PF (1979) The biosynthesis of cytokinins in crown gall tissue of *Vinca rosea*. Planta 147:97–102

Sugiyama T, Hashizume T (1980) An alternative synthesis of deuterated cytokinins. Nucleic Acid Res 8:27–31

Sugiyama T, Iwasawa H, Hashizume T (1980) Synthesis of deuterated N^6-(o-hydroxybenzyl)adenosine-d_3. Agric Biol Chem 44:1057–1060

Sugiyama T, Suye S, Hashizume T (1983) Mass spectrometric determination of cytokinins in young sweet potato plants using deuterium labelled standards. Agric Biol Chem 47:315–318

Summons RE, MacLeod JK, Parker CW, Letham DS (1977) The occurrence of raphanatin as an endogenous cytokinin in radish seed. FEBS Lett 82:211–214

Summons RE, Duke CC, Eichholzer JV, Entsch B, Letham DS, MacLeod JK, Parker CW (1979 a) Mass spectrometric analysis of cytokinins in plant tissues. II. Quantitation of cytokinins in Zea mays kernels using deuterium labelled standards. Biomed Mass Spectrom 6:407–413

Summons RE, Entsch B, Parker CW, Letham DS (1979 b) Mass spectrometric analysis of cytokinins in plant tissues. III. Quantitation of the cytokinin glycoside complex of lupin pods by stable isotope dilution. FEBS LEtt 107:21–25

Summons RE, Entsch B, Letham DS, Gollnow BI, MacLeod JK (1980) Regulators of cell division in plant tissues. XXVIII. Metabolites of zeatin in sweet-corn kernels: purification and identification using high-performance liquid chromatography and chemical-ionisation mass spectrometry. Planta 147:422–434

Summons RE, Palni LMS, Letham DS (1983) Determination of intact zeatin nucleotide by direct chemical ionisation mass spectrometry. FEBS Lett 151:122–126

Sweeley CC, Elliott WH, Fries I, Ryhage R (1966) Mass spectrometric determination of unresolved components in gas chromatographic effluents. Anal Chem 38:1549–1553

Tao G-Q, Letham DS, Palni LMS, Summons RE (1983) Cytokinin biochemistry in relation to leaf senescence. I. The metabolism of 6-benzylaminopurine and zeatin in oat leaf segments. J Plant Growth Regul 2:89–102

Tay SAB, MacLeod JK, Palni LMS, Letham DS (1985) Detection of cytokinins in a seaweed extract. Phytochemistry 24:2611–2614

Tay SAB, MacLeod JK, Palni LMS (1986) On the reported occurrence of *cis*-zeatin riboside as a free cytokinin in tobacco shoots. Plant Sci 43:131–134

Taylor JS, Koshioka M, Pharis RP, Sweet GB (1984) Changes in cytokinins and gibberellin-like substances in *Pinus radiata* buds during lateral shoot initiation and the characterization of ribosyl zeatin and a novel ribosyl zeatin glycoside. Plant Physiol 74:626–631

Thompson AG, Horgan R, Heald JK (1975) Quantitative analysis of a cytokinin using single ion current monitoring. Planta 124:207–210

Tsui C, Shao LM, Wong CM, Tao G-Q, Letham DS, Parker CW, Summons RE, Hocart CH (1983) Identification of a cytokinin in water chestnuts (corms of *Eleocharis tuberosa*). Plant Sci Lett 32:225–231

Upper CD, Helgeson JP, Kemp JD, Schmidt CJ (1970) Gas-liquid chromatographic isolation of cytokinins from natural sources. 6-(3-Methyl-2-butenylamino)purine from *Agrobacterium tumefaciens*. Plant Physiol 45:543–547

von Minden DL, McCloskey JA (1973) Mass spectrometry of nucleic acid components. N,O-permethyl derivatives of nucleosides. J Am Chem Soc 95:7480–7490

Vreman HJ, Schmitz RY, Skoog F, Playtis AJ, Frihart CR, Leonard NJ (1974) Synthesis of 2-methylthio-*cis*- and *trans*-ribosylzeatin and their isolation from *Pisum* tRNA. Phytochemistry 13:31–37

Vreman HJ, Thomas R, Corse J, Swaminathan S, Murai N (1978) Cytokinins in tRNA obtained from *Spinacia oleracea* L. leaves and isolated chloroplasts. Plant Physiol 61:296–306

Waller GR, Dermer OC (1980) Biochemical applications of mass spectrometry. Wiley-Interscience, New York

Wang TL, Horgan R (1978) Dihydrozeatin riboside, a minor cytokinin from the leaves of *Phaseolus vulgaris* L. Planta 140:151–153

Wang TL, Thompson AG, Horgan R (1977) A cytokinin gluoside from the leaves of *Phaseolus vulgaris* L. Planta 135:285–288

Wang TL, Cove DJ, Beutelmann P, Hartmann E (1980) Isopentenyladenine from mutants of the moss *Physcomitrella patens*. Phytochemistry 19:1103–1105

Wang TL, Horgan R, Cove DJ (1981) Cytokinins from the moss *Physcomitrella patens*. Plant Physiol 68:735–738

Wang TL, Wood EA, Brewin NJ (1982) Growth regulators, *Rhizobium* and nodulation in peas. Planta 155:350–355

Watanabe N, Yokota T, Takahashi N (1978 a) Identification of zeatin and zeatin riboside in cones of the hop plant and their possible role in cone growth. Plant Cell Physiol 19:617–625

Watanabe N, Yokota T, Takahashi N (1978 b) *cis*-Zeatin riboside: Its occurrence as a free nucleoside in cones of the hop plant. Agric Biol Chem 42:2415–2416

Watson JT (1985) Introduction to mass spectrometry. 2nd edn. Raven, New York
Whenham RJ (1983) Evaluation of selective detectors for the rapid and sensitive gas chromatographic assay of cytokinins, and application to the analysis of cytokinins in plant extracts. Planta 157:554–560
Wiesner J (1892) Die Elementarstruktur und das Wachstum der lebenden Substanz. A. Holder, Vienna
Wilson MM, Gordon ME, Letham DS, Parker CW (1974) Regulators of cell division in plant tissues. XIX. The metabolism of 6-benzylaminopurine in radish cotyledons and seedlings. J Exp Bot 25:725–732
Wood GW (1982) Some recent applications of field ionisation/field desorption mass spectrometry to organic chemistry. Tetrahedron 38:1125–1158
Yokota T, Murofushi N, Takahashi N (1980) Extraction, purification, and identification. In: MacMillan J (ed) Molecular aspects of plant hormones. Springer, Berlin Heidelberg New York, pp 113–201 (Hormonal regulation of development, vol 1)
Yost RA, Fetterolf DD (1983) Tandem mass spectrometry (MS/MS) instrumentation. Mass Spectrom Rev 2:1–45
Young H (1977) Identification of cytokinins from natural sources by gas-liquid chromatography/mass spectrometry. Anal Biochem 79:226–233
Zelleke A, Martin GC, Labavitch JM (1980) Detection of cytokinins using a gas chromatograph equipped with a sensitive nitrogen-phosphorus detector. J Am Soc Hortic Sci 105:50–53

GC-MS Method for Volatile Flavor Components of Foods

H. Kameoka

1 Introduction

Involved in what we know as flavor are odor, taste, and "feel to the tongue". Odor is, of course, due to the volatile substances present. The substances giving taste are not always volatile components, and are thus called involatile flavor substances. The "feel to the tongue" of foods, e.g., liquidity, viscosity etc., is an important factor of flavor. The flavor of foods is a sensation that involves the three factors of volatile flavor substances, involatile flavor substances, and physical stimulus in a mouth.

The chemical study of the flavor of foods mainly involves volatile substances. The volatile substances of various kinds of food have been studied for details of the aromatic components, and the pathway of production of volatile components is an important subject of investigation. This chapter describes analysis of volatile flavor components of foods by GC-MS.

2 GC-MS Methods

GC-MS now represents one of the most advanced techniques in instrumental analysis. Complex mixtures are separated by GC, and the components are presented individually to MS for identification. The older routine of fractionation and isolation is still important, but sample mixtures which previously were not suitable for conventional separation and isolation techniques because of sensitivity to hydrolysis, oxidation etc., or sample size may now be analyzed by GC-MS.

High resolution GC-MS is one of the most powerful analytical methods used in the flavor chemistry, its advantages are numerous, one of the principal being its superior separating power. Complex mixtures, such as flavors containing as many as 100 components, can be resolved with little difficulty. GC uses both polar and nonpolar capillary columns, such as 0.28 mm i.d. × 50 m glass SCOT columns coated with Carbowax, which permit the use of higher temperature and reproducible retention patterns. A second advantage of the method is the analysis for volatile flavor components in the headspace from foods. A third advantage is the coupling to a modern computer data system which can process a huge amount of data generated in a single experiment. These data are analyzed by computer, and each component recorded and/or computer-processed, so that the search far a data base for identification has become almost routine practice. In such an analysis, a typical flavor is identified by searching a data base containing MS of authentic compounds. If a compound is unknown, a considerable amount of effort is needed to identify it by interpreting the fragmentation pattern in electron impact and/or molecular weight information.

2.1 Preparation Methods of Flavor Samples

Flavor samples obtained by, for examples, distillation or extraction from foods can yield flavor concentrates. Steam distillation is an effective way to obtain volatile components, and will accordingly be dealt with further here.

When preparing a distillate containing volatile components, particularly if it involves steam distillation yielding diluted aqueous solutions, concentration is necessary. This may involve extraction, freezing concentration, and adsorption with active carbon. A Likens and Nickerson-type apparatus (Schutz et al. 1977) uses a flavor concentration method of simultaneous distillation-extraction (Fig. 1).

A mixture of chopped fresh foods and water is placed in the large round-bottomed flask, with added solvent in the another small round-bottomed flask. The contents of both flasks are brought to the boil, and volatile flavor compounds are simultaneously steam-distilled and extracted with solvent.

For the headspace vapors containing, for example, volatile flavor compounds foodstuffs are chopped and added to a round-bottomed flask to which is attached a column packed with absorption agent such as Porapak Q. Nitrogen (or helium) streams over the foodstuffs, and the flask is warmed to 40 °–60 °C. The volatile flavor compounds in the headspace vapors are absorbed on the absorption agent in the column. After the column is taken away, the moisture on the absorption agent is removed by injecting N_2 gas (or He gas), and the column is heated and kept at 80 °–100 °C after being connected to a U-shaped glass capillary. Elution is performed under N_2 (or He) stream. The volatile flavor compounds which are

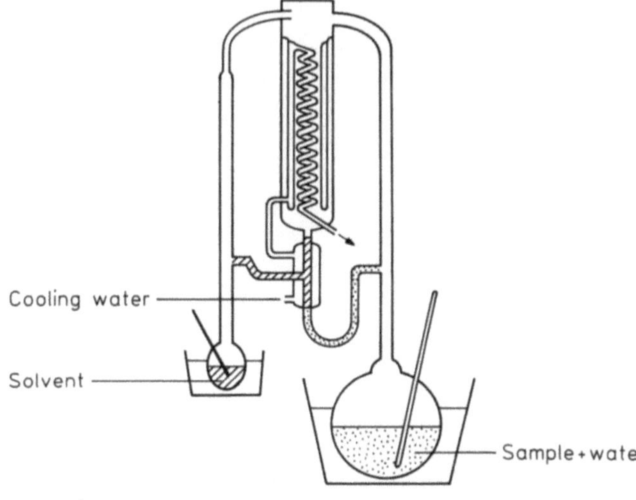

Fig. 1. Simultaneous distillation-extraction apparatus. (Yajima et al. 1981)

collected under these conditions are trapped in a U-shaped glass capillary chilled with liquid N_2. Yajima et al. (1981), using this method, have collected and analyzed headspace vapors containing volatile flavor compounds of *Tricholomamatsutake* (Ito et Imai) Sing.

In recent years, super critical gas extraction methods have been used for extraction. Using critical gas such as carbon dioxide for flavor extraction from foodstuffs, flavor extracts obtained by liquid carbon dioxide can be warmed to room temperature. Accordingly, this method is very useful, but the solubility of flavor compounds in gases at critical temperature is problematic. Today, carbon dioxide is utilized as a super critical gas. Furthermore, the resulting distillates or extracts are treated with basic and acidic solutions, separated into neutral, basic, acidic parts and/or placed on a column containing silicagel or alumina, and divided into fractions (e.g., hydrocarbon, oxygen-containing compounds etc.) by elution with solvents (nonpolar or polar). These separated flavor samples are then analyzed by GC-MS.

2.2 Operational Methods

Generally, the flavor compounds are analyzed by GC-MS (using both electron-impact and chemical ionization mass spectrometry). These GC-MS instruments link on-line to a data-processing system computer.

When using GC-MS, it is most important to select the right GC columns. A number of columns are employed, including packed columns and fused silica capillary columns (both with PEG 20M or OV 101 stationary phases), and bonded-phase fused silica capillary columns containing BPI (equivalent to OV 101), BP 20 (equivalent to PEG 20M) stationary phase, WCOT fused silica capillary columns, and SCOT column coated with PEG 20M, etc. The length and type of these columns differs according to the subject.

3 Volatile Flavor Components

3.1 Fruits

Citrus. Shinoda et al. (1970) reported constituents of Yuzu (*Citrus junnos*) oil. Yuzu oil was analyzed by fast-scan GC-MS using 0.5 mm × 45 m capillary column coated with PEG 2000. Like other citrus oils, limonene is the major component in Yuzu oil, with a content of nearly 80%. α-Pinene, β-pinene, myrcene, γ-terpinene, terpinolene, p-cymene, and p-isopropyl toluene were found in monoterpene hydrocarbon fractions. The most important fraction in the characteristic flavor of Yuzu oil is the oxygenated compound fraction, where the predominant compounds are linalool, α-terpineol, terpinen-4-ol, and thymol.

Yamanishi et al. (1980) studied the chemical composition of the essential oil and aroma characteristics of *Citrus iyo*. Since the oxygenated fraction had the typical aroma of *C. iyo*, analysis and characterization of the components were carried out on the carbonyl and noncarbonyl fractions. The quantities of main aroma components (i.e., 8 carbonyls and 14 noncarbonyl compounds) in the essential oil were determined. The compositional characteristics of the essential oil from *C. iyo* showed large amounts of linalool and smaller amounts of decanal, perillaldehyde, geranyl acetate, and neryl acetate. However, the key substance contributing to the aroma characteristics of *C. iyo* could not be identified, as it is apparently present in very small amounts. Uchida et al. (1983) also carried out qualitative and quantitative analyses of volatiles from peel of *C. iyo* by the use of a GC-MS and a GC-FID with a computing integrator. Among 53 oxygenated compounds, 2 aldehydes, 19 esters, and 9 alcohols were reported for the first time (1980) as components of *C. iyo*. By comparing the volatile compositions in peel among the fruits at four different stages, i.e., at harvest, pre-treatment, ripening, and overripening, it was found that the contributory aroma constituents such as C_{10}-, C_{11}-aldehydes, and C_6-, C_8-, neryl-, geranyl-, citronellyl-acetates increased during storage, up to ripening.

GC-MS has been used to study flavor characteristics of Satsuma mandarin orange juice extracted by a new type of screw press extraction system (Ohta et al. 1982). Sawamura et al. (1983) report the aroma components of the peel and juice of Satsuma mandarin orange grown in vinyl houses (VH) compared with those grown in open field (OF). The major common component of the peel oil from both sources was limonene, amounting to 86–90% of the total peak area. The oxygenated compound contributing to the characteristic aroma amounted to 1.84–3.33% in VH as against 1.28% in OF oranges. The characteristic components of these compounds in VH were geranyl acetate, terpinolene, terpinen-4-ol, nonanol, α-terpineol, decanal, perillaldehyde, and β-ionone.

Uchida et al. (1984) reported composition of oxygenated compounds in the peel oil of Fukuhara oranges. Forty-one oxygenated compounds were identified,

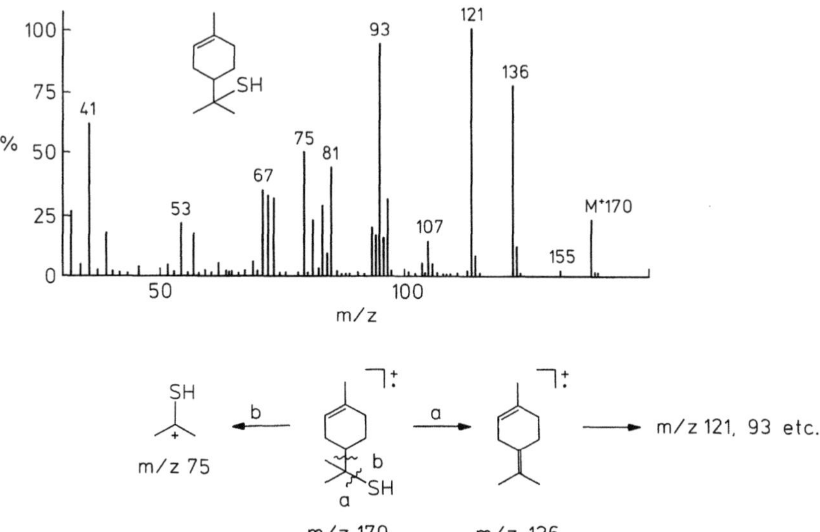

Fig. 2. MS of 1-p-menthene-8-thiol and its fragmentations. (Demole et al. 1982)

the main components being linalool (26%), octanol (14%), octanal (12%), and decanal (12%), the aliphatic aldehydes being important for the sweet aroma of the Fukuhara orange.

Koyasako and Bernhard (1983) investigated volatile constituents of the essential oil from Kumquat. In this study 120 compounds were found and 71 volatile compounds were identified in essential oil of Kumquat. A simultaneous distillation-extraction technique was used to obtain the essential oil from the fruit. Limonene was the most abundant compound, comprising 93% of the whole oil.

The powerful flavor constituent of grapefruit juice (*Citrus paradisi* Macfayden) was investigated by Demole et al. (1982) and Demole and Enggist (1983) using GC-MS etc. 1-p-Menthene-8-thiol is shown to be a potent character-donating constituent of grapefruit juice, in which it occurs at the ppb level or below, and apparently the most powerful flavor compound ever found in nature. Mass spectra and major fragmentation modes under EI-MS of 1-p-methene-8-thiol are shown in Fig. 2. Fourteen sesquiterpeneketones pertaining to the valencane (7 kinds) and eudesmane (7 kinds) groups have been identified for the first time as grapefruit juice flavor. In particular, (+)-8,9-didehydronootkatone has a powerful flavor with good grapefruit juice character.

Grapes. The volatile constituents of the muscadine grape (*Vitio rotundifolia*) (Welch et al. 1982) have been studied by GC-MS using a 2 mm (i.d.) by 3 m glass column packed with 5% Carbowax 20M on acid-washed DMCS chromosorb W. Forty-nine compounds were identified and confirmed, the main components being ethyl-, butyl acetate, butanol, isoamyl alcohol, limonene, hexanol, benzaldehyde, 2-phenyl ethanol, and 2-methoxy-1-phenyl-2-hydroxyethane. The aroma and flavor of the muscadine grape are attributable to the above compounds.

Shimizu (1982) used GC-MS to identify and characterize the volatiles in four varieties of grape musts (Koshu, Zenkoji, Muscat of Alexandria and Muscat of Humbourg grown in Japan). Forty-four volatiles identified in Koshu must included all the 39 volatile components found in Zenkoji must. Terpinen-4-ol and linalool (monoterpenoids) were detected in Koshu must, while only terpinen-4-ol was found in Zenkoji must. The content ratios of these monoterpenoids to the total volatiles were about 3% in Koshu and 0.8% in Zenkoji. On the other hand, 18 volatile monoterpenes were found in each of the muscat grapes, although terpinen-4-ol was not detected. The ratio of monoterpenes to the whole volatiles was approximately 33% in Muscat of Alexandria and 35% in Muscat of Humbourg grapes. Large differences of the monoterpenoid constituents were recognized in the flavor of grape musts and suggest the characteristic of the grape varieties.

In 1984, Horvat and Senter studied volatile constituents from sucppernong berries (*Vitis rotundifolia*). Scuppernong berries are grown commercially in the southern United States as a fresh fruit for home use and for wine production. Steam volatile oils were obtained from three cultivars of scuppernong berries at three levels of maturity and analyzed by GC-MS using both a 50 m × 0.5 mm glass open-tubular column coated with Superox 4 and a 50 m × 0.5 mm glass column coated with OV-1. Thirty-eight compounds consisting of aldehydes, substituted furans, ketones, alcohols, aromatic and aliphatic hydrocarbons, terpenes, and esters were identified. Major components from ripe berries were 2-pentanone, hexanal, toluene, furfural, 2-hexenal, xylene, benzaldehyde, geraniol, β-phenylethanol formate, λ-terpinene, benzylalcohol, phenylethanol, linalool, isomer of nonadienal, and β-ionone.

Prunus Mume. The constituents of the volatile oil obtained from the flesh of the fruits of *Prunus mume* Sieb. et Zucc. have been studied (Kameoka and Kitagawa 1976) using GC-MS etc. Amyl- and isoamyl alcohol, p-cymene, cis-3-hexen-1-ol, trans- and cis-linalool oxide, furfural, linalool, 5-methyl-2-furfural, α-terpineol, valeric- and isovaleric acid, 2,3-dimethyl-maleic anhydride, caproic acid, benzyl alcohol, guaiacol, o- and p-cresol, eugenol, and C_{14}-C_{24} aliphatic hydrocarbons were identified.

Apples. The volatile flavor components of Cashew (*Anacadium occidentale*) and Kogyoku apple have been the subject of investigation. An essence of fresh Cashew apple, obtained by well-established procedures and possessing the characteristic aroma of that fruit, was analyzed by GC and GS-MS, using both the EI and the CI techniques (MacLeod and Gonzales de Troconis 1982). Five aldehydes comprised ca. 26% of the sample, but terpene hydrocarbons (ca. 38%) provided the major group of compounds, consisting of four monoterpenes (ca. 34%) and three sesquiterpenes (ca. 4%). Important constituents of the essence included hexanal, car-3-ene, limonene, trans-hex-2-enal, and benzaldehyde. Car-3-ene (24.3%) was the major constituent. A steam distillate of the juice of Kogyoku apple (Yajima et al. 1984), was extracted with ethyl ether. The extract was separated into its acidic and neutral fraction and the neutral fraction was further separated into five fractions by column chromatography. All these fractions were analyzed by GC and GC-MS. Table 1 shows volatile flavor components from the juice of Kogyoku apple.

Starfruit. Starfruit (*Averrhoa carambola* L.) is popular in tropical and subtropical zones. Its aroma was analyzed with GC-MS; aroma constituents were mainly ethyl anthranilate (11.07%), ethyl benzoate contained diethyl succinate (7.59%), ethyl cinnamate (6.73%), ethyl N-methyl anthranilate (4.57%), tricosene (4.52%), tricosane (4.03%), ethyl caproate (3.67%), and nonanal (3.52%). Ethyl N-methylanthranilate was detected for the first time as a component in natural botanic products (Siota 1984).

Quince Fruit. Steam-distilled oil of quince fruit (*Cydonia oblonga* Mil.) was analyzed (Schreyen et al. 1979; Tsuneya et al. 1983). In the quince oil obtained by steam distillation and subsequent continuous ether extraction, 79 components were identified. The essential oil, obtained by headspace condensation, had a pleasant natural quince flavor and showed a much simpler composition. Aromagrams indicated many organoleptic important esters as the base of the fruity flavor, and ethyl 2-methyl-2-butenoate was recognized as an important contributor to the typical quince flavor. A total of 62 compounds, 2 hydrocarbons, 13 esters, 11 alcohols, 11 aldehydes, 11 ketones, 5 lactones, and 9 miscellaneous compounds was identified. Of these, the chemical structures of two new oxide com-

Table 1. Volatile flavor compounds identified from the juice of Kogyoku apples. (Yajima et al. 1984)

Peak no.	Compound	Peak no.	Compounds
		94[a]	(Z, Z)-3,5-Octadien-1-ol
12'[a]	Toluene	95'	1-Decanol
37[a]	Tridecane	112	2-Phenylethanol
95	(E, E)-α-Farnesene	133[a]	Octane-1,3-diol
45'[a]	Tetradecane	133[a]	(E)-5-Octene-1,3-diol
6	Ethanol	3	Acetaldehyde
9	1-Propanol	4	Butanal
17	2-Methylpropanol	18	Hexenal
20	1-Butanol	27	(Z)-3-Hexenal
28	2-Methylbutanol	30	(Z)-2-Hexanal
32	1-Pentanol	44'	Nonanal
44	1-Hexanol	64	Decanal
49	(Z)-3-Hexenol		
50	3-Octanol	44'[a]	6-Methyl-5-hepten-2-on
52	(Z)-2-Hexenol	66[a]	Camphor
57	1-Heptanol	82	Acetophenone
59[a]	6-Methyl-5-hepten-2-ol		
73	Linalool	9	n-Propyl acetate
74	1-Octanol	11	2-Methylpropyl acetate
79[a]	(Z)-5-Octenol	15	n-Propyl propionate
89	α-Terpineol	85'[a]	(Z, Z)-3,5-Octadien-1-yl-acetate
16	n-Butyl acetate	88[a]	n-Butyl 3-hydroxybutyrate
19	2-Methylbutyl acetate	89'	Benzyl acetate
21	n-Butyl propionate	103	2-Phenylethyl acetate
25	n-Pentyl acetate	123[a]	3-Acetoxyoctan-1-ol
29	n-Butyl n-butyrate	125[a]	3-Hydroxyoctan-1-yl acetate
31[a]	n-Butyl 2-methylbutyrate	126[a]	3-Hydroxy-(Z)-5-octen-1-yl acetate
35	n-Hexyl acetate		
39	(Z)-3-Hexenyl acetate	56	Linalool oxide
43	(Z)-2-Hexenyl acetate	85	2-Methylbutyric acid
44'	n-Hexyl propionate	85'[a]	Methylchavicol
54	n-Hexyl n-butyrate	86	γ-Hexalactone
55	n-Hexyl 2-methylbutyrate	117[a]	Benzothiazole
61	n-Octyl acetate	122[a]	Methyleugenol
80	n-Hexyl hexanoate		
84[a]	(E, Z)-3,5-Octadien-1-yl acetate		
92[a]	(E, Z)-3,5-Octadien-1-ol		

[a] Newly identified. Hitachi Model RMU-6MG mass spectrometer, glass SCOT column (0.28 mm i.d. × 50 m) coated with PEG 20M, column temp. 50–170 °C, 2 °C min^{-1}, ionizing volt 70 eV, ion source temp. 200 °C.

pounds, trans- and cis-3-methyl-5-[(*E*)-3'-methyl-1',3'-butadien-1'-yl]tetrahydrofuran were elucidated by instrumental analyses.

Apricot. The volatile components of apricot variety Rouge du Roussillon were isolated by vacuum distillation and fractionated on silica gel. Analysis by GC-MS led to identification of several compounds identified for the first time from this product (Chairote et al. 1981). These compounds included p-hydroxybenzalde-hyde, p-methoxybenzaldehyde, (3-phenyl propyl)-propyl or isopropyl ketone, 3-nonene-2-one, damascenone, β-ionone, dihydroactinidiolide, 2-phenylethanol, cis rose oxide and nerol oxide.

Blueberry. Three types of blueberry, i.e., low bush dwarf, high bush blueberry and rabbiteye blueberry, are grown commercially in the United States. Steam volatile oils were obtained from rabbiteye blueberries (*Vaccinium ashei* Reade, cv. Tift-blue) in amounts of ca. 25 ppm of the berry and analyzed by capillary GC-MS (Horvat et al. 1983). Major components identified were ethyl acetate, limonene, hexanol, cis-2-hexenol, heptanol, cinerolone, β-ionone, terpinene-4-ol, 2-undeca-none, α-terpineol, 1-carveol, nerol and eugenol. Of the 42 compounds identified, 29 have not been previously reported as constituents of blueberry volatiles.

Papaya. Volatile components of fresh papaya fruit were concentrated by several different methods. The concentrates were examined by GC-MS, and a total of 106 compounds was identified (Flath and Forrey 1977). Linalool (67.69%) is the ma-jor component of these concentrates, followed by benzyl isothiocyanate (13.11%), cis- and trans-2,6-dimethyl-3,6-epoxy-7-octen-2-ol (8.24%, 4.86%). The relative proportions of the major components are shown to be dependent upon the treatment received by the fruit tissue before and during volatile concen-tration.

In 1983, MacLeod and Pieris studied the aroma volatiles. Fresh papaya fruits from Sri Lanka were extracted and concentrated to a valid essence by well-estab-lished techniques. Analysis by GC and GC-MS showed at least 50 components (total about 97 μg kg^{-1} of fruit), of which 18 were identified for the first time as papaya volatiles.

Pineapple. Berger et al. (1983) isolated the volatile constituents of pineapple fruit under enzyme inhibition, enriched by liquid-liquid extraction, and fractionated on silica gel. Analysis of the nonpolar fraction by GC and GC-MS showed at least 20 sesquiterpene hydrocarbons to be present. α-Copaene, β-ylangene, α-patchou-lene, γ-gurjunene, germacrene D, α-muurolene, and δ-cadinene were identified. One of the minor compounds of the fraction seems to be responsible for the fra-grant odor (reminiscent of fresh-cut pineapple).

Guava. An essence of fresh guava fruit (*Psidium guajava* L.) was analyzed by GC-MS using both EI and CI techniques (MacLeod and Gonzalez de Troconis 1982 a). Esters comprised over 55% of the essence and an extensive series of the even carbon-numbered carboxylic acids was observed (over 52% of the sample). Two monoterpene hydrocarbons and five sesquiterpene hydrocarbons were also identified, myrcene being the major terpene.

Mango. Mango (*Mangifera indica* L.) is one of the most popular and best-known tropical fruits. MacLeod and Gonzalez de Troconis (1982 b) examined an essence of fresh Venezuelan mango fruit and characterized the aroma of the fruit. Impor-tant constituents are α-pinene, car-3-ene, limonene, γ-terpinene, α-humulene, β-

selinene, acetophenone, benzaldehyde, and dimethylstyrene. Car-3-ene (26%) was the major constituent. The composition of two mango varieties (Alphonso and Baladi) was investigated by means of standard-controlled distillation-extraction, GC and GC-MS (Engel and Tressl 1983). This combination led to the characterization of a spectrum of 114 components, of which 81 were identified for the first time as mango constituents.

Mangosteen. A representative sample of the aroma volatile of mangosteen, a tropical fruit, was obtained, using previously devised procedures. Components of the sample were identified as far as possible by GC-MS using both EIMS and CIMS (MacLeod and Pieris 1982). The fruit produced a relatively small quantity of aroma components (about 3 μg kg^{-1} fresh fruit), less than that obtained from similar fruit, and this partly explains its delicate flavor. The most important aroma components were hexyl acetate, cis-hex-3-enyl acetate and cis-hex-3-en-1-ol.

Jack Fruit. Jack fruit trees abound in tropical countries as a cultivated crop or as wild-growing trees, bearing very large fruits. Sword et al. (1978) studied volatile constituents of jack fruits (*Arthocarpus heterophyllus*). Among these constituents, 16 esters (methyl-, ethyl-, propyl-, isobutyl-, and butyl butyrate, and ethyl-, propyl-, isobutyl-, butyl-, and isoamyl isovalerate, butyl- and isoamyl hexanoate etc.) and four aliphatic alcohols (hexanol, octanol etc.) were identified. The total absence of terpenoids in jack fruit is unique in the plant kingdom.

Loquat Fruit. Shaw and Wilson (1982) investigated volatile constituents of loquat (*Eriobotrya japonica* L.) fruit. The loquat, also known as the Japanese plum (Biwa), is an important food crop in Japan and Israel. The fresh fruit has a mild and weak acid flavor. Major components phenyl ethyl alcohol, 3-hydroxy-2-butanone, phenylacetaldehyde and hexen-1-ols, and minor components, ethyl acetate, methyl cinnamate, and β-ionone probably contribute to the fruity-floral flavor of the fruit.

Kiwi Fruit. The aroma of the kiwi fruit was also analyzed with GC-MS for the first time by Shiota (1982). The major components were methyl butyrate, trans-2-hexenal, trans-2-hexenol, ethyl caproate, methyl benzoate, and ethyl benzoate. β-Damascenone, trans-hotrienol, methyl furoate, and ethyl furoate were regarded as interesting constituents.

Litchi Fruit. The fruit of the litchi (lychee) (*Litchi chinesis* Sonn.) has a white, juicy flesh which is surrounded by a reddish, prickly leather-like skin and contains a shiny brown, usually large seed. The volatile constituents of litchi have been investigated by the combined GC-MS technique with 42 components being identified and confirmed (Johnston et al. 1980). Of these, β-phenethyl alcohol, its derivatives, and terpenoids comprised the major portion of the volatiles.

3.2 Vegetables

Carrot Roots. The volatile flavor of fresh carrots root is known to be due to terpenoids and aliphatic alcohols, aldehydes etc. Nine additional components of the
steam volatile oil of fresh roots were identified using direct GC-MS and other instruments (Buttery et al. 1979). These components were geranyl 2-methyl butyrate, geranyl isobutyrate, β-ionone, geranyl acetone, p-cymene-8-ol, elimicin,
eugenol, p-vinylguaiacol, and 4-methyl isopropylbenzene. A compound with a
raw carrot-like aroma was isolated, but could not be positively identified.

Tomato. Chung et al. (1983) investigated volatile components in neutral, basic,
phenolic, and acidic fractions of the steam-distillate and in headspace gas trapped
from to varieties of fully ripe tomato fruits and their juice, purees, and pastes were
analyzed by GC and GC-MS using a glass capillary column (50 m × 0.28 mm i.d.)
coated with PEG 20M. One hundred and fifty-four compounds were identified.
In tomato purees and pastes, 14 components, which were detected in tomato
fruits and juices, disappeared, and 11 components, pyrrole, 2,5-dimethylpyrrole,
2-formylpyrrole, 2-methylpyrazine, 2,6-dimethylpyrazine, dibenzofuran, o-phe-
nylanisole, indole, and diphenylamine, were newly produced. 2-Acetylpyrrole, 2-
formylpyrrole and the pyrazines given above, which all have a burnt odor, are
considered to be important for the characteristic heated aroma of tomato puree
and paste.

Potato. In 1980, Coleman and Ho examined the volatile flavor that was isolated
from 540 lb of Idaho Russet Burbank baked potatoes. Extensive GC analysis
yield 420 fractions. The odor of each chromatographic fraction was evaluated,
and the fractions were analysed by IR and GC-MS. Thirty-one pyrazines and
three thiazoles were identified. The results of this work indicate that a natural
baked potato flavor is not due to a single compound, but is the result of the mixture of a number of components.

Allium Genus. Kameoka et al. (1981) studied volatile flavors of allium genus of
several kinds such as *Allium fistulosum* L. var. *caespitosum* Makino, *A. schoenoprasum* (Kameoka and Hashimoto 1983; Hashimoto et al. 1983), *A. tuberosum*
Rottl. (Iida et al. 1983), *A. grayi* Regal (Hashimoto et al. 1984) and *A. chinense*
(Kameoka et al. 1984).

   With the *A. fistulosum*, a neutral fraction of steam volatile oil was examined
by GC-MS etc., and 8 compounds identified, dipropyl disulfide, dipropyl trisulfide, trans- and cis-3,5-diethyl-1,2,4-trithiolane, 2-undecanone, 2-tridecanone, 2-
undecanol, and 2-tridecanol. Figure 3, and Table 2 show the GC of the total
flavor compounds from chive (*A. schoenoprasum* L.) oil and identification of volatile components found in the neutral oil in chive. Further, two disulfides were
found from volatile oil of *A. schoenoprasum*, and their structures established as
methyl pentyl disulfide and pentyl hydrodisulfide. These compounds, which smell
like sweet onions, are important aroma constituents.

   Next, the flavor components of Nira (*A. tuberosum* Rottl.) have been studied
by GC-MS. Twenty-nine compounds were identified in the oil obtained from ex-

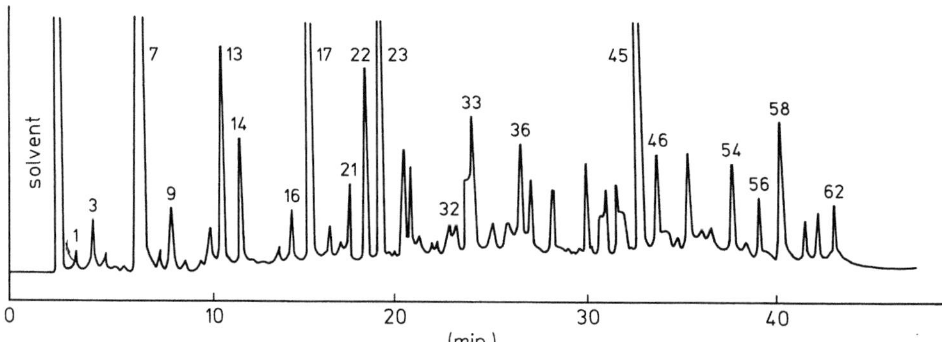

Fig. 3. Preliminary GC of the total flavor compounds from chive oil (*Allium schenoprasum* L.). GC; Shimazu GC-4C PTF type, 40 m × 0.3 mm i.d. SCOT glass capillary column coated with Thermon-600T, column temp. 80 °–210 °C at 4 °C min^{-1} injector and detector temp. 250 °C, carrier gas N_2 40 ml min^{-1}. (Hashimoto et al. 1983)

Table 2. Identification of volatile components found in the neutral oil in chive (*A. schenoprasum* L.). (Hashimoto et al. 1983)

Peak no.	Compound	Paek area %
2.	Pentanethiol	0.3
3.	Methyl 2-propenyl disulfide	0.8
7.	Dipropyl disulfide	7.8
9.	Di-2-propenyl disulfide	0.9
12.	Linalool	2.2
13.	Methyl pentyl disulfide	1.6
14.	Methyl 2-propenyl trisulfide	0.9
16.	2-Undecanone	0.6
17.	Pentyl hydrodisulfide	2.5
18.	2-Undecanol	0.8
22.	trans-3,5-Diethyl-1,2,4-trithiolane	1.9
23.	cis-3,5-Diethyl-1,2,4-trithiolane	3.0
25.	2-Tridecanone	1.5
27.	Guaiacol	0.8
32.	2-Phenylethyl alcohol	2.2
34.	2-Tridecanol	2.1
35.	2,4-Dihydroxy acetophenone	1.7
36.	Methyl myristate	3.0
37.	α-Methyl benzyl alcohol	0.1
41.	m-Ethyl phenylacetate	1.5
42.	Thymol	1.9
45.	Methyl palmitate	4.1
50.	Dihydro benzofuran	1.6
55.	Methyl stearate	1.1
56.	Methyl oleate	3.9
57.	Vanilin	0.2
58.	Methyl linoleate	3.8
60.	o-Nonyl phenol	1.5
61.	p-Nonyl phenol	1.6
62.	m-Nonyl phenol	1.8

traction of the steam distillate of Nira. The identified compounds include 7 sulfides, 2 ketones, 18 alcohols, and 2 esters. The main volatile components were dimethyl disulfide and dimethyl trisulfide. Of the volatile flavor components of *A. grayi* Regal, from the more than 90 component peaks observed by capillary GC of the oil obtained by ether extraction of the steam distillate, 47 compounds were positively identified. They included 11 sulfur compounds, 5 alcohols, 7 aldehydes, 3 ketones, 2 furanones and 19 miscellaneous compounds. One after another, the constituents of the steam volatile oils from two kinds of *A. fistulosum*, *A. fistulosum* var. *caespitosum*, and *A. chinense*, have been investigated by GC and spectral techniques. The compounds identified from the neutral fraction of each volatile oil included sulfides, thiolanes, alcohols, aldehydes, ketones, furanones, and others. Among the sulfur compounds, dipropyl disulfides make up ca. 28% of *A. fistulosum* oil, ca. 23% of *A. fistulosum* var. *caespitosum* oil, and ca. 30% of *A. chinense* oil. *A. fistulosum* oil was characterized by a large quantity of tridecan-2-one (ca. 52%) and 2,3-dihydro-2-octyl-5-methylfuran-3-one (ca. 16%). A large amount of 2,3-dihydro-2-hexyl-5-methylfuran-3-one (ca. 20%) was also isolated from *A. chinense* oil.

Cruciferae Family. Glycosides of mustard oil exist in *Brassica juncea* Czern. et Coss., *Raphanus sativus* L. var. *hortensis* Backer, *B. rapa* L. var. *laciniitolia* Kitam., *B. rapa* L. var. *nippoleifera* Kitam. and *Wasabia japonica* Matsum. etc., and

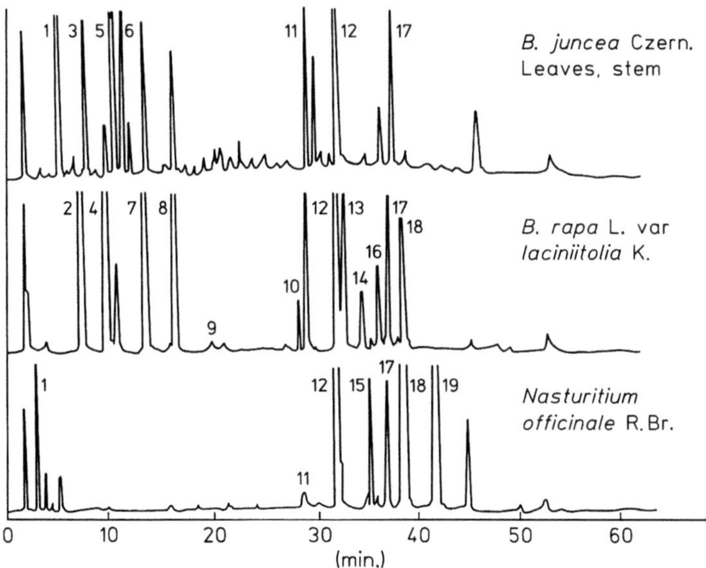

Fig. 4. GC of the steam volatile oil from Cruciferae Thermon-600T SCOT glass capillary column 40 m × 0.3 mm i.d., C.T. 80 °–200 °C at 4 °C min^{-1}. *1* 3-Butenonitrile; *2* 4-Pentenonitrile; *3* n-Butyl isothiocyanate; *4* 5-Hexenonitrile; *5* Allyl isothiocyanate; *6* Dimethyl trisulfide; *7* 3-Butenyl isothiocyanate; *8* 4-Pentenyl isothiocyanate; *9* n-Hexyl isothiocyanate; *10* 5-Methylthio pentanonitrile; *11* Phenylacetonitrile; *12* 3-Phenylpropiononitrile; *13* 6-Methylthio hexanonitrile; *14* 4-Methylthiobutyl isothiocyanate; *15* 7-Methylthio heptanonitrile; *16* 5-Methylthiopentyl isothiocyanate *17* 2-Phenylethyl isothiocyanate; *18* 8-Methylthio octanonitrile; *19* 9-Methylthio nonanonitrile. (Kameoka 1984)

glycosides of these plants, after hydrolysis with myrosinase (thioglucosidase), produce various volatile isothiocyanates, so-called mustard oil, and gave rise to the characteristic pungent odour. These volatile components are well known as follows: allyl-, β-phenylethyl-, p-hydroxy benzyl-, n-butyl-, 3-butenyl-, and sec-butyl isothiocyanate etc.

Kameoka et al. reported on the volatile flavor constituents identified from fresh *B. juncea* Czern. et Coss. (Kameoka and Hashimoto 1980a), fresh *B. rapa* L. var. Kitam. (Kameoka and Hashimoto 1980b), seed of *Brassica* genus (Ka-

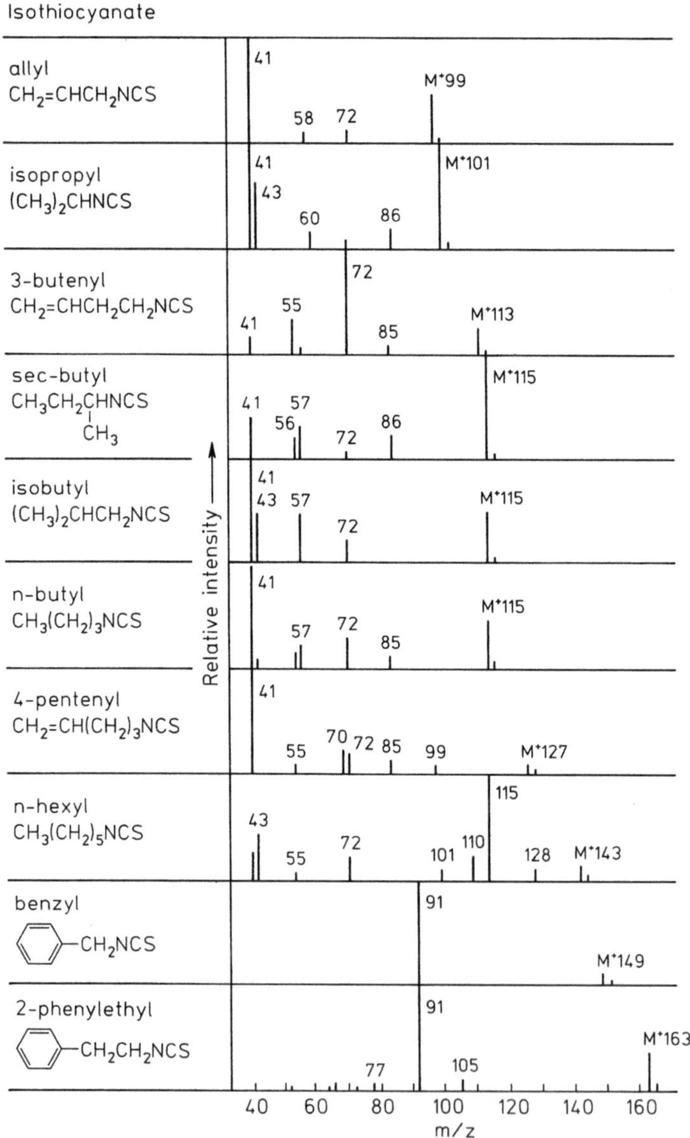

Fig. 5. MS of isothiocyanates from Cruciferae . (Kameoka 1984)

meoka and Hashimoto 1980c), fresh *B. rapa* L. sub var. *Hiroshimana* Kitam. (Hashimoto et al. 1982), *W. japonica* Matsum. and *Nasturitium officinale* R. Br. (Kameoka and Hashimoto 1982). They studied especially the MS of isothiocyanates, methylthioalkyl isothiocyanates, and methylthioalkyl nitriles (Kameoka 1984). Twenty-four sulfur and nitrogen compounds were found in these volatile oils; 10 isothiocyanates, 5 nitriles, 3 methylthioalkyl isothiocyanates, and 6 meth-

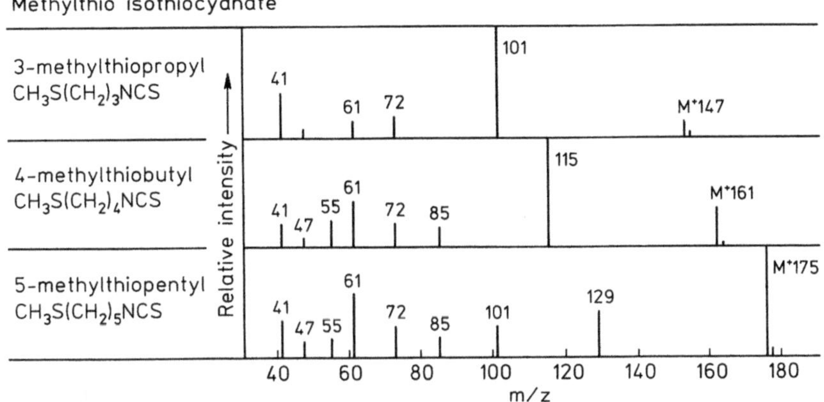

Fig. 6. MS of methylthioalkyl isothiocyanate from Cruciferae GC-MS was carried out on a Hitachi RUM-6 mass analyser connected with a Hitachi M-5201 GC; the operation parameters were as follows: column Thermon-600TS WCOT glass capillary 30 m × 0.25 mm i.d., column temperature 60 °–180 °C, ionization voltage 70 eV, accelerative voltage 3500 V. (Kameoka 1984)

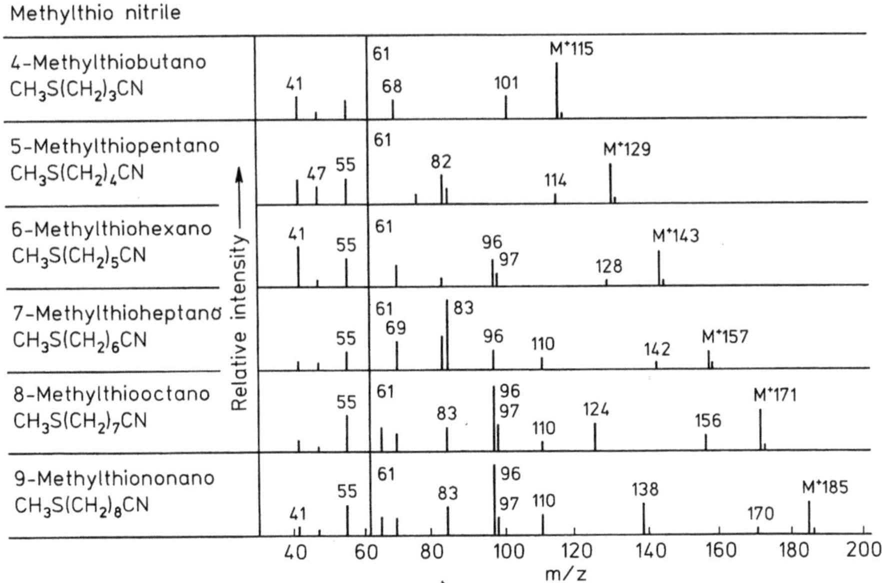

Fig. 7. MS of methylthioalkyl nitrile from Cruciferae. (Kameoka 1984)

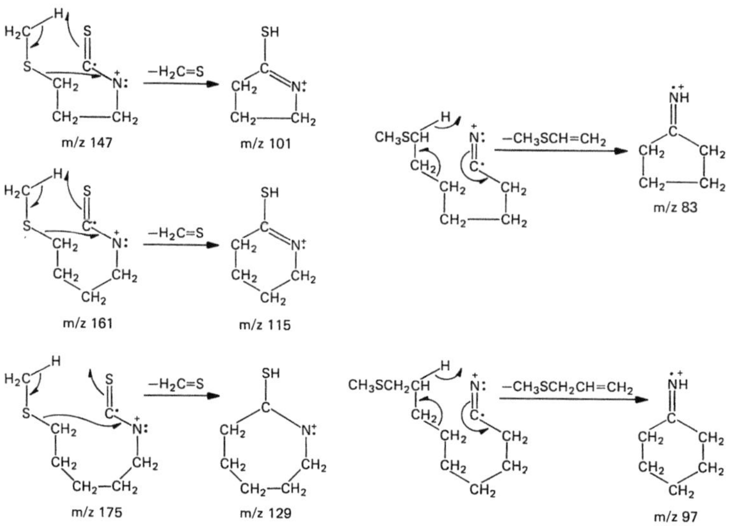

Scheme 1. Formation mechanism of m/z 101, 115, 129 and m/z 83, 97. (Kameoka 1984)

ylthioalkyl nitriles, Figure 4 shows the GC of the steam volatile oils from fresh Cruciferae of several kinds, Figs. 5–7 show the MS of isothiocyanates, methylthioalkyl isothiocyanates and methylthioalkyl nitriles, respectively. Some characteristic fragmentations occur. Especially, m/z M^+-46 (m/z 101, 115, 129) from methylthioalkyl isothiocyanates, and m/z 61 (100%), m/z 83, 97 from methylthioalkyl nitriles are useful for analyses. Scheme 1 shows the formation mechanism of these fragmentations.

Ginger. The essential oil of ginger (*Zingiber officinale*) from Fiji was analyzed by GC-MS (Smith and Robinson 1981), and a number of sesquiterpenes not previously reported in ginger oil were identified, including α-copaene, β-bourbonene, α-bergamotene, α-selinene, calamenene, and cuparene. The composition of the oil was unusual in having a much higher neral and geranial content than oils reported from India, Australia, Japan, and Africa.

Two types of Sri Lankan ginger were examined (Sidda and Chinese varieties). The essential oils from both fresh and dried samples were prepared by standard procedures (MacLeod and Pieris 1984). Both varieties yielded relatively high percentages of oil (between 1.8 and 4.3%) and total aroma volatiles (ca. 5 mg g^{-1} for dried samples). Analysis by GC and GC-MS showed terpenes to be the main aroma components (ca. 99% for all samples).

3.3 Mushrooms

The volatile flavor constituents of various mushroom species has been extensively studied. For example, Pyysalo (1976) reported an investigation of the volatile constituents of the cultivated mushroom *Agaricus bisporus,* and summarized past research on mushroom flavor, and its analysis by GC and GC-MS of aroma concentrates characteristic of both raw and cooked flavors. The major volatile constituents of raw mushroom were 3-octanone, 3-octanol, 1-octen-3-ol, benzaldehyde, octanol, and 2-octen-1-ol. The major volatile constituents of cooked mushrooms were the above six compounds plus 1-octen-3-one. Furthermore, about 50 volatile compounds were identified in each of the seven edible fresh mushroom *Cantharellus cibarius*, *Gyromitra esculenta*, *Boletus edulis*, *Lactarius trivialis*, *Lactarius torminosus*, *Lactarius rufus*, and *Agaricus bisporus* (Pyysalo 1976).

Of the flavor components of dried *Lentinus edodes* Sing., lentionine and its related compounds are well known. Kameoka and Higuchi (1976) have studied further the constituents of the steam volatile components of fresh *Lentinus* substances and the following compounds were identified: 1-octen-3-ol, octanol, 3-octanol, cis-2-octen-1-ol, furfuryl alcohol, linoleic acid, palmitic acid, myristic acid, and alkanes C_{18}-C_{28}, dimethyltrisulfide, 1-methylthio dimethyldisulfide, and 1,2,4-trithiolane. The sulfides and C_8-alcohols seem to influence the odor of fresh substances.

Tricholoma matsutake (Ito et Imai) Sing. is considered as having a most pleasant and appetizing aroma/flavor, and in Japan has become very popular in a variety of dishes. Volatile flavor compounds in the headspace vapors of fresh *T. matsutake* were absorbed on Poropak Q column, then flushed out by heating the column and collected in a cold trap, then analyzed directly by GC and GC-MS. The aroma concentrate obtained by simultaneous distillation-extraction with a Linkens and Nickerson-type apparatus was also analyzed (Yajima et al. 1981). Many C_8-aliphatic oxygenated compounds such as 3-octanone, 1-octen-3-one, 3-octanol, 2-octenol, 1-octen-2-ol, octanol, and cis-2-octenol were found unexpectedly in the headspace vapors of *T. matsutake*. Figure 8 shows a GC of the aroma concentrate. Major components were 3-octanone, 1-octen-3-one, 3-octanol, octanol, cis-2-octenol, and methyl trans-cinnamate, all, with the exception of methyl trans-cinnamate, being identical to the C_8-aliphatic oxygenated compounds found in the headspace vapors of *T. matsutake*. Tressl et al. (1982) investigated the formation of volatile C_8 components and less volatile C_{10} (C_{11}) oxo

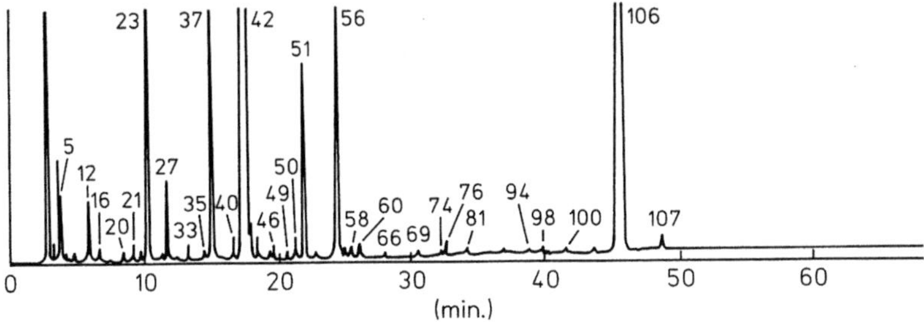

Fig. 8. GC of aroma concentrate of Matsutake-*Tricholomamatsutake* (Ito et Imai) Sing. GC-MS: Hitachi Model RMO-6MG mass spectrometer, 0.28 mm i.d., × 50 m glass SCOT column coated with PEG 20M, column temp. 200 °C at 2 °C min^{-1}, He, 70 eV, ion source temp. 200 °C. Peak No. *12* hexanal; *23* 3-octanone; *27* 1-octen-3-one; *33* hexanol; *37* 3-octanol; *40* 2-octanol; *42* 1-octen-3-ol; *50* linalool; *51* octanol; *56* cis-2-octenol; *106* methyl trans-cinnamate; *107* ethyl cinnamate. (Yajima et al. 1981)

and hydroxy acids in fresh mushroom (*Agaricus campestris*). Subsequently, Mac-Leod and Panchasara (1983) analyzed the aroma volatile of cooked edible mushroom (*Agaricus bisporus*), showing at least 40 components, including trace amounts of benzyl isothiocyanate, which indicated the presence of benzylglucosinolate in *A. bisporus*. Previously identified C_8 components, which are important to the characteristic flavor of mushroom, were also obtained, with 1-octen-3-ol contributing ca. 59% to the sample. A new mushroom volatile was detected, cyclo-octanol (ca. 24%), presumably formed by cyclization of 1-octen-3-ol during cooking. The C_8 compounds comprised ca. 98% of the sample.

3.4 Tea

Investigation of the aroma of green-, black-, and Chinese tea etc. has received much attention. Up to now more than 100 kinds of aroma compounds have been found. However, the aroma characteristics of green tea are still not well understood.

Kawakami and Yamanishi (1981, 1983) have studied aroma characteristics of Kabuse-cha (shaded green tea) and flavor constituents of Longjing tea. The essential oils from Kabuse-cha and Sen-cha (Typical Japanese green tea made from tea leaves grown in an open field) and their fresh leaves were analyzed by GC-MS. Remarkable differences were found between the essential oils from Kabuse-cha and Sen-cha, while no significant differences were observed in their fresh leaves. Kabuse-cha contained large quantities of the ionone series compounds, such as α-ionone, β-ionone, 2,6,6-trimethyl-2-hydroxy cyclohexan-1-one and 4-(2,6,6-trimethyl-1,2-epoxycyclohexyl)-3-buten-2-one, the latter compounds being newly identified from green tea. On the other hand, the flavor constituents of the highest quality Longjing tea, parched on a pan while the leaves were turned over by hand, were identified by GC-MS, and compared with those of Japanese Kamairi-cha. The amount of pyrazine, linalool oxides, carboxylic acids, lactones, geraniol, 2-phenylethanol, and ionone compounds was greater, while the amount of cis-3-hexenol, cis-jasmone, nerolidol, indole, and benzyl cyanide was much smaller in Longjing tea than in Japanese Kamairi-cha.

Takeo (1983) reported variations in the aroma compound content of semi-fermented and black tea. The volatile contents of semi-fermented and black tea were compared from the same tea cultivar. Black tea had a higher level of trans-2-hexenal, cis-3-hexenol, trans-2-hexenyl formate, monoterpene alcohols, and methyl salicylate. On the contrary, semi-fermented tea had a high content of cis-jasmone, β-ionone, nerolidol, jasmine lactone, methyl jasmonate, and indole. These variations in volatile compounds may explain the characteristic differences in flavor between teas.

Hara and Kubota (1982) used GC and GC-MS to investigate the changes in aroma components of green tea after storage for 4 months at 5 °C and 25 °C. 1-Penten-3-ol, trans-2, cis-4-heptadienal, cis-2-penten-1-ol and trans-2, trans-4-heptadienal increased remarkably during storage at 25 °C. Larger quantities of these compounds were formed during storage of low-grade tea and Ban-cha than high and medium-grade tea.

3.5 Beans and Nuts

Coffee. The aroma components in roasted coffee have been studied in detail by many investigators, and more than 580 components identified.

In 1981, Tressl and Silwar, by means of distillation-extraction, adsorption chromatography, and capillary GC-MS, identified 23 sulfur components (mercaptans, sulfides, and di- and trisulfides) in roasted coffee. Fifteen components were identified for the first time and confirmed by synthesis (among them are sulfur-containing furans and two dithiolanes).

Cocoa Beans. The aroma volatiles of raw, fermented and roasted cocoa beans were analyzed by GC and GC-MS, and showed at least 84 components, 13 of which were identified for the first time as cocoa volatiles (Gill et al. 1984). The most abundant groups of volatile from fermented beans were alcohols (ca. 40%) and esters (ca. 32%), while those from roasted beans were pyrazines (ca. 40%) and aldehydes (ca. 23%). Trimethyl- and tetramethyl pyrazine were also detected in fermented beans, and it is suggested that they contribute to the noticeable cocoa/chocolate aroma of fermented unroasted beans. Phenyl acetonitrile, benzyl isothiocyanate, and benzyl thiocyanate were all identified among cocoa volatiles, together showing the presence of precursor benzyl glucosinolate in cocoa.

Coconut. The volatile components of roasted and unroasted dried coconut shreds, isolated by steam distillation, were analyzed by GC and GC-MS (Saittagaroon et al. 1984). A JEOL Model JMS-D100 mass spectrometer connected with a JEOL Model JGC-20K gas chromatograph equipped with a 2.0 mm i.d. $\times$ 1 m glass column of Carbowax 20M on chromosorb W, 100–120 mesh, was used. Seventeen compounds were identified from the unroasted coconut, and nine of them were newly identified in coconut meat aroma. Saturated δ-C_8-C_{10}-, C_{12}-lactones were determined as the main components giving the characteristic mild, sweet, and pleasant coconut flavor. The roasted coconut gave the strong characteristic sweet and nutty aroma, and the GC-MS indicated the saturated δ-lactones as main components, and the 6 pyrazines, 2 furans, and 2 pyrroles also found seemed to contribute greatly to the nut-like aroma of roasted coconut.

3.6 Grains

Rice. In 1980 (Tsugita et al.) four kinds of rice milled to different degrees (92, 85, 75, and 50% milled rice) were subjected after cooking to odor evaluation tests and to GC and GC-MS analyses by use of the Tenax GC trap and injection techniques. Forty volatile components were identified in cooked rice, and there were remarkable differences in both the odor of cooked rice and in the quantity of volatile components between 92% milled rice and the other three degrees of milling. The main volatile components of cooked rice were shown as follows: acetone, butanal, pentanal, hexanal, 2-pentyl furan, nonanal, and 2,6-dibutyl-4-methyl phenol etc. Furthermore, according to Tsugita et al. (1983), rice stored at 40 °C and 80% relative humidity for 60 days, and which was sensorially evaluated as old rice, showed a significant difference in cooking quality and texture when cooked, from rice stored at 4 °C. Free phenolic acids were detected in a larger amounts in the rice stored at 40 °C, and it has been suggested that the increase of phenolic acids during storage may partly contribute to the cooking properties of old rice.

Buttery and Ling (1982) identified 2-acetyl-1-pyrroline as an important aroma component of cooked rice.

Rice Bran. Neutral and basic fractions obtained from the volatile concentrate of rice bran were analyzed by glass capillary GC and GC-MS (Tsugita et al. 1978). In addition to the 7 phenols and 17 acids reported previously, a total of 146 compounds (alkanes, aromatic hydrocarbons, alcohols, aldehydes, ketones, esters, lactones, furans, pyridines, pyrazines, quinolines, thiazoles, thiophenols, etc.) were newly identified in the neutral and basic fractions from the volatile concentrate of rice bran. 2-Acetyl thiazole and benzothiazole in the concentrate were considered to be the key compounds of the rice bran odor, and other compounds, for example lactones, were also estimated to be necessary for reproduction of the odor.

Soybeans. Soybeans were roasted at 200 °C for 10, 20, and 30 min, and their headspace volatiles trapped by Tenax GC were analyzed by GC and GC-MS, and compared with those of raw soybean (Doi et al. 1980). Fifty compounds were identified, and the quantity of important compounds for beany flavor, such as hexanal and 1-hexanol, decreased with longer roasting. However, the decrease was not remarkable, especially between 10 and 20 min roasting (in other words between 110 ° and 150 °C), while pyrazines, furans, and pyrrole were newly formed in this period and increased with roasting.

3.7 Jams

The characteristic sweet aroma of apple jam was collected during jam processing, and analyzed by GC-MS (Abe et al. 1980). This sweet aroma was considered to be produced while apple and sugar were cooked together. Nine components already known as apple volatile, nerolidol, and 4 furan compounds were identified, and 7 components were tentatively identified. The sweet aroma of apple jam consisted of alcohols and esters in apple volatiles, and volatile components formed during the cooking of jam. Especially the latter components containing 5-methyl-2-furfural were considered to contribute to the sweet aroma of jam.

Sugawara (Abe) et al. (1982) examined sweet aroma components in three kinds of jam. Aroma concentrates were obtained from distillates during the processing of grape, strawberry, and blueberry jams. The slurry aroma concentrates were obtained from the distillates of fruit slurries without the addition of sucrose. Each aroma concentrate was analyzed by GC-MS. The aroma concentrates of the jams had a sweeter aroma than those of the slurries in organoleptic teste. The sweet aroma of the jams was considered to be produced from the interaction of sucrose and organic acids in the fruits. 2-Furyl hydroxymethyl ketone and 5-hydroxymethyl-2-furfural were found in the jam aroma concentrates exclusively, and were considered to contribute to the characteristic sweet aroma of the jams. The presence of 2,5-diemthyl-4-hydroxy-3(2 H)-furanone in the aroma concentrate of strawberry jam explains why it has the sweetest odor of the three jams.

3.8 Fermentation Products

The many flavor compounds of beers, wines, sakes, and brandies has been thoroughly investigated. Accordingly, only recent investigations will be described.

Grape Wines. Volatile components of the neutral and phenolic fractions of wines from Koshu and Zenkoji grapes (*Vitis vinifera orientalis*) have been investigated by GC-MS (Shimizu and Watanabe 1981, 1982). The odor of dichloromethane extracts was very similar to that of original wines. The neutral fraction of the extracts was especially close to the original wine flavor. The fraction from Koshu wine was composed of 13 volatile phenols, 11 fatty acids, 8 lactones, 4 ketones, and many alcohols. The volatile compounds identified in the fraction from Zenkoji wine showed only slight differences from Koshu wine. The phenolic volatile compound 2-methoxy-5-vinylphenol was newly found as a component in both wines. A large amount of terpinen-4-ol and a trace of linalool were found in Koshu wine, while a small amount of terpinen-4-ol was detected in Zenkoji wine. Only slight differences in constituents of esters, alcohols, and hydrocarbons were recognized in the flavor of both wines. The phenolic fraction had no wine-like flavor, but possesed a phenolic and disagreeable odor. Further, volatiles in grape musts and characteristic terpenoid constituents were identified.

Brandies. Schreier et al. (1978, 1979) have studied the neutral volatile components of apple and grape brandies by GC and GC-MS. Quantitative determinations of 97 aroma components have shown that the aroma of apple brandy is characterized by components produced as a result of the yeast fermentation and technological steps such as mashing and heating rather than by the genuine apple-aroma components. Furthermore, 119 neutral volatile constituents in different grape brandies (French and German grape brandies, French cognacs) were determined.

Whisky. Volatile sulfur compounds of malt whiskies were analyzed by Masuda and Nishimura (1981). Sulfur compounds identified by GC-MS and by matching GC retention times were dimethyl sulfide, dimethyl disulfide, dimethyl trisulfide, 3-(methylthio)propanal, 3-(methylthio)propanol, 3-(methylthio)propyl acetate, ethyl 3-(methylthio)-propanoate, dihydro-2-methyl-3(2 H)-thiophenone, 2-thiophene-carboxyaldehyde, 5-methyl-2-thiophene-carboxyaldehyde, benzothiophene, and benzothiazole.

Soy Sauce. Nunomura et al. (1980) reported that 93 volatile components, such as n-butanol, isoamyl alcohol, ethyl lactate, furfuryl alcohol, etc., were identified in soy sauce flavor and 4-hydroxy-2-(or 5)-ethyl-5-(or 2)-methyl-3(2 H)-furanone (tautomers ratio; ca. 3 : 2) which exhibited a weak acidic property was a main and important constituent of the characteristic soy sauce flavor.

References

Abe E, Ito T, Odagiri S (1980) Flavor of apple jam. Nippon Nôgeikagaku Kaishi 54:761–764
Berger RG, Drawert F, Nitz S (1983) Sesquiterpene hydrocarbons in pineapple fruit. J Agric Food Chem 31:1237–1239
Buttery RG, Black DR, Haddon WF, Ling LC, Teranishi R (1979) Identification of additional volatile constituents of carrot roots. J Agric Food Chem 27:1–3
Buttery RG, Ling LC (1982) 2-Acetyl-1-pyrroline: An important aroma component of cooked rice. Chem Ind 4 Dec:958–959
Chairote G, Rodriquez F, Crouzet J (1981) Characterization of additional volatile flavor components of apricot. J Food Sci 46:1898–1902
Chung T-Y, Hayase F, Kato H (1983) Volatile components of ripe tomatoes and their juices, purees, and pastes. Agric Biol Chem 47:343–351
Coleman EC, Ho C-T (1980) Chemistry of baked potato flavor I. Pyrazines and thiazoles identified in the volatile flavor of baked potato. J Agric Food Chem 28:66–68
Demole E, Enggist P (1983) Further investigation of grapefruit juice flavor components (*Citrus paradisi* Macfayden) valencane- and eudesmane-type sesquiterpene. Helv Chim Acta 66:1381–1391
Demole E, Enggist P, Ohloff G (1982) 1-p-Menthene-8-thiol: A powerful flavor impact constituent of grapefruit juice (*Citrus paradisi* Macfayden). Helv Chim Acta 65:1785–1794
Doi Y, Tsugita T, Kurata T, Kato H (1980) Changes of headspace volatile components of soybeans during roasting. Agric Biol Chem 44:1043–1047
Engel K-H, Tressl R (1983) Studies on the volatile components of two mango varieties. J Agric Food Chem 31:796–801
Flath RA, Forrey RR (1977) Volatile components of papaya (*Carica papaya* L., Solo variety). J Agric Food Chem 25:103–109
Gill MS, MacLeod AJ, Moreau M (1984) Volatile component of cocoa with particular reference to glucosinolate products. Phytochemistry 23:1937–1942
Hara T, Kubota E (1982) Changes in aroma compounds of green tea after storage. Nippon Nôgeikagaku Kaishi 56:625–630
Hashimoto S, Miyazawa M, Kameoka H (1982) Volatile flavor sulfur and nitrogen constituents of *Brassica rapa* L. J Food Sci 47:2084–2085
Hashimoto S, Miyazawa M, Kameoka H (1983) Volatile flavor components of chive (*Allium schoenoprasum* L.). J Food Sci 48:1858–1859
Hashimoto S, Miyazawa M, Kameoka H (1984) Volatile flavor components of *Allium grayi* Regal. J Sci Food Agric 35:353–356
Horvat RJ, Senter SD (1984) Identification of the volatile constituents from scuppernong berries (*Vitis rotundifolia*). J Food Sci 49:64–66
Horvat RJ, Senter SD, Dekazos ED (1983) GC-MS analysis of volatile constituents in rabbiteye blueberries. J Food Sci 48:278–279
Iida H, Hashimoto S, Miyazawa M, Kameoka H (1983) Volatile flavor components of Nira (*Allim tuberosum* Rottl.). J Food Sci 48:660–661
Johnston JC, Welch RC, Hunter GLK (1980) Volatile constituents of litchi (*Litchi chinesis* Sonn.). J Agric Food Chem 28:859–861
Kameoka H (1984) Analysis of sulfur and nitrogen compounds from *Cruciferae* by gas chromatography-mass spectrometry. Koryo 142:19–29
Kameoka H, Hashimoto S (1980a) The constituents of the steam volatile oil from *Brassica juncea* Czern. et Coss. Nippon Nôgeikagaku Kaishi 54:99–103
Kameoka H, Hashimoto S (1980b) The constituents of steam volatile oil from *Brassica rapa* L. var. *laciniitolia* Kitamura. Nippon Nôgeikagaku Kaishi 54:865–869

Kameoka H, Hashimoto S (1980 c) Constituents of steam volatile oils from seeds of various varieties of *Brassica* of various districts. Nippon Nôgeikagaku Kaishi 54:535–539

Kameoka H, Hashimoto S (1982) Volatile flavor components from wild *Wasabia japonica* Matsum. (Wasabi) and *Nasturitium officinale* R. Br. Nippon Nôgeikagaku Kaishi 56:441–443

Kameoka H, Hashimoto S (1983) Two sulfur constituents from *Allium schoenoprasum.* Phytochemistry 22:294–295

Kameoka H, Higuchi M (1976) The constituents of the steam volatile oil from *Lentinus edodes* Sing. Nippon Nôgeikagaku Kaishi 50:185–186

Kameoka H, Kitagawa C (1976) The constituents of the fruits of *Prunus mume* Sieb. et Zucc. Nippon Nôgeikagaku Kaishi 50:389–393

Kameoka H, Demizu Y, Iwase Y (1981) Constituents of neutral fraction of steam volatile oil from *Allium fistulosum* Linn. var. *caespitosum* Makino. Nippon Nôgeikagaku Kaishi 55:315–318

Kameoka H, Iida H, Hashimoto S, Miyazawa M (1984) Sulphides and furanones from steam volatile oil of *Allium fistulosum* and *Allium chinense*. Phytochemistry 23:155–158

Kawakami M, Yamanishi T (1981) Aroma characteristic of Kabuse-cha (Shaded green tea). Nippon Nôgeikagaku Kaishi 55:117–123

Kawakami M, Yamanishi T (1983) Flavor constituents of Longjing tea. Agric Biol Chem 47:2077–2083

Koyasako A, Bernhard RA (1983) Volatile constituents of the essential oil of Kumquat. J Food Sci 48:1807–1812

MacLeod AJ, Gonzalez de Troconis N (1982) Volatile flavor components of cashew "Apple" (*Anacardium occidentale*). Phytochemistry 21:2527–2530

MacLeod AJ, Gonzalez de Troconis N (1982 a) Volatile flavor components of guava. Phytochemistry 21:1339–1342

MacLeod AJ, Gonzalez de Troconis N (1982 b) Volatile flavor components of mango fruit. Phytochemistry 21:2523–2526

MacLeod AJ, Panchasara SD (1983) Volatile aroma components, particularly glucosinolate products of cooked edible mushroom (*Agaricus bisporus*) and cooked dried mushroom. Phytochemistry 22:705–709

MacLeod AJ, Pieris NM (1982) Volatile flavor components of mangosteen, *Garcinia mangostana*. Phytochemistry 21:117–119

MacLeod AJ, Pieris NM (1983) Volatile components of papaya (*Carica papaya* L.) with particular reference to glucosinolate products. J Agric Food Chem 31:1005–1008

MacLeod AJ, Pieris NM (1984) Volatile aroma constituents of Srilankan ginger. Phytochemistry 23:353–359

Masuda M, Nishimura K (1981) Changes in volatile sulfur compounds of whisky during aging. J Food Sci 47:101–105

Nunomura N, Sasaki M, Yokotsuka T (1980) Shoyu (soy sauce) flavor components: Acidic fractions and the characteristic flavor components. Agric Biol Chem 44:339–351

Ohta H, Tonohara K, Watanabe A, Iino K (1982) Flavor specificities of Satsuma mandarin juice extracted by a new-type screw press extraction system. Agric Biol Chem 46:1385–1386

Pyysalo H (1976) Identification of volatile compounds in seven edible fresh mushrooms. Acta Chem Scand B 30:235–244

Saittagaroon S, Kawakishi S, Namiki M (1984) Aroma constituents of roasted coconut. Agric Biol Chem 48:2301–2307

Sawamura M, Yanogawa K, Hattori M (1983) Aroma components of Satsuma mandarin orange grown in vinyl houses and in open fields. Nippon Nôgeikagaku Kaishi 57:863–871

Schreier P, Drawert F, Schmid M (1978) Changes in the composition of neutral volatile components during the production of apple brandy. J Sci Food Agric 29:728–736

Schreier P, Drawert F, Winkler F (1979) Composition of neutral volatile constituents in grape brandies. J Agric Food Chem 27:365–372

Schreyen L, Dirinck P, Sandra P, Schamp N (1979) Flavor analysis of quince. J Agric Food Chem 27:872–876

Schutz TH, Flath RA, Mon TR, Eggling SB, Teranishi R (1977) Isolation of volatile components from a model system. J Agric Food Chem 25:446–449
Shaw PE, Wilson III CW (1982) Volatile constituents of loquat (*Eriobotrya japonica* L.) fruit. J Food Sci 47:1743–1744
Shimizu J (1982) Identification of volatiles in grape musts and characteristic terpenoid constituents. Agric Biol Chem 46:2265–2274
Shimizu J, Watanabe M (1981) Neutral volatile components in wines of Koshu and Zenkoji grapes. Agric Biol Chem 45:2797–2803
Shimizu J, Watanabe M (1982) Volatile components identified in the phenolic fractions of wines from Koshu and Zenkoji grapes. Agric Biol Chem 46:1447–1452
Shinoda N, Shiga M, Nishimura K (1970) Constituents of Yuzu (*Citrus junnos*) oil. Agric Biol Chem 34:234–242
Shiota H (1982) Kiwi fruit. Koryo 137:59–64
Siota H (1982) Starfruit. Koryo 143:37–42
Smith RM, Robinson JM (1981) The essential oil of ginger from Fiji. Phytochemistry 20:203–206
Sugawara (Abe) E, Ito T, Odagiri S (1982) Sweet aroma components in three kinds of jam. Nippon Nôgeikagaku Kaishi 56:101–108
Sword G, Bobbio PA, Hunter GLK (1978) Volatile constituents of jack fruit (*Arthocarpus heterophyllus*). J Food Sci 43:639–640
Takeo T (1983) Variations in the aroma compound content of semi-fermented and black tea. Nippon Nôgeikagaku Kaishi 57:457–459
Tressl R, Silwar R (1981) Investigation of sulfur-containing components in roasted coffee. J Agric Food Chem 29:1078–1082
Tressl R, Bahri D, Engel K-H (1982) Formation of eight-carbon and ten-carbon components in mushroom (*Agaricus campestris*). J Agric Food Chem 30:89–93
Tsugita T, Kurata T, Fujimaki M (1978) Volatile components in the steam distillate of rice bran: Identification of neutral and basic compounds. Agric Biol Chem 42:643–651
Tsugita T, Kurata T, Kato H (1980) Volatile components after cooking rice milled to differents degrees. Agric Biol Chem 44:835–840
Tsugita T, Ohta T, Kato H (1983) Cooking flavor and texture of rice stored under different conditions. Agric Biol Chem 47:543–549
Tsuneya T, Ishihara M, Shiota H, Shiga M (1983) Volatile components of quince fruit (*Cydonia oblonga* Mill.). Agric Biol Chem 47:2495–2502
Uchida K, Matsumoto M, Kobayashi A, Yamanishi T (1983) Composition of oxygenated compounds in peel oil from *Citrus iyo* and its variation during storage. Agric Biol Chem 47:1841–1845
Uchida K, Kobayashi A, Yamanishi T (1984) Composition of oxygenated compounds in the peel oil of Fukuhara oranges. Nippon Nôgeikagaku Kaishi 58:691–694
Welch RC, Johnston JC, Hunter GLK (1982) Volatile constituents of the muscadine grape (*Vitis rotundifolia*). J Agric Food Chem 30:681–684
Yajima I, Yanai T, Nakanura M, Sakakibara H, Hayashi K (1981) Volatile flavor compounds of Matsutake-*Tricholoma matsutake* (Ito et Imai) Sing. Agric Biol Chem 45:373–377
Yajima T, Yanai T, Nakanura M, Sakakibara H, Hayashi K (1984) Volatile flavor components of Kogyoku apples. Agric Biol Chem 48:849–855
Yamanishi T, Fukawa S, Takei Y (1980) Aroma characteristics of *Citrus iyo*. Nippon Nôgeikagaku Kaishi 54:21–25

GC-MS Methods for Tobacco Constituents

H. KODAMA

1 Introduction

Gas-chromatography mass spectrometry (GC-MS) has contributed to the advance of tobacco chemistry. Although the structure of unknown compounds cannot be determined by GC-MS alone, it is to great use to the identification of the tobacco constituents. Recently, mass spectra of tobacco isoprenoids have been reviewed by Enzell et al. (1984).

Tobacco volatiles consist of a large number of minor compounds. Up to the present, more than 800 volatile compounds are known to exist in tobacco leaves (Fujimori and Kaneko 1979; Enzell 1976; Enzell et al. 1977). Most of these compounds are thought to be generated by oxidative degradation of terpenoids, such as cembranoids, labdanoids, and carotenoids, during the curing and aging process. These compounds, generally obtained by steam distillation, have a representative tobacco aroma. Their mass spectra are not encountered in standard libraries (Stenhagen et al. 1974), because these compounds are relatively peculiar to the tobacco constituents. Therefore it is important for GC-MS analysis of tobacco constituents to have its own compilations of data bases of the mas spectra by tobacco constituents.

2 Cembranoids and Their Degraded Compounds

Most abundant cembranoids in tobacco are (1 S, 2 E, 4 R, 6 R, 7 E, 11 E)- and (1 S, 2 E, 4 S, 6 R, 7 E, 11 E)-2,7,11-cembratriene-4,6-diols (1), (2), which differ only in the absolute configuration at C-4. They cannot be distinguished from each other by the spectrum in which the base peak occurs at m/z 43 and other fragment peaks are of low abundance (Fig. 1). The detailed fragmentation has been proposed (Enzell et al. 1984). Generally, these compounds cannot be identified by mass spectra, because the characteristic fragment ions are absent in their spectra. Further, it is difficult to obtain clear separations of the compounds on GC. For these reasons, GC-MS is not so useful for the analysis of these compounds.

About 100 cembranoid-degraded compounds which possess a characteristic isopropyl group are known to exist in tobacco leaves. Solanone (3) is the most abundant cembranoid-degraded compound in tobacco, and is thought to be generated by the degradation of cembratrien-4,6-diols because its absolute configuration at C-5, where isopropyl group attaches, is identical with those of cembratrien-4,6-diols (1), (2). Its spectrum exhibits the molecular peak at m/z 194 and

                                                                    H. Kodama

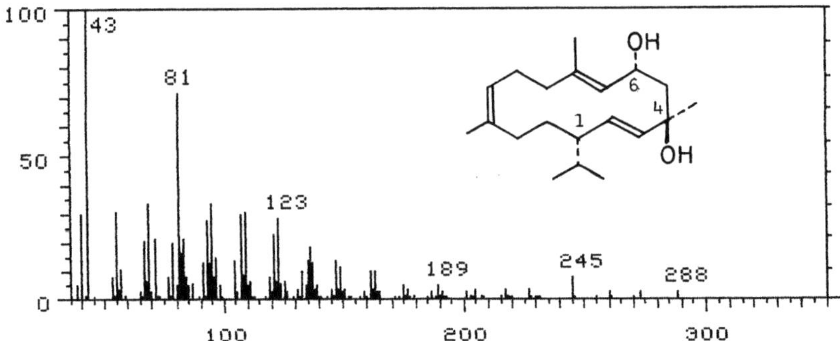

Fig. 1. EI mass spectrum of (1S,2E,4R,6R,7E,11E)-cembratrien-4,6-diol (1)

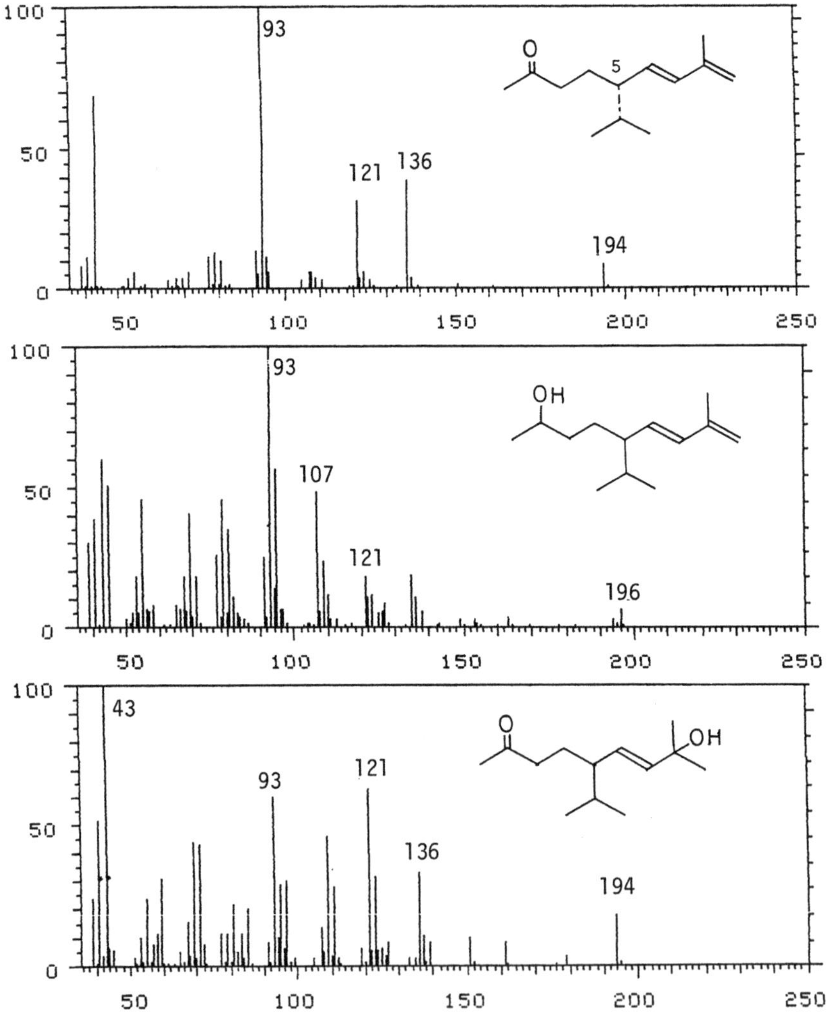

Fig. 2. EI mass spectra of solanone (3), solanol (4) and 2-hydroxy-5-isopropyl-2-methyl-3-nonen-8-one (5)

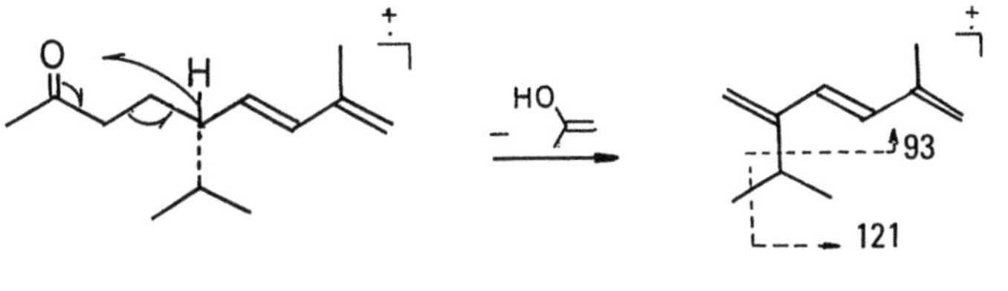

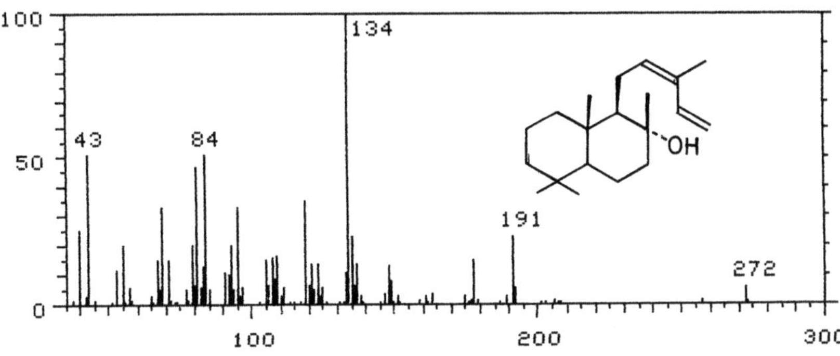

(3)

Scheme 1

characteristic fragment peaks at m/z 136, 121, and 93 (Fig. 2). The loss of 2-propanone group from the molecule generates at m/z 136 ion and the subsequent loss of methyl and isopropyl group account for the generation of the m/z 121 and 93 ions (Scheme 1). Solanol (4) possesses a hydroxyl group at C-8 instead of the carbonyl group. Its spectrum exhibits a molecular ion at m/z 196 and similar fragment ions to those of solanone (3) (Fig. 2). These similar fragment ions may arise by the initial loss of 2-propanol group. The hydroxylated derivative of solanone (3), 2-hydroxy-5-isopropyl-2-methyl-3-nonen-8-one (5), gives a spectrum in which the same fragment ions with that of solanone (3) are of relative abundance (Fig. 2). It is probable that initial loss of water affords the same species as the molecular ion of solanone (3), followed by the reaction outlined in Scheme 1.

3 Labdanoids and Their Degraded Compounds

More than 50 labdanoids and their degraded compounds are encountered in tobacco. 12Z-abienol (6), whose spectrum is shown in Fig. 3, is the most abundant labdanoid in tobacco and thought to be the precursor of the majority of tobacco labdanoids. Diagnostically important fragmentation in these compounds occurs predominantly in ring B. The detailed fragmentation pathways of these compounds have been proposed and reviewed (Enzell et al. 1984).

Fig. 3. EI mass spectrum of 12Z-abienol (6)

4 Carotenoid-Degraded Compounds

About 90 compounds which have a trimethylcyclohexane ring have been encountered in tobacco levels. These compounds are thought to be generated by oxidative degradation of carotenoids and to be key flavor components in tobacco. Most of these compounds are C-13 compounds, and separated into two types by the position of double bond in cyclohexane ring. One having the C-4,5 double bond is α-type and the other having the C-5,6 double bond is β-type. At C-3 most of the α-type have a carbonyl substituent, whereas the β-type have a hydroxyl sub-

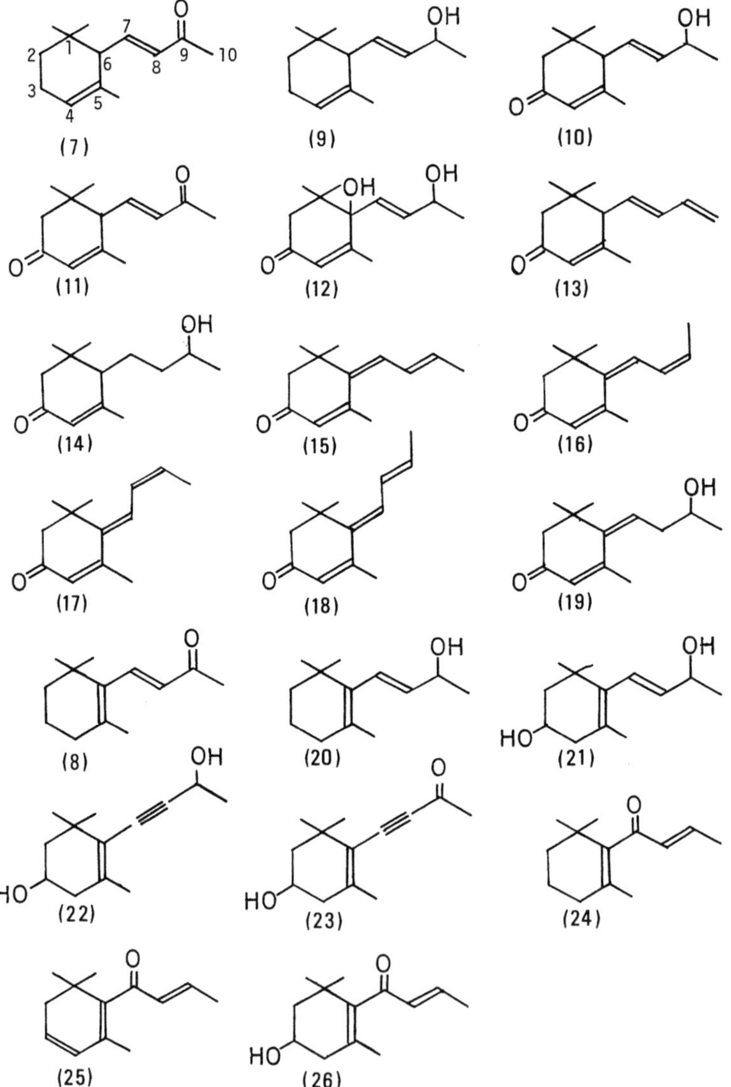

Fig. 4. Carotenoid-degraded compounds in tobacco [except for α-ionol (9)]

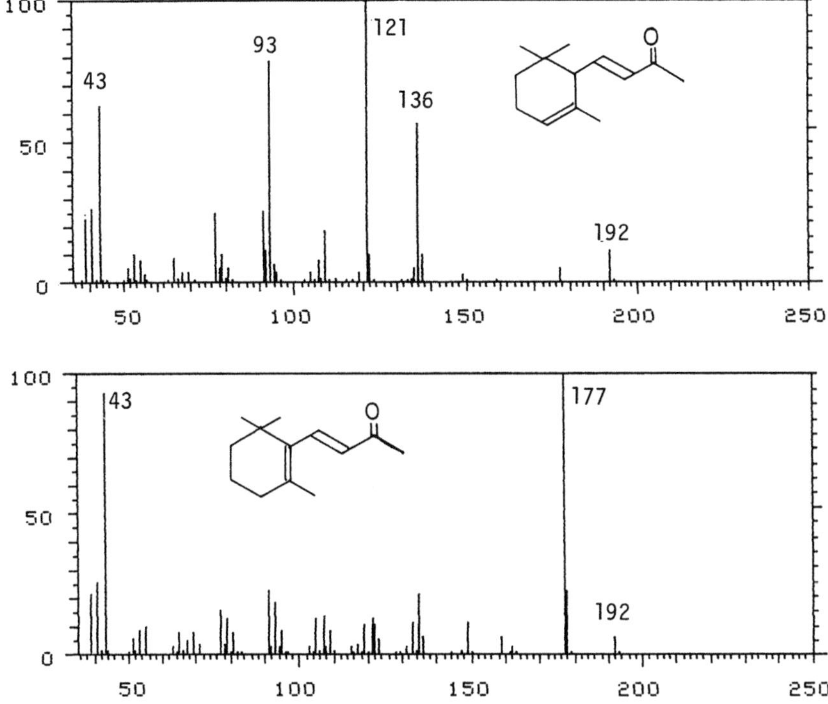

Fig. 5. EI mass spectra of α-ionone (7) and β-ionone (8)

stituent. In the spectra of these compounds, the slight difference of the position of double bond or the substituents afford the significantly dissimilar feature. Therefore, GC-MS is of great use for the identification of these compounds.

For example, the spectra of α- and β-ionone (7), (8) differ considerably from each other, although only in the position of double bond (Figs. 4 and 5). The detailed fragmentation of β-ionone has been proposed by Thomas et al. (1967). The deuteration studies on this compounds have shown that the generation of the most abundant ion at m/z 177 involves the loss of the methyl group from C-5. The spectrum of α-ionone (7), which exhibits little abundance of this ion by the shift of the double bond, contains a characteristic [M-56] ion (m/z 136). This ion is generated by retro-Diels-Alder rearrangement leading to elimination of isobutyl group. Subsequent loss of methyl and carbonyl groups yields the m/z 121 and 93 ions, respectively (Biemann 1962). Although it has not been reported as a tobacco constituent, α-ionol (9) gives a spectrum which exhibits the characteristic ion at m/z 138 (Fig. 6). The formation of this ion can be rationalized by the retro-Diels Alder rearrangement leading to the elimination of isobutyl group (Scheme 2).

The influence of the carbonyl functionality has been demonstrated in the fragmentation of 3-oxo-α-ionol (10) and 3-oxo-α-ionone (11) proposed (Aasen et al. 1973). Their spectra show the almost identical feature except for the diagnostic fragment peak at m/z 152 in the spectrum of 3-oxo-α-ionol (10), and 150 in that

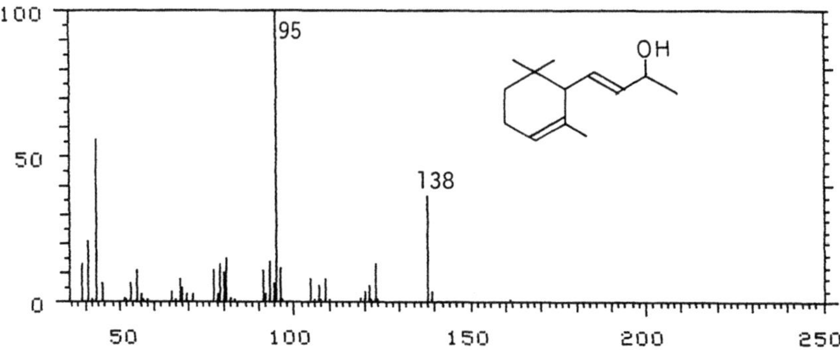

Fig. 6. EI mass spectrum of α-ionol (9)

Scheme 2

(10)

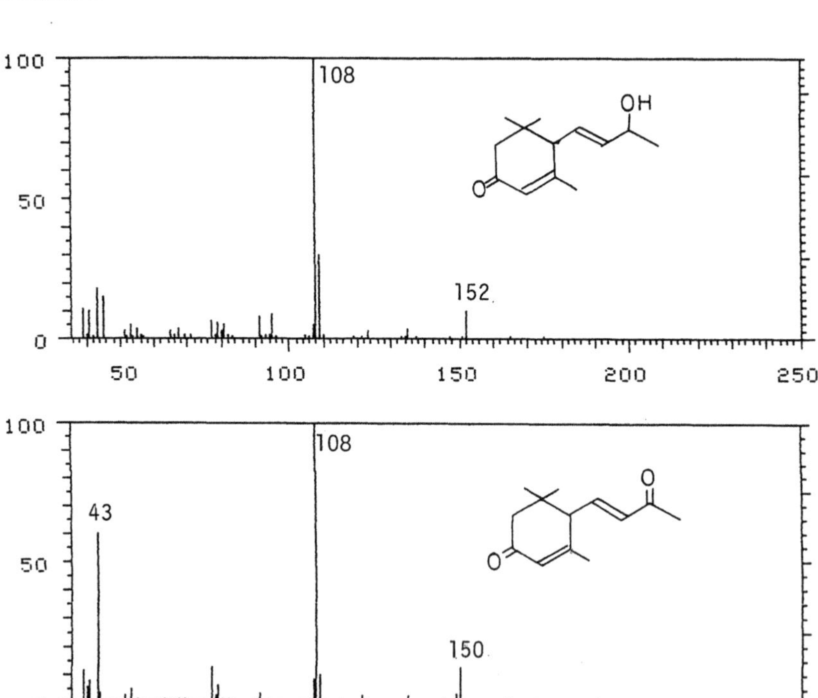

Fig. 7. EI mass spectra of 3-oxo-α-ionol (10) and 3-oxo-α-ionone (11)

of 3-oxo-α-ionone (11) (Fig. 7). This difference of the fragment peaks is due to that of C-9 substituent. High-resolution and deuterium-labeling studies allowed them to clarify the formation of these ions.

An analogous type of fragmentation may account for the generation of m/z 124 in from blumenol A (12) which is C-6 hydroxy derivative of 3-oxo-α-ionol (10) (Fig. 8). Most abundant ion at m/z 124 can be rationalized by that elimination of isobutyl group and subsequent McLafferty type of rearrangement. The plausible route for the formation is described in Scheme 3.

The base peak in the spectrum of megastigma-4,7E,9-trien-3-one (13) (Fig. 9), dehydroxyl derivative of 3-oxo-α-ionol (10), occures at m/z 134 as a result of the facile elimination of the isobutyl group which is promoted by the C-7,8 double bond. In this fragmentation, further decomposition, the simple cleavage of C-8,9 bond may be surpressed by the absence of C-9 substituent.

Blumenol C (14), 7,8-dihydroderivative of 3-oxo-α-ionol (10), gives a spectrum which exhibit diagnostic peaks at m/z 150 and 95 (Fig. 9). These ions may arise by the cleavage of C-7,8 followed by the loss of hydrogen and isobutyl group, respectively (Scheme 4).

Megastigmatrienones (15), (16), (17), (18) (4 isomers) are the most abundant carotenoid degraded compounds in tobacco. Although many fragment peaks occur in their spectra (Fig. 10), it is difficult to distinguish them on the spectra be-

Fig. 8. EI mass spectrum of blumenol A (12)

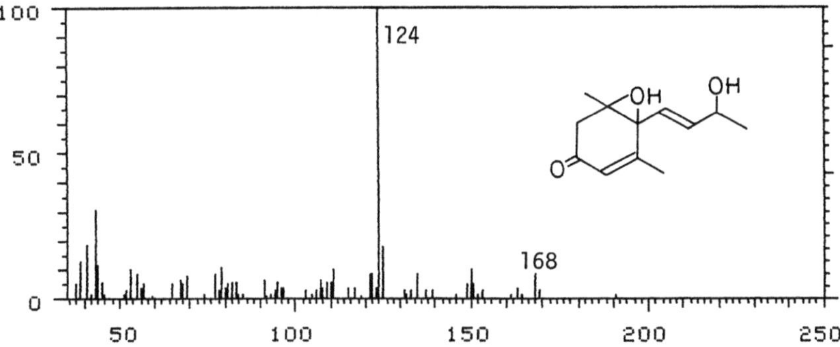

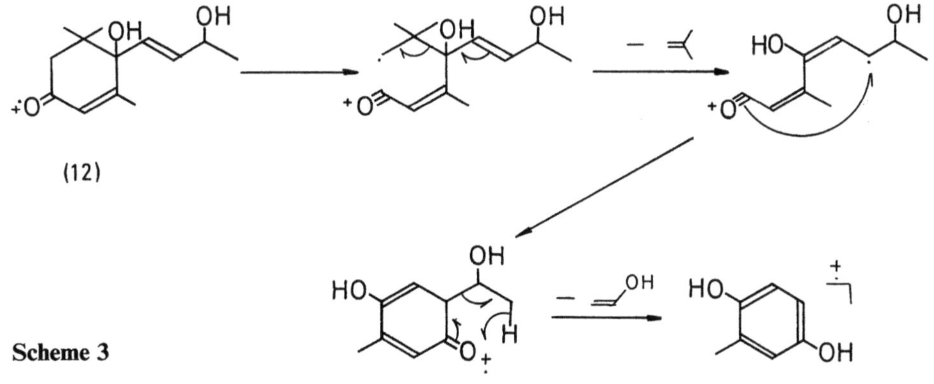

Scheme 3

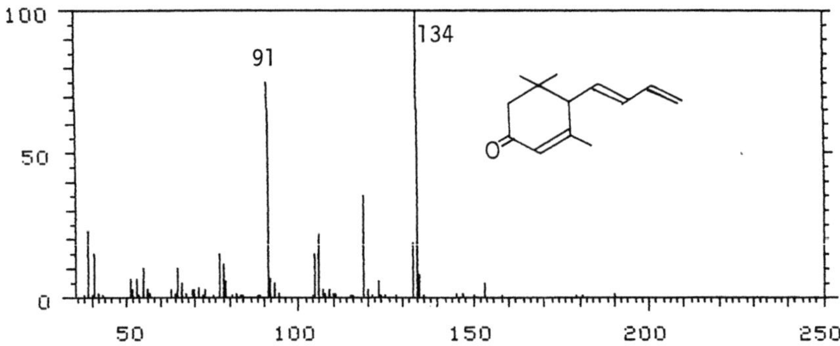

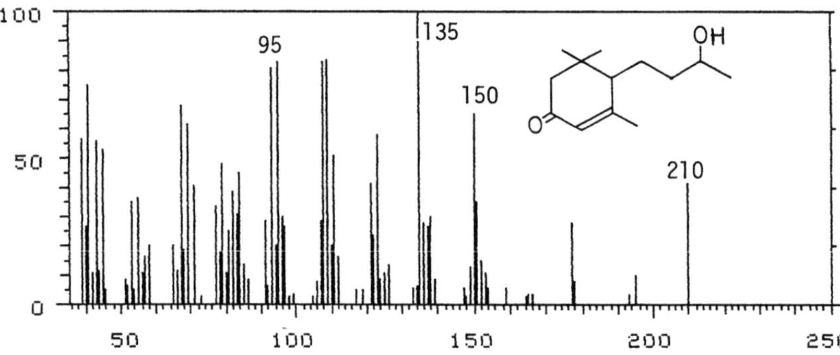

Fig. 9. Ei mass spectra of 4-(1,3-butadienyl)-3,5,5-trimethyl-2-cyclohexen-1-one (13) and blumenol C (14)

Scheme 4

cause their spectra are almost identical. They are identified by the retention times on GC equipped with capilally column coated with OV-101 [the retention times, (16) < (15) < (18) < (17)]. The most abundant megastigmatrienone among the four is (6 Z, 8 E)-isomer (15), whose spectrum shows the highly abundant molecular ion at m/z 190 stabilized by conjugated enone.

In the spectrum of 4-(3-hydroxybutylidene)-3,5,5-trimethyl-2-cyclohexane-1-one (19) (Fig. 10), which also has a C-5,6 double bond, the abundant ions at m/z 164 and 149 are accounted for by the cleavage of C-7,8 bonds with concomitant hydrogen shift and of C-6,7 bond respectively (Scheme 5). The generation of [M-

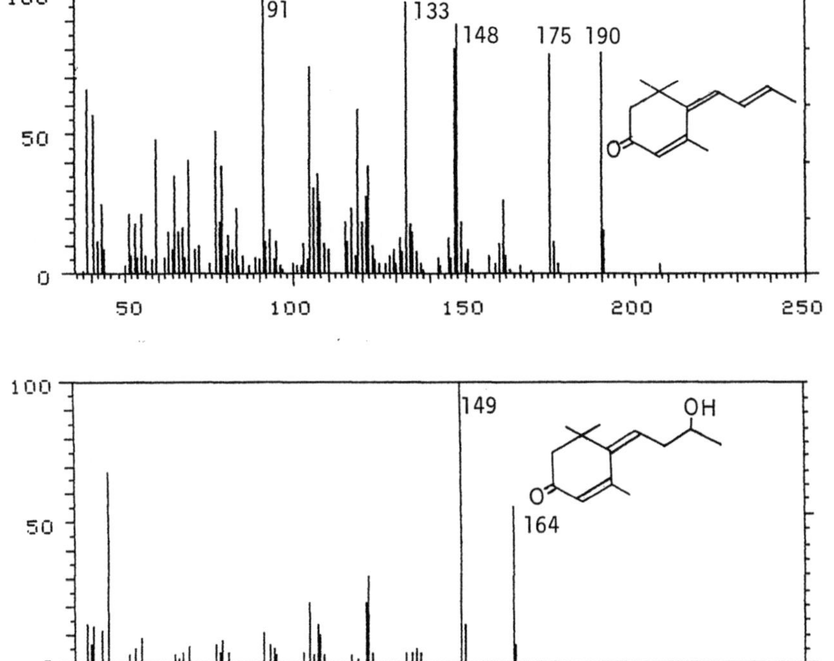

Fig. 10. EI mass spectra of megastigma 4,6Z,8E-trien-3-one (15) and 4-(3-hydroxybuty-lidene)-3,5,5-trimethyl-2-cyclohexene-1-one (19)

Scheme 5

56]$^+$ion, which arise by the elimination of isobutyl group, is supressed in the fragmentation of these compounds, which possess C-6,7 double bond.

The spectrum of β-ionol (20) (Fig. 11) shows features significantly dissimilar to those of α-ionol (9). Its spectrum exhibits a molecular ion at m/z 194, characteristic ions at m/z 179 and 176, which may arise by the elimination of methyl group and water respectively. Two fragment ions subsequently generate m/z 161 ion by the elimination of methyl group or water.

The introduction of hydroxyl group at C-3 seems to have no influence on the fragmentation of β-ionol, since the characteristic fragment ions of 3-hydroxy-β-ionol (21) are two mass lower than that of β-ionol (20) (Fig. 11). The initial loss of water followed by the same fragmentation to that of β-ionol (20) can account

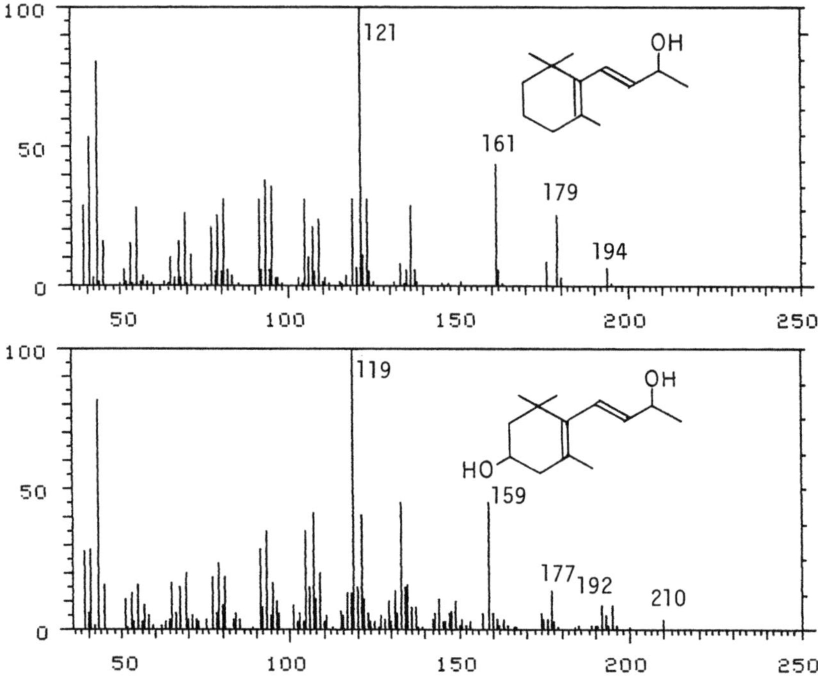

Fig. 11. EI mass spectra of β-ionol (20) and 3-hydroxy-β-ionol (21)

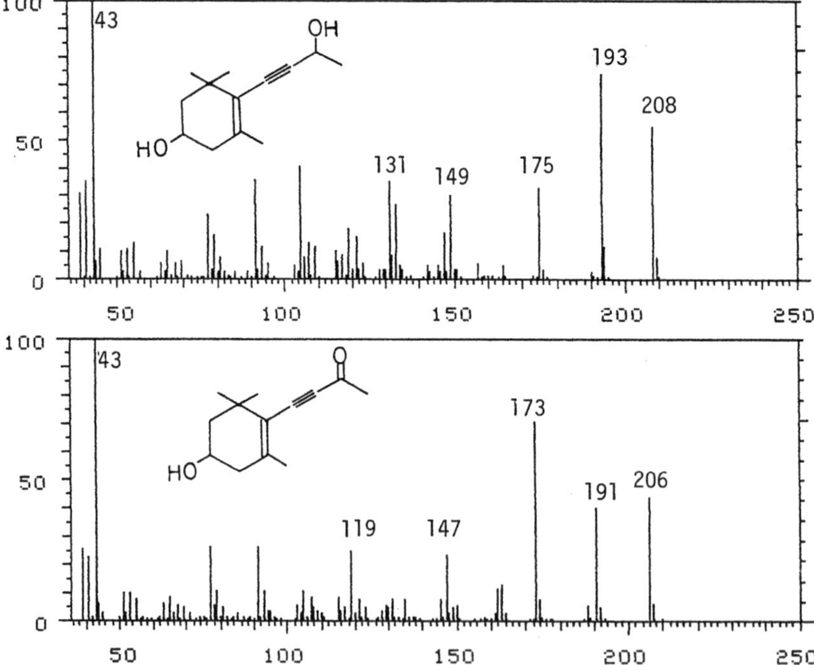

Fig. 12. EI mass spectra of 3-hydroxy-7,8-dehydro-β-ionol (22) and 3-hydroxy-7,8-dehydro-β-ionone (23)

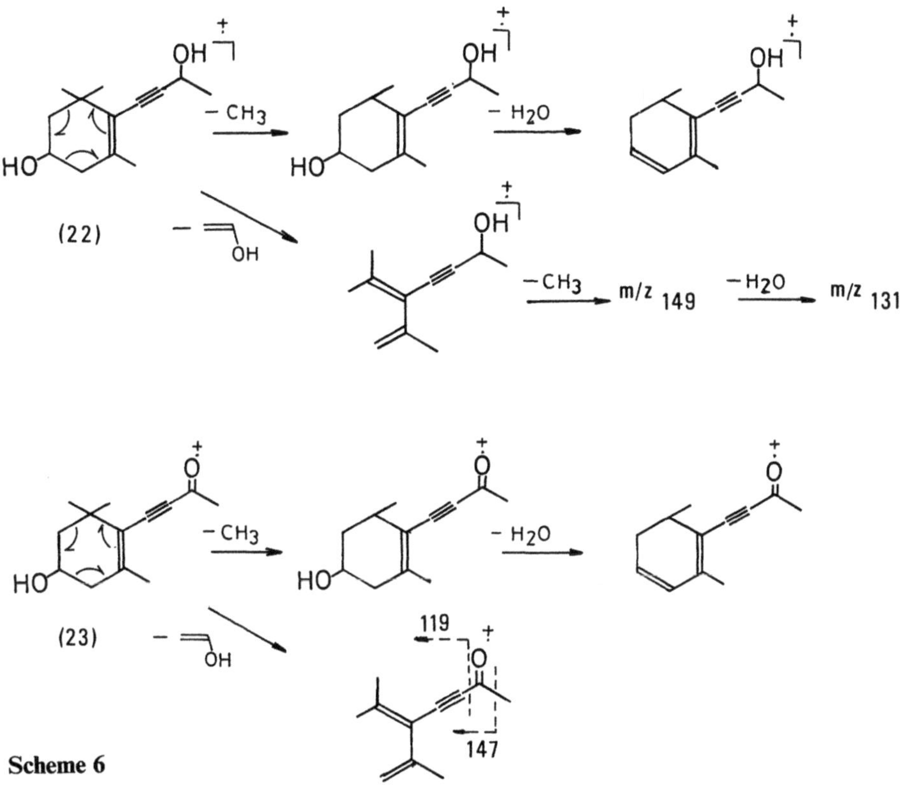

Scheme 6

for the generation of these ions, however, in the absence of labeling studies, fragmentation pathway of the most abundant ion remains uncertain.

The spectra of 3-hydroxy-7,8-dehydro-β-ionol (22) and 3-hydroxy-7,8-dehydro-β-ionone (23) afford the relatively abundant [M]$^+$, [M-15]$^+$, [M-18]$^+$ and [M-59]$^+$ ions (Fig. 12). The generation of diagnostic [M-59]$^+$ ion can be accounted for by the retro-Diels-Alder rearrangement followed by the loss of methyl group (Scheme 6). Subsequent loss of water affords the m/z 131 ion in the former, whereas the latter does not shows the corresponded ion by the absence of C-9 hydroxyl substituent. The generation of diagnostic ion at m/z 119 in the latter can be accounted for by the retro-Diels-Alder rearrangement and subsequent cleavage of C-8,9 bond.

β-Damascone (24), β-damascenone (25) and 3-hydroxy-β-damascone (26) possess carbonyl groups at C-7. The simple cleavage of C-6,7 bond is highly favored process in these type of compounds (Fig. 13). The spectrum of β-damascone (24) shows the molecular ion at m/z 192 and highly abundant ion at m/z 177 which is generated by the elimination of methyl group. The cleavage of C-7,8 bond affords the characteristic fragment ions at m/z 123 and 69. The elimination of methyl group is surpressed in the fragmentation of β-damascenone (25) which is the C-3,4 dehydro derivative of β-damascone (24). Its spectrum shows the molecular peak at m/z 190 and diagnostical peak at m/z 121 and 69. These complementary ions are generated by the simple cleavage of C-6,7 bond. 3-Hydroxy-β-

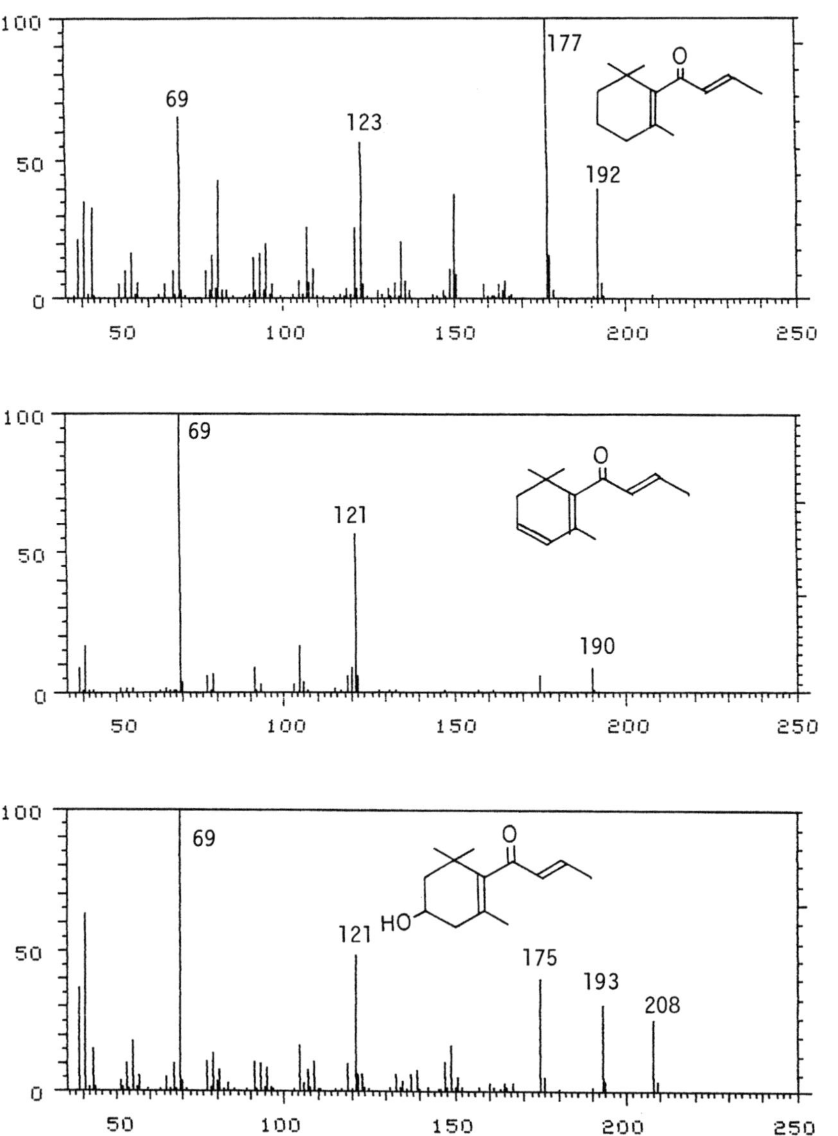

Fig. 13. EI mass spectra of β-damascone (24), β-damascenone (25) and 3-hydroxy-β-damascone (26)

damascone (26) affords characteristic fragment ions at m/z 121 and 69. Such species are also present in the spectrum of β-damascenone (25). Therefore it is reasonable to speculate that the initial fragmentation takes place by the elimination of water and subsequent cleavage of C-6,7 bond generate the m/z 121 and 69 ions (Scheme 7).

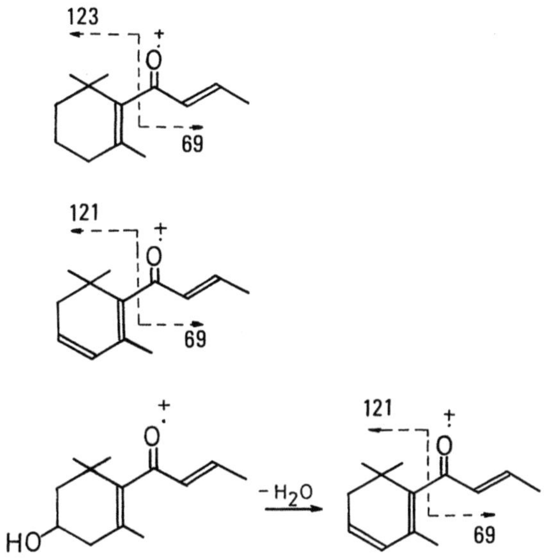

Scheme 7

5 Sesquiterpenoids

Tobacco leaves have been known to produce many structurally related sesquiterpenoid stress compounds, are shown in Fig. 14. These compounds are produced in response to infection of virus and bacteria, such as tobacco mosic virus (TMV), cucumber mosic virus (CMV) and *Pseudomonas solanacearum* (Fujimori 1984).

Solavetivone (27) gives rise to a molecular ion at m/z 218 and fragment ions at m/z 203, 190, 176, 175, 162, 147, 133 and 108 (Fig. 15). These fragment peaks also observed in the spectrum of solanascone (29), though the peak intensities are significantly different. The plausible route for their formation has been proposed (Enzell et al. 1984). Their hydroxylated derivatives, 3-hydroxysolavetivone (28) and 3-hydroxysolanascone (30) are also known to exist in tobacco leaves, however, 3-hydroxysolanascone (30) have been reported as the glycoside form. The characteristic fragment peaks in the spectra of 3-hydroxysolavetivone (28) are also observed in that of 3-hydroxysolanascone (30), although these are of low abundance (Fig. 16). It has been reported that solavetivone (27) and 3-hydroxysolavetivone (28) are converted into solanascone (29) and 3-hydroxysolanascone (30), respectively, by [2+2] photocyclization (Fujimori et al. 1978). The retro-rearrangement from solanascone (29) and 3-hydroxysolanascone (30) to solavetivone (27) and 3-hydroxysolavetivone (28) respectively may take place at least in part. An analogous type of rearrangement may account for the generation of the same fragment ion from dehydrosolanascone (31) as those from solavetivenone, however solavetivenone has not been reported as a tobacco constituent.

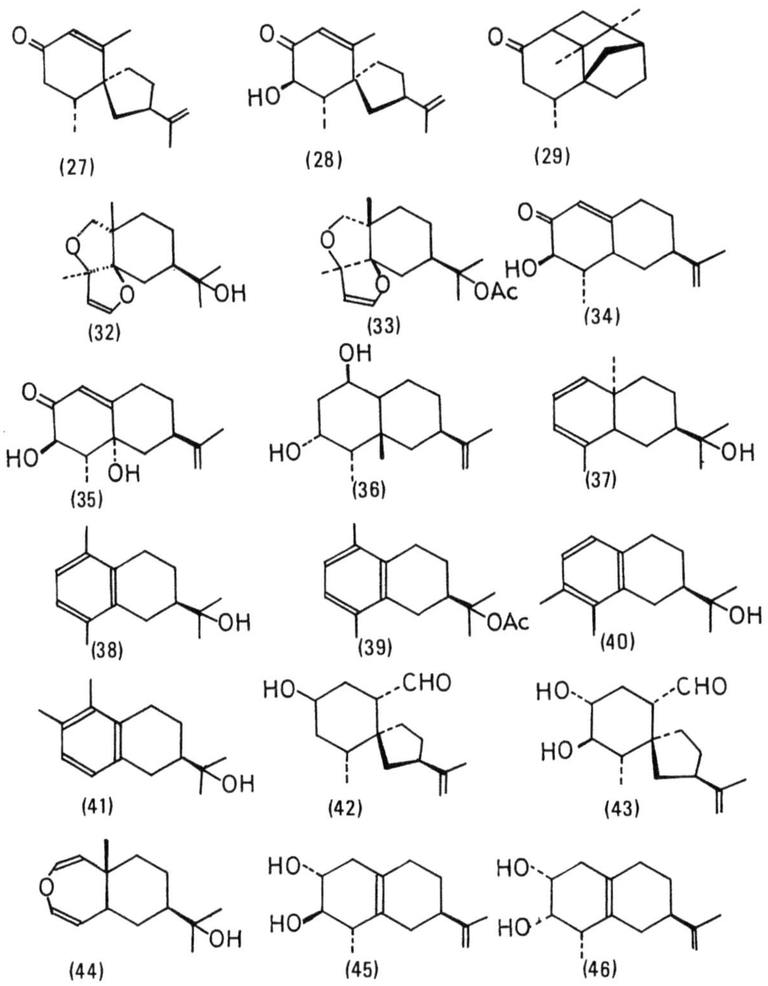

Fig. 14. Sesquiterpenoid stress compounds in tobacco

The spectra of phytuberol (32) and phytuberine (33) exhibit the most abundant ion at m/z 205 and similar fragment ions. The significant difference is the occurrence of abundant ion at m/z 207 in the former, whereas no corresponding peak is found in the latter.

In the spectrum of glutinosone (34) (Fig. 17), the abundant ions at m/z 162 and 94 are accounted for by the initial cleavage of bound between the hydroxylated carbon atom and carbonyl carbon atom, followed by the reactions outlined in Scheme 7. An analogous type of elimination accounts for the generation of the most abundant ion in the spectrum of oxyglutinosone (35).

Capsidiol (36), occidentalol (37), occidol (38), occidol acetate (39), occidol isomers (40), (41), occidenol (44), lubimin (42), and 3-hydroxylubimin (43) afford the spectra in which the fragment ions, generated by the elimination of methyl

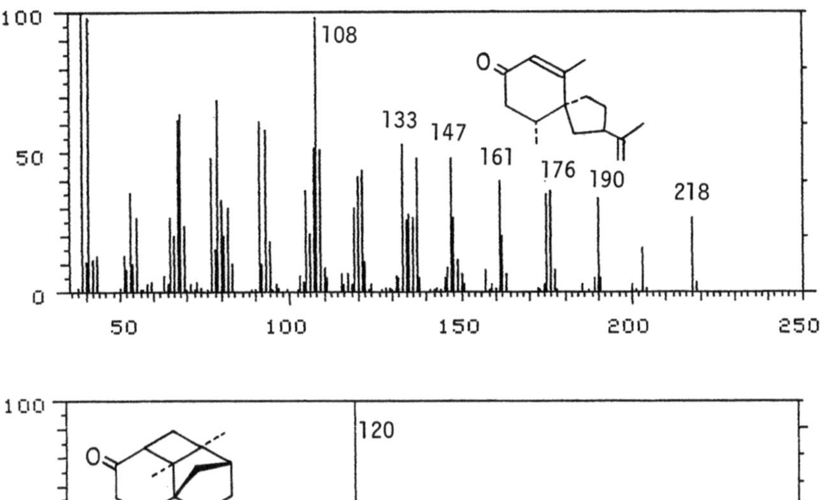

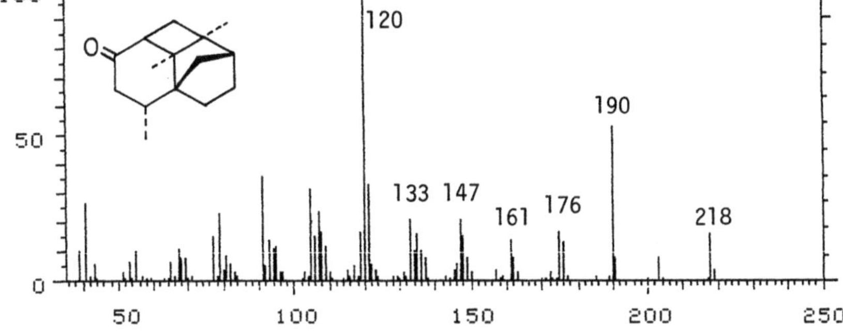

Fig. 15. EI mass spectra of solavetivone (27) and solanascone (29)

and water, are of relative abundance. The peaks at m/z 41 and 55 clearly show the presence of isopropanol and isopropenyl substituents at C-7, respectively.

Rishitin (45) and its epimer (46) have been isolated from TMV-inoculated tobacco leaves (Fuchs et al. 1983). The mass spectrum of the isomer is almost identical to that of rishitin, except for the lower intensity of the peaks at m/z 222 and 204 corresponding with the $[M]^+$ and $[M-H_2O]^+$, respectively (Fig. 18). By this susceptibility of the elimination of water in the fragmentation, the structure of the isomer has been assumed to be 2- or 3-epirishitin. This estimation is supported by the fact that this isomer is converted into acetonide more easily than rishitin.

6 Terpenoid Glycosides

During recent years, some carotenoid-degraded compounds and sesquiterpenoids have been isolated as a glycoside form from tobacco leaves. These glycosides can be analyzed as their acetyl derivatives by GC-MS. A diagnostically important fragmentation reaction in these compounds is the cleavage of the glycosidic linkage, which affords the characteristic ions at m/z 331 and [M-347]. The spectrum

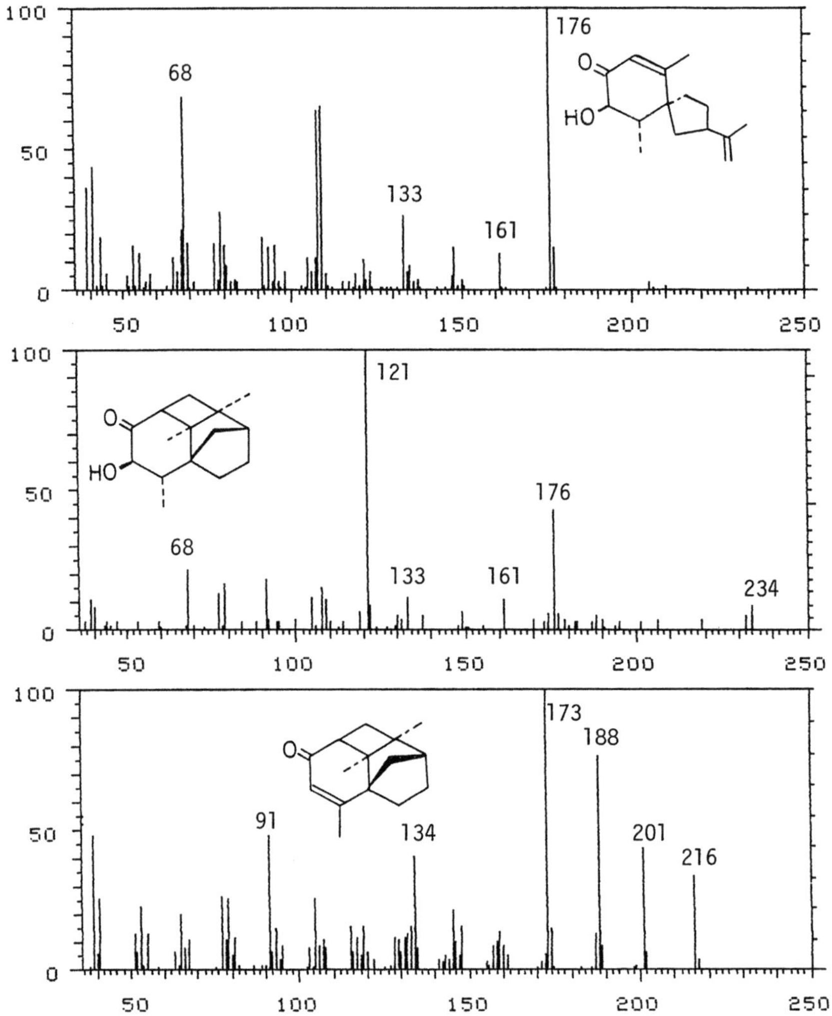

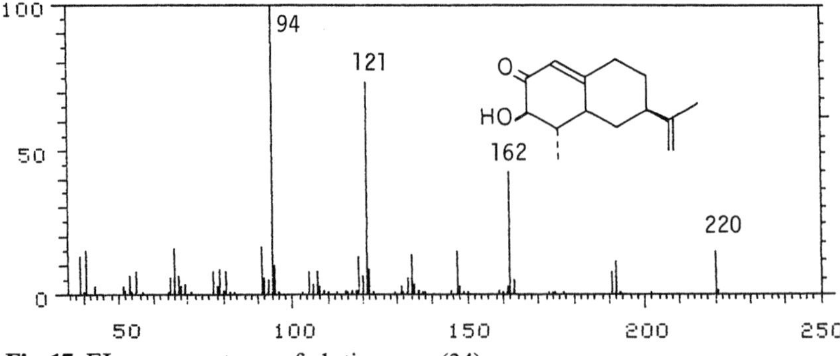

Fig. 16. EI mass spectra of 3-hydroxysolavetivone (28), 3-hydroxysolanascone (30) and dehydrosolanascone (31)

Fig. 17. EI mass spectrum of glutinosone (34)

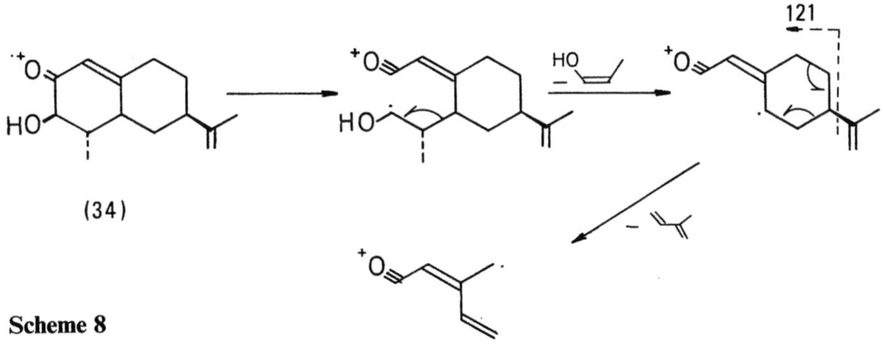

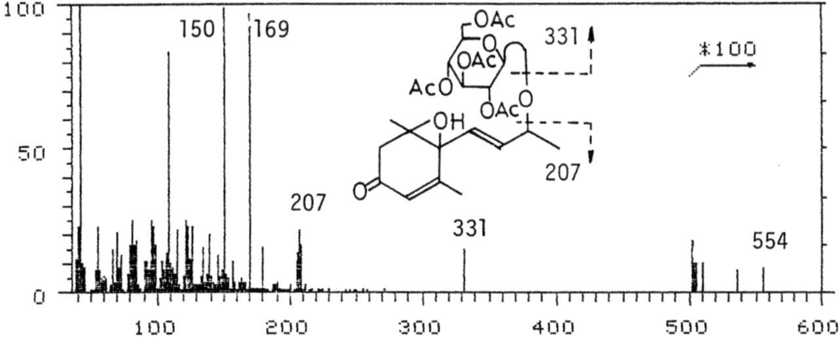

Scheme 8

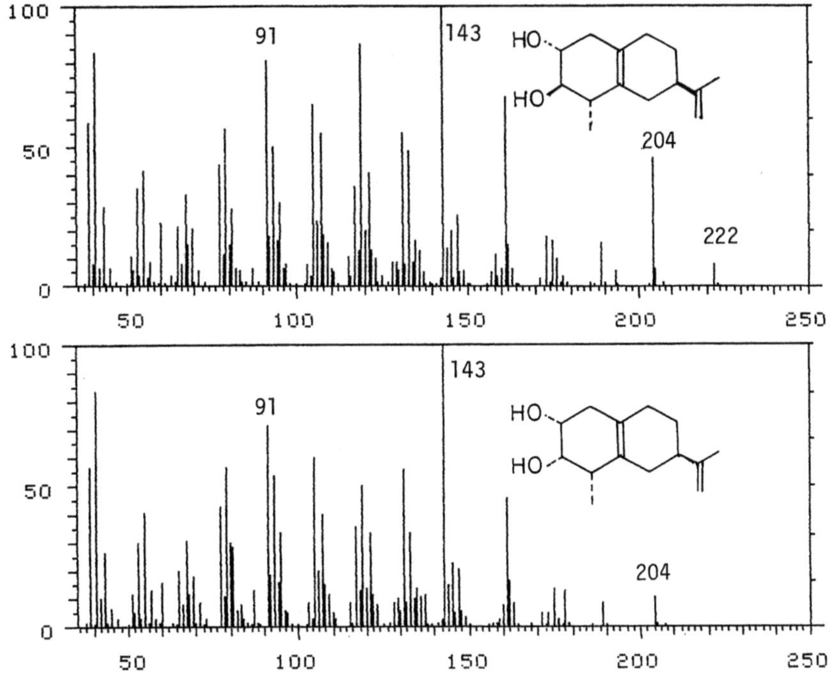

Fig. 18. EI mass spectra of rishitin (45) and epirishitin (46)

Fig. 19. EI mass spectrum of blumenol A-β-glucoside tetraacetate

of blumenol A-β-glucoside tetraacetate shows the relatively low abundance of molecular peak at m/z 554 and diagnostic fragment peaks at m/z 331 and 207 (Fig. 19). It is of interest to note that the same rearrangement to aglycone, elimination of isobutyl group (Scheme 2), from the dehydroxylated ion, can account for the generation of abundant ion at m/z 150.

7 Linked Scanning

There are some primary ions which happen to decompose into fragments while traversing the first field region. These decomposed ions, called "metastable ion", are of great use for the analysis of ion structure, fragmentation pathway and structural determination. Although the kinetic energy will be reduced to m_d/m_p (m_d and m_p; mass of daughter and parent ions respectively), these ions are accelarated as mass mp. Consequently, metastable ions are recorded as mass m*. Mass m* can be simply related to the mass of the parent ion (m_p) and the mass of the daughter ion (m_d) by the following equation.

$$m^* = m_d/m_p$$

Linked scanning is the method which detects the metastable ions by attenuating the electron field. The relation of metastable ion and daughter ion, and parent ion is indicated by the following equations.

$$m_d = m^*/r, \ m_p = m^*/r^2$$

($r = E/EO = m_d/m_p$; attenuation coefficient of electron field). Therefore, mass of metastable, daughter, and parent ion can be obtained by the mass of recorded ions and the electron field voltage.

Three methods of linked scanning, B/E, B^2/E and $B(1\text{-}E')^{1/2}/E'$ scanning, are widely available. The B/E linked scanning is the most useful among the three. The major information obtained from this scanning is that the generation of the daughter ion from the primary ion can be clearly defined.

The B/E linked scanning spectra of carotenoid degraded compounds are shown in Fig. 19. In the spectra the broad peaks of the metastable ions are exhibited as line shape and the mass of the metastable ions are corrected to those of daughter ions by computer calculation. All the spectra show the first fragmentation step from the molecule. Most of these spectra exhibit the initial elimination of water or methyl group. The spectrum of 3-hydroxy-β-ionol (21) exhibits the most abundant peak at m/z 192 which arises by the elimination of water, whereas that of 3-hydroxy-β-ionone exhibits the elimination of the methyl group instead of water by the difference of the C-9 substituent. It is of interest to note that the spectra of their 7,8-dehydro derivatives exhibit the contrary daughter ions to that of 3-hydroxy-β-ionol (21) and 3-hydroxy-β-ionone. The spectra of 3-hydroxy-7,8-dehydro-β-ionol (21) and 3-hydroxy-7,8-dehydro-β-ionone (22) exhibit the most abundant ions which arise by the elimination of the methyl group and water, respectively. The spectra of 3-oxo-α-ionol (10) and 3-oxo-α-ionone (11) exhibit

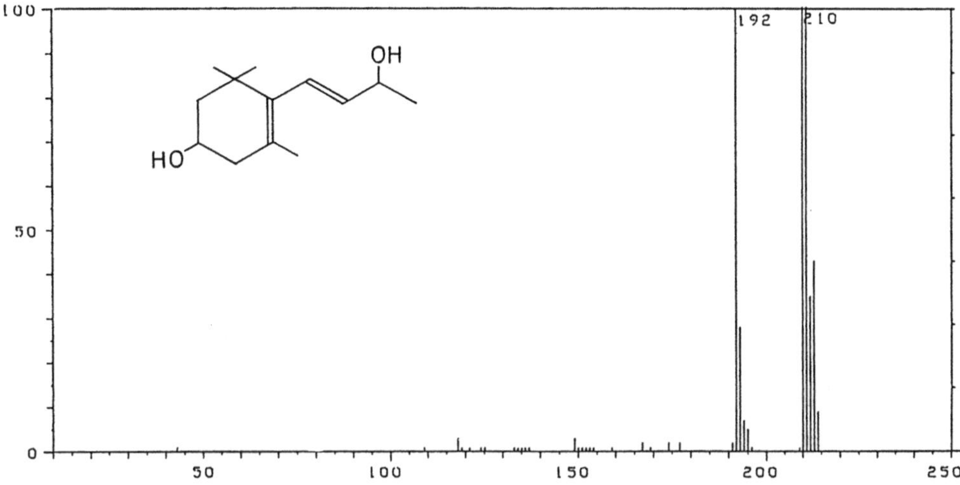

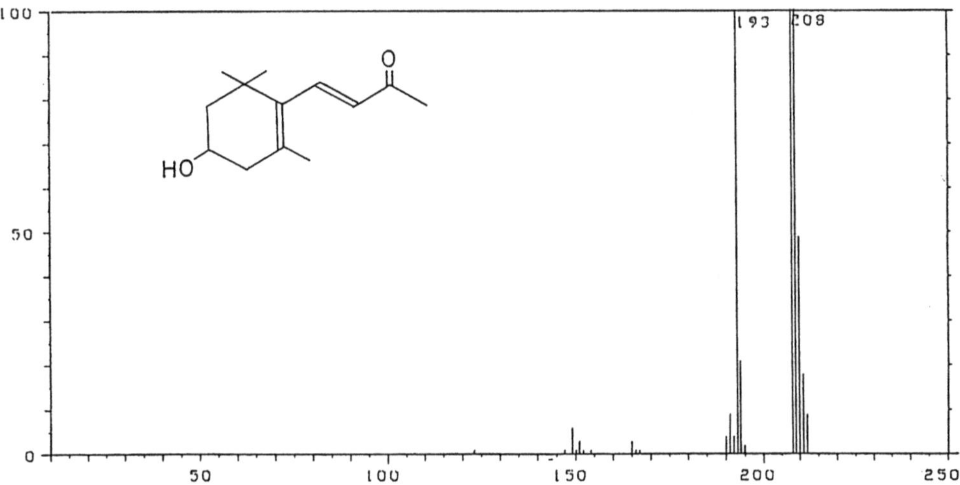

Fig. 20 a

Fig. 20 a–d. Linked scanning spectra of carotenoid-degraded compounds

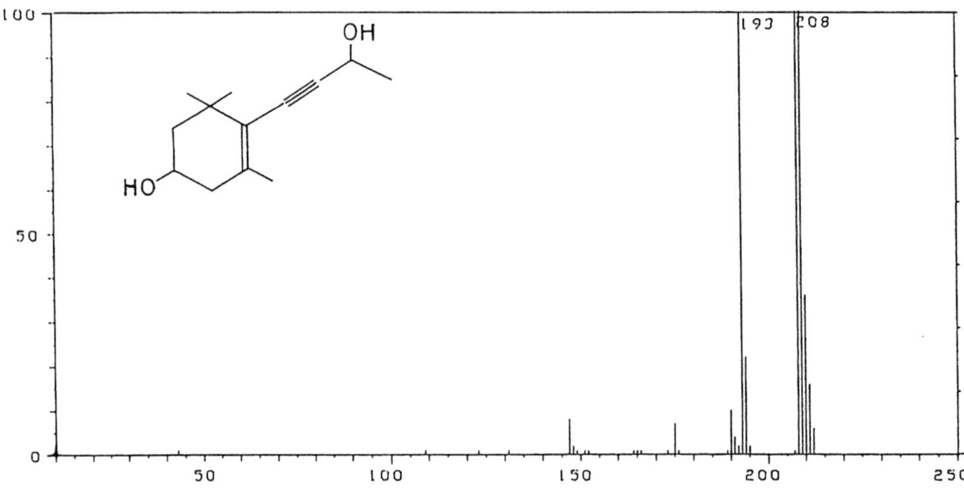

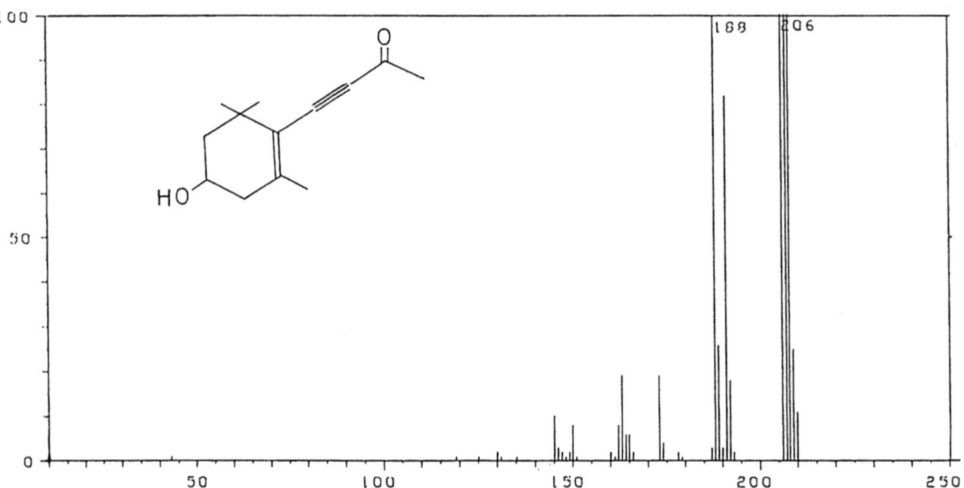

Fig. 20b (Legend see p. 295)

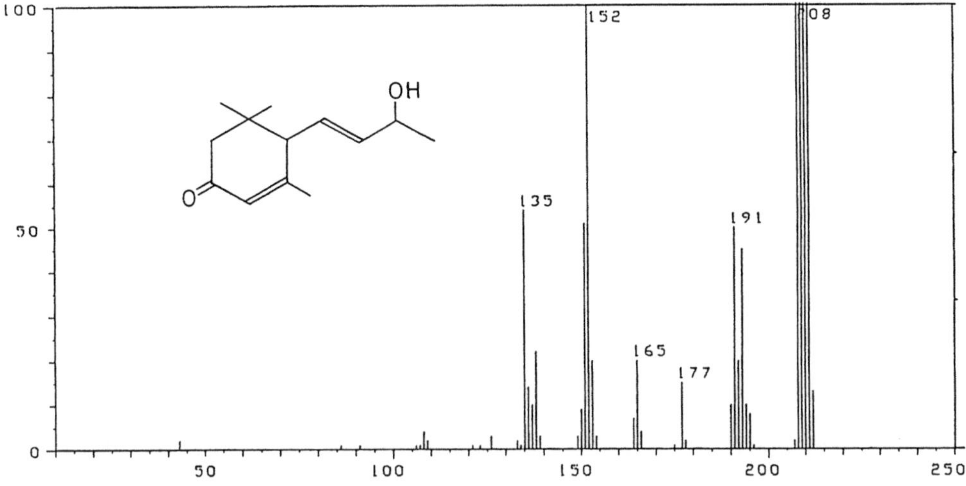

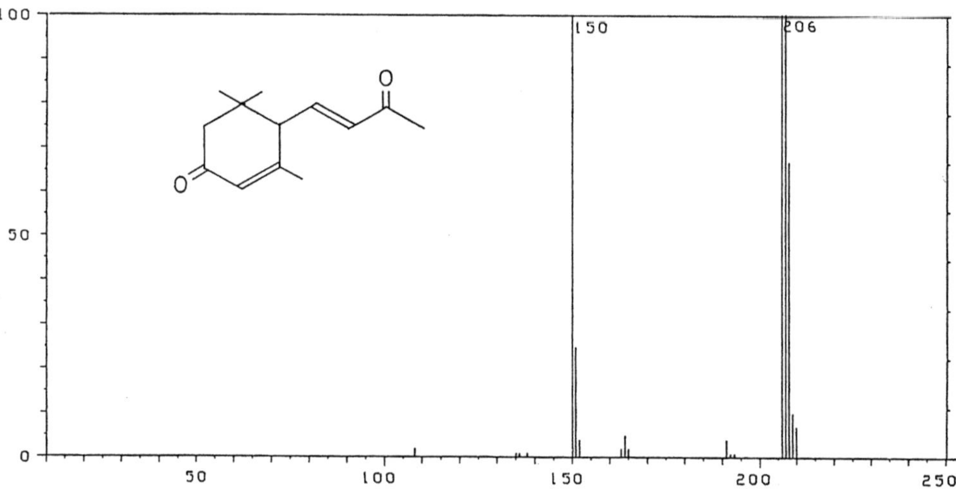

Fig. 20 c (Legend see p. 295)

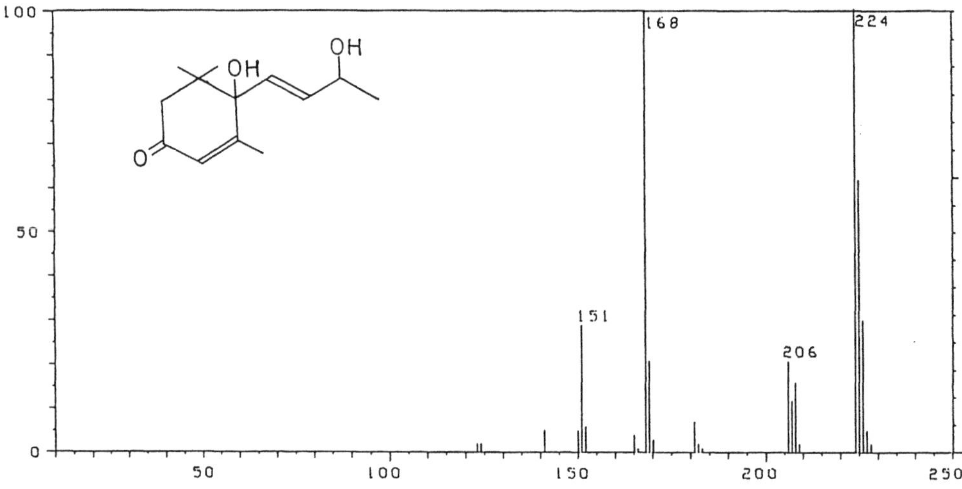

Fig. 20 d (Legend see p. 295)

the most abundant ions at m/z 152 and 150, respectively, which proves the direct generation of these ions from the molecular ions by the elimination of isobutyl group. The most abundant ion at m/z 168 in the spectrum of blumenol A (12) also proves the same fragmentation (Scheme 3).

Acknowledgment. I thank Mr. K. Shizukuishi of Hitachi Ltd. for linked scanning spectra measurement.

References

Aasen AJ, Kimland B, Enzell CR (1973) Acta Chem Scand 27:2107
Biemann K (1962) Mass spectrometry organic chemical applications. McGraw-Hill, New York
Enzell CR (1976) Res Adv Tob Sci 2:32
Enzell CR, Wahlberg I, Aasen AJ (1977) In: Herz W, Grisebach H, Kirby GW (eds) Isoprenoids and alkaloids of tobacco. Springer, Berlin Heidelberg New York
Enzell CR, Wahlberg I, Ryhage R (1984) Mass spectrometry reviews, p 395
Fuchs A, Slobbe W, Mol PC, Posthumus MA (1983) Phytochemistry 22:1197
Fujimori T (1984) Kagaku to Seibutsu 22:358
Fujimori T, Kaneko H (1979) Nippon Nogeikagaku Kaishi 53:95
Fujimori T, Kasuga R, Kaneko H, Sakamura S, Noguchi M, Furusaki A, Hashiba N, Matsumoto T (1978) J Chem Soc Chem Commun 1978:563
Stenhagen E, Abrahamsson S, McLafferty FW (1974) Registry of Mass Spectra Data. John Wiley, New York
Thomas AF, Willhalm B (1967) Tetrahedron Lett 1967:5129

Subject Index

ABA compounds, structures 15
ABA metabolites, methylesters 16
ABAGE (ABA glycosyl ester) 15
abienol, MS 279
abscisic acid (ABA) 3, 14–18
acetylating agents 125
acyl chains, cleavage 114
   positional distribution 117–118
Aesculus parviflora 190
Agaricus bisporus 268, 269, 270
Agaricus campestris 270
alcohols in apple juice 260
alditol acetates, partially methylated,
   identification 27–29
alditols 25
aldobioronic acid 36
Allium chinense 265
Allium fistulosum 263, 265
Allium schoenoprasum 263
Allium tuberosum 263
Amansiae 126
Amaranthus betacyanin test 215
Amphidium carterae 126
Anacardium occidentale 259
Anacystis 74
anthranilate 259
apple embryos, GA analysis 7, 8
apple juice, volatile flavor compounds
   260
apple volatiles 272
apricots, volatile components 260
arachidonic acid 60, 62, 63, 64
Arthocarpus heterophyllus 262
Ascophyllus nodosum 129
Asparagopsis armata 129
Aspergillus niger 12
auxin analysis, GC-MS strategy 165–176
auxin conjugates, hydrolysis 151, 159
auxin, generic term 148
auxins, derivatisation 157–160
   derivatives, structures 159
   extraction 150–152
   purification 152–153
   quantification 176–178
   reference compounds 148–150
Avena coleoptile curvature 146
Averrhoa carambola 259

barndies, volatile components 273
t-BDMS ethers, GC-MS spectrum 112
benzyladenine (BAP) 215, 221
benzylglucosinolate 270
biosynthetic relationship, GA 10
Bligh Dyer method 101
blumeol 283, 284, 293
bound auxin 189, 191
Brassica juncea 265
Brassica napus 18
brassinolide 19
brassinosteroids 18, 19
   MS data 19
BSTFA (N,O-bistrimethylsilyltrifluoro-
   acetamide) 2

cabbage pectins 27
cadinene 261
cAMP, content in maize seedlings 51
   extraction 48–50
   isotope dilution assay 51–55
   mass spectrum 50
   synthetic 59
   trisilyted, mass spectrum 50
Cantharellus cibarius 269
capillary splitter 81
carbodiimide method 25, 39
carbohydrate derivatives 23
β-carbolines 152
carotenoid degradation products 280–289
carrier gas systems 79
carrot roots, volatile flavor 263
castasterone 18, 19
cell-wall polysaccharides 36
cembranoids 277–279
cGMP, in maize seedlings 56–59
   silyted, mass spectrum 57, 58
   synthetic 59
chemical ionisation 4
chive oil, flavor compounds, GC 264
Chlorella 74
4-chloro-indole-3-acetic acid (4-Cl-IAA)
   147
cinnamate 259
Citrus iyo 256–257
Citrus junnos 256
Citrus paradisi 258

cocoa beans, aroma volatiles   271
coconut, volatile components   271
coffee, roasted, aroma   271
computer-based GC-MS systems   130
computerized data systems GC-MS   163
configuration, anomeric   23
copaene   261, 268
cross-scanning   4
Crotalus adamanteus   118
Cruciferae, isothiocyanates, MS   266, 267
   methylthioalkyl nitrile, MS   267
   steam volatile oil   265
Cucurbita pepo   147
Cyanidium   72
cyclic purine nucleotides, stability   55
Cydonia oblonga   259
cymene   256
cytokinin analysis, structural studies
   234–236
cytokinin-like activity   215
cytokinins, abbreviations, synthesis
   217–219
   biosynthesis   216
   chemical structure   214
   combined GC-MS   233–234
   derivatisation   225–229
   FAB spectra   243
   isolation of novel   230
   metabolic profiling   244
   MS   230–234
   preparative GC   230
   quantification   236–244
   separation by GC   220–230

damascone   287, 288
data-processing system computer, on-line
   with GC-MS   256
Datura crown gall, cytokinin glycosides
   241, 243
DBU (1,5-diazabicyclo[5.4.0]undec-5-ene)
   31, 33
de-esterification, pectic polysaccharides
   40
degradation, β-eliminative   27, 31
depolymerisation, partial, permethylated
   derivatives   38–40
derivatisation   2–3
desorption chemical ionisation (DCI)
   241
detector   82–83
deuteration, catalytic, of diacylglycerols,
   apparatus   116
diacylglycerols, tBDMS-derivatives   109,
   111
diazomethane, toxic   2
dinosterol   130
direct inlet   51

disulfides, in *Allium*   263
DMSO (dimethylsulphoxide)   26
double bounds, reduction by deuterium
   115–116
double focusing instruments   163
Dunaliella   72, 109, 110

ECL (equivalent chain lengths)   90–91
EI mass spectrometry   4
electron capture detection (ECD)   168
electron impact (EI)   4, 163
β-elimination reactions   30
enantiometric nature, sugar residues   24
endogenuous auxins, GC MS analyses
   176
endopolygalacturonanase   38
Eriobotrya japonica   262
esterification, artifactual, auxins   152
Eucalyptus haemastoma   17
external standardization, GC, fatty acids
   89
extraction fatty acids   68–69

FAB-MS (fast atom bombardmentmass
   spectrometry)   35
fast atom bombardment (FAB)   164
fatty acid esters, interpretation of mass
   spectra   92–94
   structure determination   90–94
   suitable capillary columns   85
   suitable packet columns   84
fatty acids, column selection   83–87
   lower plants   90
   glycolipids, lower plant   67 ff
   standard compounds   87–89
   unsaturated   60–64
fermentation products   273–274
FID (flame ionization detector)   82–83
field desorption (FD)   164
fingerprint, mass spectra   92
flame ionization detectors   82–83, 224
flavor components, volatile   256–274
flavor of food, chemical study   254
flavor samples, preparation   255–256
5-fluoro-indole-3-acetic acid methyl ester
   (5-F-IAA-Me), mass spectra   171
fragmentation, MS   2
fragmentation pattern   25, 37
fragmentography   4
free-IAA   151
fruit constituents   256–263
FSCC (fused silica capillary columns)   83

GA-conjugates   11, 13
   extraction   10–11
   identification   13–14
   structure types   10, 12

GC data, interpretation 87–89
GC results, calculation 87–89
geranial, ginger oil 268
Gibberella fujikuroi 14
gibberellins 3, 9–14
ginger, essential oils 268
Ginkgo biloba 64
Girard-T reagent 137
GLC analysis fatty acids, column types
  83–84
Gleditsia species 128
glutinosone 290, 292
Glycine max 210
glycolipid fatty acids, derivatization
  76–78
  esterification 77
glycolipid separations by TLC 72–73
glycolipids, isolation 74–76
  separation 70–74
glycosidic linkage 23, 25
  determination 24–27
  GA 13
glycosiduronic acid linkage, stability 24
glycosyl residues, neutral 30
glycosyl-sequencing technique 39
Gracilaria lichenoides 64
grape fruit juice 258
grape wine, volatile components 273
Grob type injector 223
guava, essence 261
gurjunene 261

Halopytis pinastroides 124
headspace technique 135
heptafluorobutyrylimidazole (HFBI) 160
hexamethyldisilizane (HMDS) 159
hormone, conjugated 189, 191
humulene 261
hydrocarbons in apple juice 260
hydrolysis studies, pectic polysaccharides
  29–33

IAA, calibration plot 184
  conjugates, amount in plant tissue 193
    hydrolysis 192–193
    occurrence 190
  derivatives, stability 160
  determination, internal standards
  194–195
  free, amount in plant tissue 193
  -inositols, derivatization 196–197
    discovery 189–190
    identification 191–195
    qualitative analysis 195–210
  -Me, EI spectrum 176
    fragmentation pattern 165
    memory effects 158

methyl esters (ME) 158, 169, 189
methyl esters, fragmentation pattern
  165
-myo-inositol-galactoside 191
reaction with diazomethane 158
ring-labelled 181
sugar esters 152
identification by GC-MS 1–9
indol-3-acetic acid (IAA) 146, 147, 155
  methyl ester trifluoroacetyl derivative
    (IAA-Me-TFA) 172
  pentafluorobenzyl ester (IAA-PFB),
    mass spectra 173
indole-3-carboxylic acid (ICA) 147, 155
indole glucosinolates 148
indole-3-methanol (IMeOH) 147, 155
indole-3-propionic acid (IPA) 147, 155
indoles, fragmentation mechanism 168
  mass spectra 166–168
  retention time 155
inhibitors, ABA related 17
injection 79–82
ionol 282, 283, 285, 286, 287
ionone, MS 281, 282
inositol moiety 195–196
internal standards (IS), auxin
  determination 177, 178–182
internal standardization, GC fatty acids
  89
interphase GC-MS 161
ionization, chemical (CI) 164
ionisation, suppression 10
IS, stable labelled 182
isotope dilution assay, cAMP 51–55

jack fruit, volatile constituents 262
jams, characteristic aroma 272–273

Kalanchoe blossfeldiana 64
kiwi fruits, aroma 262
Kováts indices 3, 126
Kováts retention index system 139

labdanoids 279
liagosterol 129
limonene 256, 261
linalool 258
linked scanning 294
lipid degradation 68
lipid extraction 101–104
lipid fatty acids, methyl esters 77
lipids, separation by gas chromatography
  108–110
Litchi chinensis 262
litchi fruit, volatile constituents 262
loquat fruit, volatile constituents 262

lupinic acid, direct probe mass spectrum
    235

magnetic sector instruments   162
Mangifera indica   261
mango essence   261
mangosteen, volatile aroma   262
mannans, yeast   38
mass fragmentography (MF)   169
mass spectrometry, quantitative   5–9
McReynolds constants stationary phases
    86
Me (methyl esters)   2
Medicago sativa   128
megastigmatrienones   283, 284
menthene, fragmentations, MS   257
metastable ions   294
methyl arachidonate (MA)   62, 63
methylation   2
    analysis   36
        carbohydrate polymers   24–27
        extensions   29–33
        polysaccharides   41
methyl eicosapentaenoate, mass spectrum
    93
methylene units (MU)   126
5-methyl indole-3-acetic acid methyl ester,
    electron impact mass spectrum   179
methyl palmitate, mass spectrum   93
methylthioalkyl nitrile, Cruciferae, MS
    267
MeTMS derivatives of GA   12
MID (multiple ion detection)   52
mixed spectra   4
mixture cAMP – cGMP   59
molecular ions (M)   92
monoterpenes, volatile   258
MSTFA (N-methyl-O-trimethylsilyl-
    trifluoroacetamine)   2
MTBSTFA (N-Methyl-N(tertbutyl-
    dimethylsilyl) trifluoroacetamide)   108
multiple ion monitoring (MIM)   169, 177
muurolene   261
mushrooms, volatile flavor constituents
    268–270
mustard oil, glycosides   265
myo-inositol   189, 190
myo-inositols, mass spectral fragmentation
    pattern   199–210
myrosinase   266
myrcene   256

Nasturtium officinale   265, 267
Nitzschia alba   72

octanol   269
octen   269

oligosaccharides, characterization as
    permethylated derivatives   34–37
orange juice   257
Oryza sativa   190
2-ox-indole-3-acetic acid (OxIAA)   148

Padina vickersiae   126
papaya volatile components   261
partial acid hydrolysis, pectic
    polysaccharides   29–33
patchoulene   261
peak capacity   3
peak identification by standard
    compounds, fatty acids   87–88
pectic polysaccharides, hydrolysis   24
pectin, lemon peel   38
pectins, enzymatic hydrolysis   38
    methylation, single operation   26
    partial acetolysis   37–38
    partial acid hydrolysis   33–37
    retention times   26
pentafluoropropionyl anhydride (PEPA)
    160
permethylation   3
Pharbitis purpurea
Phaseolus vulgaris   51, 182, 244
phosphatidyl-choline-derived diacyl-
    glycerol trimethylsilyl ethers, *Dunaliella
    salina*   110
phospholipase hydrolysis   104–106
phospholipids, analysis of under-
    derivatized   117
    formation of derivatives for GC-MS
        104–108
    mass spectrometry   110–117
    purification   102–104
    separation on silica gel   104
phytuberol   290
pinapple, volatile constituents   261
pinene   256, 261
Pinus contorta   182
Pinus sylvestris   147
Pisum sativum   184
plant sterol analysis, development
    123–124
plant sterols, isolation   124
PMAA (partially methylated alditol
    acetates)   24, 25, 26, 27–29
PMAA, preparation   42
polysaccharides, bacterial   30
    complex, sequencing   38–40
    pectic   23–43
polyunsaturated acyl chains, fragmentation
    114
Porphyridium   72, 74
Prorocentrum cordatum   130
Prunus mume   259

Pseudomonas solanacearum 289
Psidium guajava 261
pumpkin endosperm, GA 14
PVP (polyvenylpyrrolidone) 10
pyrroline, aroma of coocked rice 272

quadrupole mass filters 162
quantitation of molecular species by
GC-MS 113–114
quince flavor 259

raphanatin, direct probe mass spectrum
235
Raphanus sativus 265
relative retention time (RR$_t$) 125
remethylation 29
retention indices, terpenoids 140
retention time 3
hydrogen as carrier gas 110
rhamnogalacturonan 31, 33, 39
Rhodymenia palmata 129
rice bran, odor 272
rice, volatile components 271–272
Ricinus communis 182
ring form, pectic polysaccharides 23
rishitin 291, 293
Risoella verruculosa 124

saringosterol 129
scan, substraction 4
Scenedesmus 94
SCOT (support-coated open tubular)
capillary columns 83
selected ion current monitoring (SICM)
168
selected ion monitoring (SIM) 162, 169,
178
selective detectors 224
sequencing pectic polysaccharides 38–40
sequencing sugar residues 33–38
sesquiterpenoids, tobacco leaves 289–291
silylation 77–78, 125
simultaneous distillationextraction
apparatus 255
SIM (selective ion monitoring) 5–9
Sinapis alba 55
sodium dimsyl (sodium methylsulphinylmethanide) 26
preparation 40
solanol 278
solanone 277, 278, 279, 289, 291
solavetivone 289, 291
soybeans, roasted, headspace volatiles
272
soy sauce, volatile components 273
spectrum, back-ground substracted 7
spectrum libraries 4

Sphagnum 72, 74
Sphagnum magellanicum, glycolipds mass
spectrum 93
Spinacia oleracea 128
split-injection 155–157
splitless injections 81, 156–157
splitter system, all-glass 80
spots, removing from TLC plates 76
standards, authentic 4
stationary phases, fatty acid separation
86
steroids, brassinolide-related 18
sterol ring systems 122
sterol, structures 121
sterolic fractions, purification 125
sterols, characterization 125–130
characterization by GC-MS data
127–128
characterization by MS data 126
Dinoflagellates 130
extraction 124
hydroxylated, red algae 129–130
marine 123
TMS derivatives 128, 129
steryl-acetates 127–128
steryl-esters 124
stress compounds, tobacco 290
substitutents, non-carbohydrate 23
substitution pattern 25
sugar residues, configuration 24

tBDMS (tert-Butyldimethylsilyl)
derivatives, formation 107–108
tBDMS derivatives, key diagnostic
fragments 112
tea, aroma 270
terpenes, ginger 268
terpenoid glycosides 291–294
terpenoids, isolation 135–136
prefractionation 136–138
retention data 139–140
terpinenes 256, 259, 261
terpinolene 256, 259
testerone 19
Thea sinensis 128
thymol 256
TIC chromatogram 5
TLS markers 183
TMCS (trimethylchlorosilan) 2
TMS (trimethylsilyl) 2, 54, 77, 107, 125,
159
TMS ethers, storage 107
TMV virus 289
tobacco tissue, cAMP determination 51
tobacco volatiles 277
tomato, volatile components 263

total lipid extracts, separation by TLC
  72–73
Tricholoma matsutake aroma concentrate,
  GC   269
tricosane   259
trifluoroacetylanhydride (TFAA)   160
trimethylcyclohexane ring, compounds
  280–289
trimethylsilyl derivatives, formation   107
trimethylsilylation   2
tryptophane (Tryp)   147, 155
typhasterol   19

Vaccinium ashei   261
vegetable substances, analytical problems
  47–48
Vitis rotundifolia   258
Vitis vinifera orientalis, odor   273
Volvox   72
vomifoliol   16

WCOT (wall-coated open tubular capillary
  columns)   1, 3, 83, 153
wheat-germ oil, methyl ester fraction   61
whisky, volatile sulfur compounds   273

xanthoxin   16
xyloglucan, potato   29
xyloglucans, apples   38

ylangene   261

Zea mays   48, 56–59, 123, 127, 148, 182,
  184, 190, 191, 192, 210, 215
zeatin (Z)   215, 217
  calibration plot, IDMS   240
  derivatives, probe mass spectra   242
  direct probe mass spectra   235
Zingiber officinale   268

MIX
Papier aus verantwortungsvollen Quellen
Paper from responsible sources
FSC® C105338

If you have any concerns about our products,
you can contact us on
ProductSafety@springernature.com

In case Publisher is established outside the EU,
the EU authorized representative is:
Springer Nature Customer Service Center GmbH
Europaplatz 3, 69115 Heidelberg, Germany

Printed by Libri Plureos GmbH
in Hamburg, Germany